IEE TELECOMMUNICATIONS SERIES

SERIES EDITOR: PROF. J.E. FLOOD

Digital transmission systems

Previous volumes in this series

Telecommunication networks
J.E. Flood (Editor)

Principles of telecommunication-traffic engineering
D. Bear

Programming electronic switching systems
M.T. Hills and S. Kano

Digital transmission systems

P. BYLANSKI, B.Sc.(Eng.), Ph.D., D.I.C., C.Eng., M.I.E.E.
The General Electric Company Limited
Hirst Research Centre
Wembley
HA9 7PP
England
and
D.G.W. INGRAM, B.Sc.(Eng.), M.A., C.Eng., F.I.E.E.
University of Cambridge Department of Engineering
Trumpington Street
Cambridge CB2 1PZ
England

PETER PEREGRINUS LTD.
on behalf of the
Institution of Electrical Engineers

Published by Peter Peregrinus Ltd.,
Southgate House, Stevenage, Herts. SG1 1HQ, England

First published 1976
© 1976: Institution of Electrical Engineers
Second impression 1978
Third impression 1979

ISBN: 0 901223 78 6

Typeset by H Charlesworth & Co Ltd, Huddersfield, England
Printed in England by Unwin Brothers Limited,
Old Woking, Surrey

Preface

In recent years we have become increasingly aware of two aspects associated with our work on digital communication systems. First, there has been an ever-increasing flood of technical papers in the field, and very little of this work has been critically assessed and integrated into unified presentations. Secondly, a large proportion of the published material has originated from academic sources, and as a natural consequence it has tended to emphasise the intellectually stimulating areas of the subject, rather than those of practical importance to communication engineers. When we commenced the preparation of this text we felt that we ought to attempt to counteract this situation. This book is therefore intended as a reference and review work, rather than as a tutorial textbook. It is aimed mainly at engineers and scientists engaged in the research, development and planning of digital communications, and, in particular, at those moving into the subject or working in associated fields, who require a digested summary of the current state of the art. It should also provide a useful background for academic staff involved in telecommunication courses and, although we cannot recommend this book to the undergraduate student, we hope that he may find time to read sections of it and get some idea of the fascinating problems encountered in trying to build large complex communication systems.

Economic factors restrict the size of the work, and as a result we have had to rely extensively on cited references to fill in the detail. This we regret, but we can see no way of avoiding it. We have tried to restrict ourselves to readily available sources; apart from references to books and periodicals we have included a number of references to reports and documents. In the UK the reports cited are available through the British Lending Library; except the documents of the International Telecommunications Union; but we feel that these, although not widely available, are reasonably accessible within the telecommunication industry.

In considering the subject we have used three basic approaches – theoretical, planning and practical. We have found that what we felt was adequate theoretical coverage occupied rather more space than originally expected. However, the theory is oriented towards the really relevant characteristics of practical systems; we have not, therefore, concerned ourselves overmuch with mathematical formalism. Those mathematical developments which offer little practical insight are avoided by the addition of suitable references.

We felt that the planning aspect had hitherto received insufficient attention. We have tried to put some of the recommendations agreed by international bodies, such as the various committees of the ITU, in perspective and to provide a guide to the underlying thinking.

We have attempted to maintain the practical approach throughout, even in the theoretical parts of the book. Detailed circuit considerations, however, have had to be trimmed.

The book is organised to follow a sequence of basic themes: historical background, Chapter 1; the services to be transmitted, their performance requirements and the planning involved, Chapters 2–5; the channel media, Chapters 6–8; design factors, Chapters 9–13; realisation and practice, Chapters 14–16.

This work in large measure reflects the efforts of our colleagues throughout the telecommunication industry. This includes not only those with whom we have closely collaborated at the Hirst Research Centre of the General Electric Company Limited and in the British Post Office, but others whom we have never met and whom we only know through their published work. Because of this, we feel a little hesitant in naming individuals, but we consider that we must pay tribute to the following who have been particularly influential and helpful: D.J. Cleobury, R.M. Dorward and J.C. Vines of GEC Telecommunications Ltd., Coventry; F.M. Clayton, K.V. Lever and P. Wells of the Hirst Research Centre; G. Marshall formerly of the Centre; K.S. Chung of Cambridge University, and not least Prof. J.E. Flood of Aston University, Birmingham, for encouraging us to undertake this work.

P. Bylanski
D.G.W. Ingram

Ealing, London
Coton, Cambridge

August 1975

Contents

Notation and abbreviations

a.s.k., f.s.k., p.s.k., q.s.k.	= amplitude, frequency, phase, quadrature shift keying
a.s.k.n. etc.	= a.s.k. with n states etc., e.g. a.s.k.4
c.a.s.k. etc.	= coherent a.s.k. etc.
d.p.s.k.	= differential p.s.k.
dBm	= signal level relative to 1 mW
dBm0	= signal level in dBm at an agreed reference point in the system
dBm0p	= as above, after psophometric weighting, usually referring to noise
d.s.v.	= digital sum variation
f.d.m.	= frequency division multiplex
f.e.x.t.	= far-end crosstalk
h.r.c.	= hypothetical reference circuit
n.e.x.t.	= near-end crosstalk
P_e	= probability of error
p.c.m.	= pulse code modulation
pW0	= signal level in picowatts at an agreed reference point in the system
pW0p	= as above, after psophometric weighting
sinc x	= $(\sin x)/x$
s.d.r.	= signal/distortion ratio
s.n.r.	= signal/noise ratio
t.d.m.	= time division multiplex
σ	= r.m.s. value

$$\bar{\Phi}\left(\frac{h}{\sigma}\right) = \tfrac{1}{2}\,\mathrm{erfc}\left(\frac{h}{\sigma}\right) = \frac{1}{\sqrt{2\pi}}\int_{\frac{h}{\sigma}}^{\infty} e^{-x^2/2}\,dx$$

Chapter 1

Introduction and historical background

In this book we shall consider the transmission of digital signals by electromagnetic transmission systems. We will not consider other types of transmission, for example, acoustical or mechanical schemes, and we will only take into account other aspects of digital communications, such as analogue/digital conversion, to the extent that these impinge on the problems of transmission. The restrictions are necessary if the book is to be kept to a manageable size, and even with these limitations we shall still have to ask the reader to make extensive use of the references cited to fill in details.

1.1 Introduction of digital communications

Most inventions arise from the need to solve problems and, paradoxically, most inventions lead in turn to the discovery of new and hitherto unsuspected problems. In addition, inventions can rarely be considered in isolation because their successful implementation is dependent on an adequate supporting technology. These factors have been just as important in the development of digital transmission systems as they have been in other fields, and so we find that the early history of our subject cannot be described simply in terms of a list of eminent inventors making their discoveries. It is necessary to take into account the political, economic, and social forces that led these men to think about particular problems, and the existence of the technology that made it possible for them to translate their ideas into something of lasting value.

The early systems suffered from many shortcomings that we now recognise to be due to fundamental limitations found in the theory of transmission systems. This, however, was not known to the original inventors who usually thought that the faults in their inventions were due to minor imperfections in design that could be eliminated by further work. It is easy to see the nature of their difficulties in hindsight, and not at all easy for us to appreciate how difficult the step of recognising a new fundamental problem proved to be at the time. We are now surprised by the very long time taken to identify these problems, and the many mistakes and incorrect interpretations of results that were made by the early workers. The same sort of situation is seen to occur time and time again, and we should realise that it is probably still happening. It is quite likely that some of the limitations we experience in our present systems are the symptoms of fundamental principles of which we are not aware, and that our present explanations of these effects are just as wildly incorrect as those made by some of the earlier workers in this field.

The early history of digital transmission is linked with the development of the electric telegraph. Nonelectric signalling systems date back over 2000 years. The Greek general Polybius is known to have used a scheme based on an array of 10 torches in 300 B.C.,[1] and the Roman armies made extensive use of a form of semaphore signalling. (A representation of a Roman signalling station appears in the carvings on Trajans Column in Rome.) The earliest known proposal for an electric telegraph system appeared in a letter published in the Scots' Magazine of the 17th February 1753, from a contributor signing himself C.M. (For a facsimile of this letter see Garratt.[2]) It is not known whether C.M. actually built a system along the lines he proposed, although judging from the detailed description he gave, and his obvious appreciation of the practical problems, one is led to suspect that he must have made some experiments. He proposed a transmission path of 26 parallel wires, supported by insulators at 20 yard intervals. The transmitter was an electrostatic generator, and at the receiving end a pith ball was suspended on a thread from each wire. Below each pith ball was a slip of paper with the appropriate letter written on it. When the transmitter was connected to a wire the paper slip rose to touch the corresponding pith ball. As an alternative, C.M. proposed an acoustic receiver in which an array of bells was excited by sparks from the ends of the wires. However, at that time no need for new communication systems was felt and there was no incentive to try to solve the difficulties inherent in C.M.'s proposals.

By the end of the 18th century a different situation existed. The turmoil of the French Revolution had produced a need for faster communications which was recognised by Claude Chappe, who initially made further experiments on electrostatic telegraphs. These were unsuccessful, the main reason, as we now know, being that the technology of that time was not capable of producing transmission lines with adequate insulation for high-voltage low-current operation. Chappe abandoned the electric telegraph and turned his attention to a visual semaphore system. This proved highly successful and became widely used in the new French Empire. By chance, at about the same time the necessary supporting inventions that would lead to a low-voltage telegraphy system were being made. The Voltaic pile was invented in 1800, and the same year it was discovered that an electric current would decompose water, thus providing a means of detecting the current. One of the first attempts to apply these discoveries to a telegraph system was made by S. T. Sömmering in 1809. The background to this experiment is worth relating since it gives an insight into the military pressures that were encouraging the development of better communication systems at that time.

In the spring of 1809 Austrian troops crossed the River Inn to invade Bavaria. The monarch, King Maximillian, fled before this invasion to Dilligen, where he was surprised, and no doubt relieved, by the unexpected arrival of a French army headed by Napoleon. This was due to the use of the Chappe telegraph by the French, which had enabled Napoleon to receive news of the Austrian invasion much sooner than had been expected. As a result, Munich, which had been captured by the Austrians on the 16th of April, was retaken by Napoleon only six days later, and Maximillian was able to return to his capital within the month. Impressed by this demonstration of the utility of fast communications, the king decided to install his own telegraph system, and through his minister, Montgelas, asked the Bavarian Academy of Sciences to submit proposals. It was no doubt the intention that these proposals should be for a semaphore system, but Sömmering was led to think of an electric scheme and by the 18th of August 1809 had successfully demonstrated transmission over a path of 2000 feet.

A drawing of Sömmering's apparatus is shown in Fig. 1.1. His system, to a large extent, was an adaption of C.M.'s ideas to low-voltage transmission. The same array of wires, one for each character, was again used, although Sömmering increased their number to 35 to include numerals. The signal source was now a voltaic pile, connected to these wires by means of flexible leads and plugs. The detector was an array of gold pins in a trough of water, receipt of a signal being registered by a formation of gas bubbles at a particular pin. An interesting accessory was a ringing

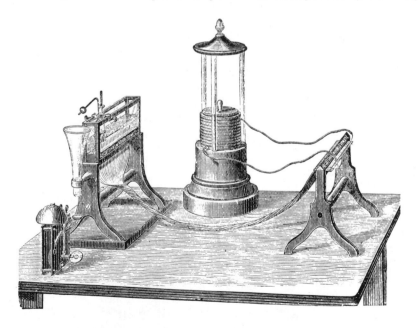

Fig. 1.1 Sömmering's telegraph, 1809

device, which in detail matched the best efforts of Heath Robinson. A spoon-shaped glass vessel was placed in the receiving trough so as to collect the gas bubbles. An arm from this held a lead ball which was released by the lifting of the spoon when a message began to arrive. This ball then fell through a funnel on to the trigger of a clockwork mechanism, which finally rang a bell.

Sömmering's equipment suffered from two shortcomings, the cost and complexity of the transmission path, and the slow response of the electrochemical detector. His system was not developed for practical use. The detector problem was, in principle, solved in 1819 when H. C. Oersted discovered the magnetic effect of an electric current, and the following year Ampere specifically suggested the use of a magnetic detector in a telegraphy system. Various other scientists dabbled in telegraphy, and, as a result of their efforts, the proposals were refined and solutions to the transmission problem were found. Notably Baron Schilling who, while on the staff of the Russian Embassy in Munich had become acquainted with Sömmerings work, invented a system using only six wires to operate five magnetic needles, and Gauss and Weber constructed a telegraph using only one pair of wires in which letters were represented by a sequential code of positive and negative signals. These systems represent some of the earliest applications of coding schemes in digital transmission, and it is interesting to note that they demonstrate the two major classes of codes that are still in use, namely, sequential and parallel encoding.

However, the end of the Napoleonic Wars had removed most of the incentive to improve communications, and the electric telegraph remained a laboratory curiosity.

A new stimulus was needed, and this appeared with the development of railways from about 1830 onwards. Even from their earliest days, railway trains were capable of running at up to 50 miles/h, and the need for a communication system that could match the speed of the trains soon became apparent. In 1837 Wheatstone and Cooke installed a demonstration system between Euston and Camden Town on the London and Birmingham railway. This was essentially an improved copy of Schilling's telegraph which Cooke had seen in Heidelberg the previous year, and used the same arrangement of six wires and five magnetic needles. A drawing of the receiver for this system is shown in Fig. 1.2. The London and Birmingham Railway did not take up the telegraph, but the following year a system was installed in Germany on the Nurnberg-Furth Railway. This installation used an improved version of the Gauss-Weber scheme, developed by C. A. von Steinheil of Munich, and hence was the first serious attempt to use a time-sequential-coding arrangement. In England, the Cooke-Wheatstone system was adopted by the Great Western Railway, and a 40 mile route between Paddington and Slough was in operation by 1840. After 1840 installation of telegraph systems proceeded with great speed, and in just over 30 years a worldwide system had been established.

Thus, we may take 1837 to be the 'moment of lift-off' of the electric telegraph. From this time the various subsidiary technologies of digital transmission began to develop independently, and it is convenient to consider these separately.

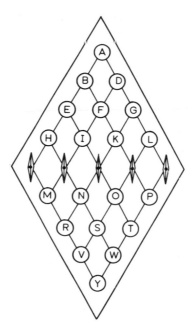

Fig. 1.2 Decoder for Cooke and Wheatstone's five-needle telegraph

1.2 Early ideas about transmission

It is important to realise that we have only reached a reasonably complete understanding of electric transmission processes during the present century (the theory
of semiconductors has been established within the last 25 years) and the early
workers had no idea at all of the nature of the forces they were dealing with. They
vaguely pictured electricity to be some sort of elusive fluid, which had the ability to
flow through metals and some other solid materials. Thus it was natural to explain
the weakening of an electric signal during transmission by leakage of this fluid out
of the conductor, and it followed that it would be a good idea to confine it by
some surrounding barrier or coating on the wire. These ideas are clearly evident in
some of the statements in C.M.'s original letter, and for many years it was generally
thought that the only limitations on transmission over long distances were caused
by imperfections in the insulation of the lines. It was also recognised that the
electric fluid flowed very rapidly, but no one had any real idea of the speed of
transmission. The first serious attempt to measure propagation velocity appears to
have been made by Sir Francis Ronalds[3] in 1816. He installed a wooden framework
in the garden of his house at Hammersmith, on which he suspended eight miles of
iron wire. Using spark gaps and other detectors he attempted to measure the
propagation delay over this length of line. He established that, as far as the human
observer was concerned, transmission was instantaneous.

In 1832 Wheatstone carried out a similar experiment using a shorter line but with much more sophisticated detection equipment[4] consisting of two spark gaps and a rotating mirror mounted on a clock mechanism. He estimated that the transmission velocity was 288 000 miles/s. This overestimate was probably due to the fact that his wires were arranged in a zig-zag stack with appreciable inductive coupling between the ends, and apparently Wheatstone later corrected this error, since in 1840 he stated the value to be 192 000 miles/s.[5] Meanwhile, the work of Ohm in 1827 had established the relation between potential, line resistance and current, and it was beginning to be realised that a limit would be set on long-distance transmission by the sensitivity of the detectors and the maximum potential that could be applied to the lines. One solution to this problem was to introduce some form of regenerator at intermediate points along the line, and using the technology available at that time such a regenerator could be realised in the form of a galvanometer operating a local mechanical switch. This is a good example of an inevitable invention, because both the requirement and the technology were clearly defined, and it is not surprising that a number of people, including Wheatstone and Morse, constructed such devices at about the same time. The prior claims must however go to a little known English inventor, A. Davey, in 1837.[5] For obvious reasons the device was called a relay, and the term later came to be applied to a wide range of electromagnetic switches beyond its original field of application. Rather surprisingly, very little further development in regenerators occurred until well into the 20th century.

As telegraph systems developed it became apparent that this was not the end of the story. It was soon discovered that some of the electric fluid appeared to 'stick' in the line, and was only slowly dissipated, thus setting a limit to the speed at which signals could be sent. It was found that it was desirable to connect a battery of reverse polarity to the line between signal elements to 'sweep out' this residual charge. Then, from about 1850 onwards, submarine cables began to appear which, as we now know, have much higher capacitances and lower characteristic impedances than the overhead-wire lines that were common before that time. The engineers of 1850 were not aware of this, but it soon became apparent to them that the characteristics of these cables were very different from those of land lines. Not only were the propagation speeds so low that it was actually possible to measure the delays on some of the longer cables (the cable installed between Varna and Balaclava in the Black Sea in 1856 was found to have a delay of 0·1 s for a 741 km route; see Grivet[6]) but it was also found that the signals at the receiving ends of these cables were noticeably distorted. While the potential transitions were applied almost instantaneously at the transmitting end, the received signals only changed leisurely from one state to another, thus setting a limit to the signalling speed. The laying of a transatlantic cable was already being considered and the sponsors of this project therefore sought scientific advice on the feasibility of their proposals. The problem was passed to the young Professor of Natural Philosophy at the University of Glasgow, W. Thomson (later Lord Kelvin) and the result of his investigations were presented to the Royal Society in May 1855 in a paper which may be considered to be the starting point of modern transmission-line theory.[7]

Kelvin considered the line as a combination of distributed capacitance and resistance as shown in Fig. 1.3. Since the laws governing the charging of capacitors and current flow in resistors had by this time been formulated he was able to set down the differential equations for this system, namely

$$\frac{\partial e}{\partial x} = -Ri \tag{1.1}$$

$$\frac{\partial i}{\partial x} = -C\frac{\partial e}{\partial t} \tag{1.2}$$

Using Fourier's methods of analysis he was then able to obtain a series solution for the received current as

$$i_r = \frac{2E}{Rl}\left(\tfrac{1}{2} - e^{-ut} + e^{-4ut} - e^{-9ut} + e^{-16ut}\ldots\right) \tag{1.3}$$

where

$$u = \frac{\pi^2}{CRl^2} \tag{1.4}$$

A plot of this function, which Kelvin called an arrival curve, is given in Fig. 1.4. The important feature of this curve is its bottom bend. No significant current is received for a finite time interval and thereafter it increases rapidly. This led to Kelvin's famous *KR* Law (Kelvin used *K* for capacitance) which stated that the signalling speed of a cable was inversely proportional to the product of its capacitance per unit length, its resistance per unit length and the square of its length.

This appeared to be a solution to the question of the speed of transmission of electric signals, but as we now know, this was not so. Remarks in Kelvin's paper reveal that he appreciated that the propagation speed of the signal wavefront was

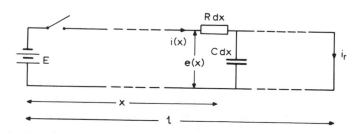

Fig. 1.3 Submarine cable as analysed by Kelvin
   R = resistance per unit length   *l* = total length
   C = capacitance per unit length

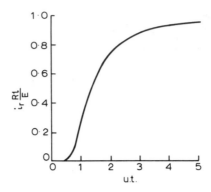

Fig. 1.4 Arrival curve for a submarine cable

not directly related to the propagation speed of the 'electric fluid', and that his solution for the submarine cable was only one special case. Three years later Kirchoff published an analysis of propagation along a telegraph wire. In effect, Kirchoff calculated the self inductance and capacitance of the wire, and showed that the propagation velocity was given by the square root of the product of these quantities. Other workers, notably Weber, had already derived the numerical value of this as about 3.1×10^8 m/s. Kirchoff pointed out that this value was in good agreement with the estimates of the velocity of light and suggested that the velocity of light and the velocity of propagation of electric disturbances along wires were the same. The relation between light and electricity had therefore been established so that, when some eight years later C. H. Maxwell[8] published his theoretical analysis of the electromagnetic field and showed that electrical signals could be described as a wave phenomenon, the inference that light was also an electromagnetic wave followed immediately. Despite this, the mathematical analysis of transmission lines in terms of travelling waves was not carried out for another 25 years.[9] This apparently was because Kirchoff's work had been based on the rather artificial concept of a completely isolated wire with the return path infinitely distant, and the relation between this and practical transmission problems, was not obvious to the telegraph engineers of that period.

1.3 Development of coding

The use of a code arises from the need to convert the signal into a form that matches the characteristics of the transmission path. Early electrical signalling systems dealt almost exclusively with the transmission of written messages. The characteristics of the telegraph limited the transmission to a very restricted set of signals (e.g. positive, negative and space) but equally, the source also originated from a restricted character set (the set of letters and numbers). Thus, the problem appeared to be bounded and well defined, but even within these restrictions a surprisingly large range of alternatives was found to be possible. Now codes are closely related to languages and share many of their characteristics. A code has its elementary symbols (corresponding to letters or phonemes in a language) and these are combined to make up code words. The way in which this is done, and the sequence in which code words can appear, is defined by rules comparable to the grammatical rules of a language, and finally different codes, like different languages, have their own special peculiarities.

All of this, of course, was not apparent to the early telegraphists, who produced their codes on an *ad hoc* basis by common sense or guesswork. As we have already seen, the earliest systems used ·very cumbersome codes, with words of 26 or more digits using only one mark in each word, thus giving a one-to-one correspondence between the code words and the source symbols. By contrast, the Cooke and Wheatstone receivers used five magnetic needles, the deflection of any two of these defining the transmitted symbol, as shown in Fig. 1.2. The system therefore used fixed-length code words of five digits, every code word contained two marks and three zeros, with parallel transmission being used. This is one of the earliest examples of a constant-disparity code, that is, a code in which the relative content of marks and zeros is fixed.

The earliest sequential transmission schemes used codes that were direct adaptations of the original parallel codes, for example, a sequence of 15 marks signified the 15th letter in the alphabet. In a 2-level system the zero then denoted the space between marks. It was clearly not necessary to build up to a fixed-length word by adding a string of terminating zeros, it being sufficient to use a longer space (two or three successive zeros) to indicate the gap between code words, and so at an early stage variable-length code words began to appear. The next step was to attempt to speed up transmission by associating the shortest code words with the most frequently used letters. Baron Schilling appears to have been among the first to adopt this in an experimental telegraph of 1825.[1] The work on variable-length sequential codes culminated in the development of the Morse code (Fig. 1.5) in 1843 and the fact that it is still in widespread use is a striking testimonial to the effectiveness of this coding scheme. It should, strictly speaking, have been called the Vail code, since most of the development was carried out by Morse's colleague A. Vail. Its most noted feature was the serious attempt to match the code-word length to the frequency of occurrence of letters in normal English, and the story of how Vail estimated this by counting the numbers of letters in the type boxes at the local newspaper office is well known. When, a century later, a mathematical theory of communication was developed, it could be shown that the Morse code came within 15% of the theoretically attainable limit. The lesson in this is clear, we should never despise the common-sense solution to theoretically difficult problems, common sense can be very effective. Another feature of the Morse code was that it used three distinct symbols: the dot, the dash and the space, and so could in one sense be regarded as one of the earliest examples of a sequential ternary code.

Letter	Morse symbol	Vail's type box count
E	·	12000
T	—	9000
A	·—	8000
I	··	8000
N	—·	8000
O	———	8000
S	···	8000
H	····	6400
R	·—·	6200
D	—··	4400
L	·—··	4000
U	··—	3400
C	—·—·	3000
M	——	3000
F	··—·	2500
W	·——	2000
Y	—·——	2000
G	——·	1700
P	·——·	1700
B	—···	1600
V	···—	1200
K	—·—	800
Q	——·—	500
J	·———	400
X	—··—	400
Z	——··	200

Fig. 1.5 Morse code, 1843

In parallel with his work on the needle telegraph, Wheatstone also developed a letter telegraph, the ABC system.[5] While most historians have regarded this as a minor contribution to the subject, its coding scheme is of interest. The arrangements are shown in Fig. 1.6. At the transmitter a pointer dial was rotated to the letter to be transmitted. During rotation this actuated a contactor that sent impulses to the line. At the receiver these operated an escapement that allowed the receiving pointer to move in step with the transmitter to the corresponding letter. The important point was that when the next letter was transmitted it was merely necessary to continue rotating the pointer to this next letter. The transmitted codes were then dependent on the difference between successive letters, and were no longer in a one-to-one correspondence with the transmitted signal. Wheatstone's

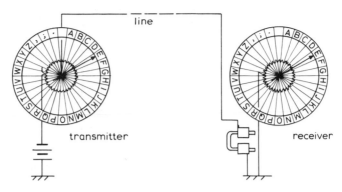

Fig. 1.6 Simplified diagram of Wheatstone's ABC system

system was therefore one of the earliest examples of a differential coding scheme, and the forerunner of modern techniques such as delta modulation and differential pulse-code modulation.

Another possibility is the use of alternative modes in the transmission code. In these schemes, the meaning of a given code word depends on the mode which the code is in at that time. One of the earliest examples of this appeared in the Hughes Printing Telegraph[10] in which the transmitting code included a mode-changing signal. In one mode the code words represented letters, and in the other the same words represented numbers or other symbols. The French engineer Baudot also used mode shifting in his later scheme.

As telegraphy developed it began to be realised that codes could be organised to exhibit other desirable properties. Most of the early workers in this area were concerned with the development of codes to ease the design of their apparatus, in particular, to meet the needs of the high-speed automatic transmitters that began to appear after about 1850. E. Gray invented the code named after him, the Gray code, some time before 1878.[11] This is a fixed-length code with the interesting property that only one digit changes between each adjacent code word in the alphabet. Nowadays, the main interest in this code is in its applications in certain classes of analogue/digital convertors.

In one of his later systems Baudot introduced a chain code. This is shown in Fig. 1.7. The interesting feature of this code is that the sequence of marks and zeros in each position of the code word are identical, but are displaced relative to one another. This means that in an automatic transmitter the code may be generated by using a number of wiper contacts on a single ring of fixed segments.

1.4 Shared transmission paths and multiplexing

We have already seen that the first digital systems used parallel coding with a separate transmission path for each element in the code word, and that at an early

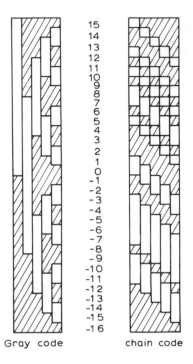

15
14
13
12
11
10
9
8
7
6
5
4
3
2
1
0
-1
-2
-3
-4
-5
-6
-7
-8
-9
-10
-11
-12
-13
-14
-15
-16

Gray code   chain code

Fig. 1.7 Baudot's chain code

stage sequential coding schemes began to be introduced to reduce the complexity of the transmission path. The next logical step was to use a single transmission path for more than one signal, and schemes employing a single line for both directions of transmission soon began to appear. So long as only one direction was in use at any one time this did not introduce any real technical problems. The only requirement was to arrange some operating procedure whereby each terminal could indicate when it had finished transmitting and was ready to receive a response from the other end. Systems of this type (known as simplex systems) are still used in some radio networks where only a single channel is shared by a number of stations.

A rather more subtle idea was to use one line for simultaneous transmission in both directions (now known as duplex transmission), and it was several years after the first introduction of the telegraph before it was realised that this was possible. The essential requirement was to provide some network that would separate the incoming and outgoing signals at each terminal, and all the early workers apparently failed to realise that this problem had in principle been solved in 1834 when Wheatstone invented his famous bridge. As far as we are aware, the first proposal for a duplex system was made in 1853 by William Ginzl, the Director of the Austrian Telegraph service, and the first account in the English literature appeared the following year as Patent No. 2308 issued to R. S. Newall. Newall's system is shown in Fig. 1.8. It consists of a straightforward proposal to cancel the effect of the transmitting key on the local receiver by use of a differential relay. Clearly, static balance is achieved if the currents in the two windings of this relay are equal, and Newall proposed to use a variable resistor to set up this condition, with a differential galvanometer included to detect balance. The fault in this arrangement, which Newall failed to realise, was that the dynamic balance condition when signalling was not the same as the static balance, and as a result current 'kicks' occurred in the receiving relay whenever the transmitting key was operated. This failure appears to have been inherent in most of the early proposals for duplex systems, and it is interesting to note that even Lord Kelvin failed to see the relation between the duplex requirements and his earlier analysis of cable characteristics when he made a proposal for a differential relay a few years later.[5]

The use of a Wheatstone Bridge as a hybrid circuit was finally proposed by Maron in 1863, and the final step, the introduction of a proper balancing network or artificial line, appears to have been made independantly by various workers round about 1870. These included J. B. Stearn in America and A. Muirhead in England.

In the meantime, other engineers had been trying to develop multiplex systems that would allow simultaneous transmission of a number of signals over one line in the same direction. We now know that the basis of any multiplexing scheme is to modulate a set of orthogonal waveforms by the signals to be transmitted, and that there are a number of different sets of orthogonal signals that are suitable for this purpose. However, at a time when all signalling was by simple binary codes, the only arrangement that occurred to workers in this field was time-division multiplexing. The basis of this method is shown in Fig. 1.9a. At the transmitting terminal a rotary switch connects the various signal paths to the transmission line in turn, and at the receiving end a similar switch, running in synchronism, routes the signals to the appropriate outgoing lines. The first system of this type was proposed in 1853 by an American, M. B. Farmer.[12]

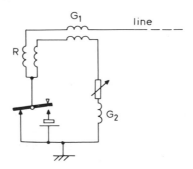

Fig. 1.8 Newall's duplex system
R = differential relay
G_1 = differential galvanometer
G_2 = simple galvanometer

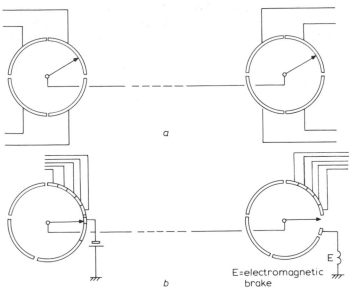

Fig. 1.9 Multiplex arrangements
   a Basic multiplex
   b Baudot system

There are a number of interesting and important principles involved in multi-plexing, the first of these being the question of synchronisation. There are in fact two approaches to this. In the first approach, the two rotating switches are initially synchronised by some agreed setting up procedure, and then, every time a signal is received, a correction is made to the position of the receiving commutator. Clearly, so long as the rate of drift of the receiving switch relative to the transmitter is slow, and there are no long gaps in the transmission, this scheme will be adequate to keep the two terminals in synchronisation. In modern terminology this is known as digit synchronisation, and its important feature is that it uses the timing information contained in the transmitted signal with no need for any additions or alterations to this signal. The first application of this type of synchronisation appears to have been made by D. E. Hughes in the printing telegraph which he invented in 1855.[12]

The second approach involves the insertion of a fixed synchronising signal once during each revolution of the switch. One of the earliest schemes of this type was introduced by J. M. Baudot, a young engineer in the French telegraph service, in 1875. Baudot's scheme, as applied to a 4-channel system, is shown in Fig. 1.9*b*. The transmitter consists of a rotary switch which scans four groups, each of five segments, corresponding to the four transmitted channels. These are followed by a 21st contact which is connected permanently to the battery, and then a final gap. Thus, after each group of 20 signal elements a fixed framing digit is transmitted. At the receiver there is a similar rotary switch with one difference; the positions of the gap and the 21st contact are reversed. The receiving synchronising contact is connected to an electromagnetic brake on the drive shaft to the switch, and the motor driving this switch is arranged to run very slightly faster than the corresponding motor at the transmitter. Thus, if the two rotating wiper contacts are initially in synchronism, the receiver tends to advance slightly on the transmitter until its rotating contact makes contact with the synchronising segment while the synchronising digit is still being transmitted. This energises the brake, and thus holds the receiver in synchronism with the transmitter. Baudot's scheme was therefore the direct ancestor of the frame-alignment arrangements used in modern digital transmission systems.

The use of groups of five segments for each channel in the Baudot system also deserves comment, since this involves another basic problem in synchronising schemes. The problem here is the need for a further level of synchronisation between the multiplex switch and the source of the digital signals. If we imagine each of the fixed-switch segments in Fig. 1.9*a* to be connected back to a telegraph key, we can see that it is highly desirable to operate this key in synchronism with the switch. Otherwise, changes in the keyed signal could occur just before the wiper loses contact with the corresponding segment, resulting in a badly distorted signal.

Baudot's solution was to use a fixed 5-unit code rather than a code of the Morse type, and to combine the coding with the multiplexing. The operators on a Baudot system used a 5-key transmitter, and were given an audible warning shortly before the wiper contact reached their segment of the switch. They then had to press the appropriate selection of keys, and a locking system then held these down until after transmission had been completed. Thus, the Baudot system also employed a crude form of feedback synchronisation to the message source, coupled with buffer storage to smooth out the residual timing irregularities. Similar features can be recognised in modern systems.

1.5 Consolidation — impact of the telephone

By 1872 most of the basic techniques of digital transmission had been discovered, and had been put into service in one form or another. In that year, Alexander Graham Bell, a young Scotsman who had recently emigrated to America, started work on an 'harmonic telegraph'. If he had completed this project it would have probably led to the first practical frequency-division-multiplex telegraphy system. In fact, various discoveries changed the course of his investigations and led to the invention of the telephone in 1876. As a means of communication the telephone was fast, personal and convenient. It needed no training in the use of codes and so made electrical communications directly accessible to the general public.

From about 1880, the telephone began to dominate the development of communications. Now telephony involves the transmission of analogue signals, and when a practical amplifying device appeared in the form of the thermionic valve this also proved to be suitable for dealing with analogue signals. Hence, after 1880, the developing communication networks were basically designed to handle analogue transmissions, and to an increasing extent digital transmission in the form of telegraphy had to be adapted to fit in with the characteristics of these networks. By 1950 the world's communication systems were almost entirely based on analogue transmission, the only exceptions being long submarine telegraph circuits and some radio systems.

However, the impact of telephony on digital techniques was not entirely detrimental. Telephony signals, being broadband, and being sensitive to the distortions caused by the frequency limitations of the transmission paths, stimulated interest in this area, and in time led to a clear understanding of the mechanisms involved in transmission distortion. This was applied to the old problems of telegraphy transmission originally studied by Kelvin, and in 1928 a fairly complete account of the fundamental theory of digital transmission was finally presented by Nyquist.[13] In parallel, an understanding of the significance of noise had been growing, and these two subjects were finally linked by Shannon.[14] A vast amount of technological support in the form of circuit theory, design methods for filters, amplifiers, transmission lines and so on was also appearing. It is impossible, in a short account, to deal with all of these, and so we must restrict our survey of work in the 20th century to a few selected topics.

1.6 Discovery of noise and crosstalk

The effects of lightning had been recognised since the earliest days of telegraphy, and when telegraph systems began to be used fairly extensively it soon became apparent that other natural phenomena, notably the aurora borealis, were also sources of interference. It was found that on long lines, and in particular on the long submarine cables, slowly fluctuating voltages appeared, and these were attributed to variations in the earth's magnetic field. (See Reference 15, pp. 97–99 for an account of early records of interference.)

These effects, however, were not considered to be a serious nuisance; lightning is not very frequent outside the tropics, and low-level signals did not seriously impair binary transmission. The advent of the telephone rapidly changed this situation. Not only was telephonic communication much more seriously disturbed by low-level interference, but the telephone receivers were considerably more sensitive than previous devices and made the effects of noise much more apparent. At the same time, electrical-power systems were beginning to be installed and man-made interference was becoming evident. The situation is illustrated by comments made at that time by W. H. Preece, the chief engineer of the British Post Office. (It is interesting to remember that these remarks were made a few years before the discovery of radio transmission.) We quote: 'Its (the telephone's) delicacy has detected the presence of current in wires contiguous to wires conveying currents, which have always been suspected, but have been evident only on wires running side by side for several miles. . . . There are other disturbing elements that are peculiar. Earth currents, which are always present in the wires, produce a peculiar crackling noise, similar to that produced by a single current battery, such as a Smee or Leclanché, not unlike the rushing of broken water. . . . When auroras are present these earth currents become very powerful and the sounds are much intensified. The effects of thunderstorms are very peculiar: a flash of lightning even though so distant as to be out of sight will still produce a sound; and if it be near enough to be only sheet lightning, it produces, according to Dr. Channing, of Providence, a sound like the quenching of a drop of molten metal in water, or the sound of a distant rocket. . . . Sometimes a peculiar wailing sound is heard, which an imaginative correspondent of mine likened to "the hungry cry of newly hatched birds in a nest".'

However, at that time it was generally thought that these were not fundamental limitations and that any transmission could always be improved by better screening and more sensitive detectors. It was only after about 1920 when high-gain valve amplifiers were becoming available that it began to be realised that there was some fundamental basic noise level that was unaffected by any attempts at screening, and this was, in 1928, put on a firm basis in a classic paper by H. Nyquist.[16] While the implications of this in terms of simple amplitude modulation or analogue baseband signals were obvious, there was at the time considerable confusion as far as more complex transmission schemes were concerned. It was at first thought that techniques such as frequency modulation succeeded in circumnavigating the noise limitation, and then, incorrectly, that they offered no advantages. It was slowly realised that there was a trade-off between noise and bandwidth, and this was put on a formal theoretical basis by the work of C. E. Shannon[14] on information theory. In the period 1940—50 a somewhat similar situation arose in the digital-transmission field, and it is only in the past twenty years that the relative merits of the various forms of digital modulation have been generally appreciated.

In parallel with this, awareness of the effects of interference between separate communication channels had been growing. The name 'cross-talk' clearly indicates that this topic originated in the consideration of coupling between audio frequency telephony channels. These couplings mainly occurred in the transmission lines, and the idea of deliberately modifying the lines to balance out this coupling seems to have occurred at a fairly early stage.[17] When paper insulated multipair cables were developed in about 1887[3] similar coupling problems appeared which were solved by controlling the stranding lay (in every-day terms, the rate of twist) of the pairs. The design of cables with low interpair crosstalk became a major preoccupation in the cable industry, but since these were designed primarily for use on audio--frequency analogue telephony systems, the fact that the crosstalk increased with frequency was not a major limitation. As we shall see later, this was a highly significant factor which has had an important influence on more recent developments in digital transmission.

When multichannel carrier systems were introduced some use was made of pair cables at higher frequencies, but the cross-talk difficulties led to a switch to coaxial cables which were specifically designed to eliminate this type of degradation. This step is now also affecting digital-transmission developments. Crosstalk, in the digital sense, has only become of major interest in the past 20 years, and most of the work in this field is recent.

1.7 Information theory and modern ideas on coding

The problems posed by the relative performance of different modulation schemes were probably the stimulus that led to further ideas on the fundamentals of communications. The first step was taken in 1928 when R. V. L. Hartley[18] published a paper on the theory of communication. The problem that Hartley was concerned with was the quantitative measurement of information. He proposed that the information content of a signal should be defined as the logarithm of the number of distinguishable values in the signal. He was then able to show that the information capacity of a channel, measured in this way, was proportional to the product of the bandwidth and time. However, this still left a lot of questions unanswered. For example, the constant of proportionality was not fixed, the role of noise was not clear, and most important, while Hartley's definition covered the case of transmission of a set of abstract symbols, it did not show how the information content of human languages and speech could be measured. All of these problems were tackled by Claude Shannon in his famous work[14] published in 1948. This included a number of revolutionary innovations. First, Shannon equated information with the reduction of uncertainty, and was thus able to measure the information content of language on a statistical basis. Put crudely, the recipient of a message can usually make a guess at what is coming next from the context of the message he has received up to the present, and the amount of information he obtains from the next symbol is dependent on its unexpectedness.

In mathematical terms, the information content of a symbol i is inversely proportional to its probability of transmission p_i, this led to Shannon's expression for the information content of a message of n symbols as

$$H = - \sum_{i=1}^{n} \log p_i \text{ bits} \tag{1.5}$$

He then went on to apply this approach to the noisy channel in which reception of a certain symbol only implied a probability that a specified symbol had been transmitted. This led to his famous expression for the information capacity of a noisy channel.

$$C = w \log (1 + S/N) \quad \text{bits/s} \tag{1.6}$$

where w is the bandwidth, S is the signal power and N is the noise power. Shannon then considered the possibility of correcting transmission errors to achieve error-free communication. In the course of this he produced two surprising and unexpected results. First, he showed that bandwidth is not a fundamental limitation, and that, providing an adequate signal/noise ratio can be achieved it is possible to transmit information at any required rate over an arbitrarily narrow band channel. Secondly, provided the attempted transmission rate is less than the limit set by the channel capacity, it is always possible to achieve perfect error-free transmission by means of some suitable coding scheme.

Shannon's work did not show how this could be achieved, he merely showed that it was possible. From the point of view of digital transmission a significant feature was that he achieved his results by expressing signals in a digital format, and that, in the course of doing this, he gave a clear insight into the relationship between analogue signals and digital signals.

An immediate consequence was renewed interest in digital coding schemes, in particular, schemes involving codes with error-detecting and error-correcting properties that might approach the theoretical performance that Shannon predicted. The first major contribution in this area was published by R. W. Hamming two years later.[19]

1.8 Digital transmission of analogue signals

In a sense, some of the earliest experiments in telephony were based on digital transmission. For example, the telephone constructed by Reis in 1861 employed a platinum make-and-break contact coupled to a diaphragm.[20] As we now know, such a device would produce dicotomised speech which, even if the instrument was working perfectly, would be severely distorted; and this limitation was the basic cause of the failure of these early schemes. After the invention of a successful analogue telephone developments started to follow the same path as that taken by telegraphy some 40 years earlier, and attempts were made to improve the utilisation of telephone lines by simultaneous transmission of more than one telephony signal. Because this work was taking place soon after the introduction of the highly successful Baudot telegraph system, it was natural that some sort of time-division-multiplex system should be the most obvious approach, and various attempts to achieve this were made. The workers at that time did not appreciate the relation between the frequency spectrum of the speech signals and the required sampling rates, and so they tried to use low-speed multiplexers similar to those employed in telegraphy. The realisation that a high sampling speed was required was probably made by Miner[21] in 1903. He employed a high-speed commutator to sample telephony signals at a rate of about 4 KHz, and reported that this gave intelligible speech. His work represented the first successful demonstration of pulse-amplitude modulation. This was, of course, only a half-way step to digital encoding of speech, since a true digital signal would have involved quantisation in both time and amplitude.

20 years later the development of pulse-duration modulation followed a somewhat similar pattern. The inventor in this case was R. A. Heising,[22] and like Miner he was not directly interested in digital transmission; his objective was to use pulse-duration modulation to improve the power efficiency of radio amplifiers. The key steps in the invention of true digital encoding of analogue signals were finally made by A. Reeves in the period 1936–37.[11,23] His first contribution was to discover yet another of the basic forms of pulse modulation, pulse-time modulation. In this, the information is transmitted by the time of occurrence of the pulse relative to some fixed time scale; see Fig. 1.10. (Note, however, that the telegraphy equivalent of p.t.m. had in fact been developed some 80 years earlier in the Hughes printing telegraph.) Reeves' next, and crucial, step was to recognise that by quantising the remaining analogue dimension of the signal, and expressing this in a binary code, he would achieve high noise immunity. This led immediately to his invention of pulse-code modulation. As far as European workers were concerned, further developments were halted soon after this by the war of 1939–45, and as a result the refinement of Reeves' basic ideas took place in America. In 1947 workers at the Bell Telephone Laboratories[24] published the first account of a fully operational p.c.m. system. At about the same time, Deloraine in France[25] invented delta modulation (the counterpart of the differential coding used in the Wheatstone ABC telegraph) and Grieg[26] described pulse-count modulation. This covered virtually all the basic forms of analogue/digital conversion, and work since then has been mainly concerned with 2nd-order schemes involving combinations of the basic approaches. Of these, the most important is differential pulse-code modulation, which combines p.c.m. and delta modulation. This was first described by Cutler[27] in 1950.

1.9 New transmission media

Up to the end of the 19th century, all practical electrical communication had been based on the use of metallic transmission lines. The introduction of radio at the beginning of the 20th century (it is not fair to describe this as an invention, since it represented an accumulation of inventions and discoveries by a number of workers, that were finally merged into a practical system by Marconi) was the first major innovation. Radio followed much the same path of development as the earlier line systems. The first radio systems were simple telegraph schemes, and when an effective method of modulation was invented radio telephony began to become the dominant mode of transmission. One significant difference was that the early radio systems were broadcast systems; they were essentially omnidirectional. As radio developed it became apparent that directional transmission was possible, but this involved aerial systems with dimensions comparable to the wavelength of the signals being transmitted; and this was therefore one of the factors that encouraged the use of higher and higher operating frequencies. By the 1930s frequencies above 100 MHz were being investigated, and it is very interesting to note that, in his account

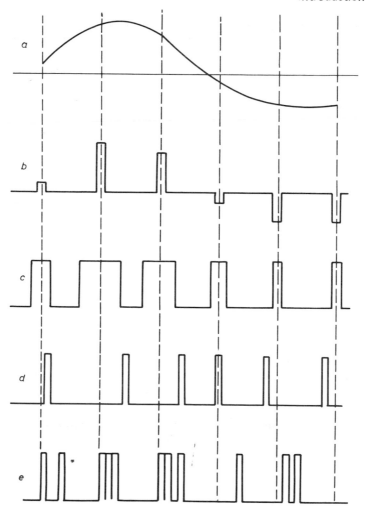

Fig. 1.10 Types of pulse modulation
    a Modulating signal
    b Pulse-amplitude modulation
    c Pulse-duration modulation
    d Pulse-position modulation
    e Pulse-code modulation

of the invention of p.c.m., Reeves indicated that the work originated in attempts to develop microwave radio systems.

Two other new transmission media have appeared during this century. The first, the waveguide, is in fact a very close relative of the convential transmission line. In 1897, Lord Rayleigh showed that electromagnetic-wave propagation could occur inside a hollow conducting tube, and over the next few years several workers investigated other similar transmission modes (see Lamont[28] for an account of these). This remained a scientific curiosity until 1936 when the problem was re-examined by scientists at the Bell Telephone Laboratories[24] and it was shown that under certain conditions waveguides could provide very low-loss transmission paths. Coupled with this, the high operating frequencies implied that extremely high transmission capacities could be made available. Since that time the waveguide has, as far as transmission has been concerned, fulfilled the role of a solution looking for a problem. Considerable research and development has been carried out and the general form of high-capacity waveguide systems has been established (see for example IEE Conference Publication 71 dealing with this.) If introduced these systems would almost certainly use digital techniques. However, the combination of demand and economic factors that would finally bring about the introduction of waveguide transmission have never occurred, and there now seems to be a possibility that the electrical waveguide will be superceded by the optical waveguide or fibre transmission system.

Since light is simply a higher-frequency form of electromagnetic radiation it had been obvious that the transmission modes discovered for waveguide transmission were in theory applicable to light. There was, however, one snag. To launch the required transmission mode into a waveguide it is necessary to have a controlled coherent signal source, and no such source was available at optical frequencies. The technological breakthrough occurred in the invention of the laser.[30] This immediately led to renewed interest in optical transmission, and within a few years low-loss glass fibres, which were potentially suitable for long distance optical transmission, had been produced.[31] As in the case of the microwave waveguide, digital transmission would be used. A great deal of work is now going on in this area, and although a number of major problems remain to be solved, it seems highly probable that optical transmission systems will be developed within the next few years.

1.10 Introduction of p.c.m.

As we discussed in Section 1.8, by 1947 the fundamental problems of pulse-code modulation had been largely solved, but another 15 years were to elapse before a commercial system (the American T1 system) was actually introduced into a communication network. This delay was caused by the same two factors that had determined the time of introduction of the first telegraph systems 120 years earlier, namely, the need for the system and the availability of an adequate supporting technology. The technological gap was the lack of a suitable low-power switching device. As previously mentioned, thermionic valves can be used as switching elements, but, in general, they are not very good switches, and they consume a lot of power. As a result, a pulse-code-modulation system based on 1947 technology was large, rather unreliable, and tended to get very hot. In fact, the key invention that altered this situation was made in the same research laboratories at about the time that the 1947 p.c.m. system was being constructed. This was the invention of the transistor. Development of the transistor took about another 10 years, but by 1957 it provided a near ideal switching device; small, very fast, reliable and with very low power consumption.

By this time a need for a digital transmission system had also appeared. This requires some explanation since the factors involved were rather complex. The primary requirements stemmed from the steady growth in the use of the telephone and the consequent increase in the loading on the cable networks, in particular, in the larger cities. Now, in theory, this problem could have been solved simply by installing more cables, but in many cities the available space under the streets was already fairly fully occupied by other services (water, gas, sewers etc.) and, in any case, the dislocation caused by widespread digging up of the streets was not to be contemplated. The next possibility was to increase the capacity of the existing cables by adapting them for multichannel carrier operation, and, in fact, this was tried on a limited scale.[32] The difficulty here was in the characteristics of the cables. As previously mentioned, these cables had been primarily designed for audio working, and at higher frequencies suffered from severe crosstalk and noise limitations. It was realised that digital transmission in the form of p.c.m. offered a solution to these problems and the development of p.c.m. systems specifically for use on these junction cable routes was initiated. This led to the first installations some five years later.[33] While the basic system design was surprisingly similar to the original 1947 scheme one innovation deserves mention. This was the use of the ternary line code. This topic will be discussed in detail in later Chapters, and for the present we will merely remark that this offered the advantages of a line signal with no d.c. component, with error-detecting capabilities and with a rather more favourable energy spectrum than that offered by a 2-level binary code. The T1 system proved to be highly successful, and junction p.c.m. systems of this general type are now in widespread use throughout the world. This, in fact, represents the starting point for modern digital communications.

1.11 Present position

The introduction of junction p.c.m. has been the turning point in the application of digital transmission techniques. These junction systems have proved to be highly successful, and are now used in large numbers. Their existence justifies the use of other digital schemes, in particular, the introduction of high-capacity, long-haul digital transmission systems, and at present, a variety of high-capacity systems are being developed for use on coaxial cables, on microwave radio channels, on waveguides and on optical fibres.

These systems are complex and highly sophisticated, and to fully appreciate their implications it is necessary to study a wide range of the supporting technologies. For example, we have to consider the nature of the signals being transmitted, and their sensitivity to various forms of impairment. We have to consider the processing of these signals before transmission, that is, the operations of coding and multiplexing. We must obviously study the transmission media, and consider any peculiarities of these media that are relevant to digital systems, and we must then find out how to design transmission equipment to match the characteristics of these media. It must also be remembered that, having built a system, it is desirable to be able to keep it running, and so the requirements for repair and fault location must not be overlooked. In all of these areas the current interest in digital techniques has led to a vast increase in activity, and technical papers on these topics appear almost continually.

In the remainder of this book we will attempt to present an account of the present situation, although it must be admitted from the outset that a comprehensive coverage of the field is impossible. We have therefore tended to concentrate on the areas where we feel that new significant information has become available, and we will give only brief mention to areas where we believe that no significant problems have been raised by the introduction of digital transmission, or where, as in the case of optical fibres, we feel that as yet insufficient information is available to make it possible to give a realistic assessment of the position.

1.12 References

1 BENNETT, W.R., and DAVEY, J.R.: 'Data transmission' (McGraw-Hill, 1965)
2 GARRATT, G.W.R.: 'The early history of telegraphy', *Philips Tech. Rev.,* 1965, **26,** pp. 265–268
3 DUNSHEATH, P.: 'A history of electrical engineering' (Faber, 1962)
4 WHEATSTONE, C.: *Phil. Trans. R. Soc.,* 1834, **124,** p. 583
5 MARLAND, E.A.: 'Early electrical communication' (Abelard Schuman, 1964)
6 GRIVET, P.: 'The physics of transmission lines at high and very high frequencies' (Academic Press, 1970) vol. 1
7 THOMSON, W.: 'On the theory of the electric telegraph', *Phil. Trans. R. Soc.,* May 1855
8 MAXWELL, J.C.: 'A dynamic theory of the electromagnetic field', *Philos. Trans. R. Soc.,* 1865
9 HEAVISIDE, O.: 'Electrical papers' 1892, vol. 2
10 FLEMING, J.A.: 'Fifty years of electricity' (The Wireless Press, 1921)
11 CATTERMOLE, K.W.: 'Principles of pulse code modulation' (Illiffe, 1969)
12 STONE, A.E.: 'Text book of telegraphy' (MacMillan, 1928)
13 NYQUIST, H.: 'Certain topics in telegraph transmission theory', *Trans. AIEE,* 1928, **47,** pp. 617–644
14 SHANNON, C.E.: 'A mathematical theory of communication', *Bell Syst. Tech. J.,* July and Oct. 1948, **27,** pp. 379–423 and 623–656
15 STILL, A.: 'Communication through the ages' (Murray Hill Books, New York, 1946)
16 NYQUIST, H.: 'Thermal agitation of electricity in conductors', *Phys. Rev.,* 1928, **32,** pp. 110–113
17 HUGHES, D.E.: *J. Soc. Telegraph Eng.,* 1879, **8,** p. 163
18 HARTLEY, R.V.L.: 'Transmission of information', *Bell Syst. Tech. J.,* 1928, **7,** pp. 535–563
19 HAMMING, R.W.: 'Error detecting and error correcting codes', *ibid.,* 1950, **26,** pp. 147–160
20 VON URBANITZKY, A.R.: 'Electricity in the service of man' (Cassel, 1890)

21 MINER, W.M.: US Patent no. 745734, 1903
22 HEISING, R.A.: US Patent no. 1655543, 1924
23 REEVES, A.H.: 'The past, present, and future of p.c.m.', *IEEE Spectrum*, May 1965
24 BLACK, H.S., and EDSON, J.O.: 'Pulse code modulation', *Trans. AIEE*, 1947, 66, pp. 895–899
25 DELORAINE, E.M.: 'Pulse modulation', *Proc. IRE*, 1949, 37, pp. 702–705
26 GRIEG, D.D.: 'Pulse count modulation', *Electr. Commun.*, 1947, 24, pp. 287–296
27 CUTLER, C.C.: US Patent no. 26055361, 1950
28 LAMONT, H.R.L.: 'Waveguides' (Methuen, 1947)
29 CARSON, J.R., MEAD, S.P., and SCHELKUNOFF, S.A.: 'Hyper-frequency waveguides – mathematical theory', *Bell Syst. Tech. J.*, 1936, 15, pp. 310–330
30 FISHLOCK D.: 'A guide to the laser' (MacDonald, 1967)
31 KAPRON, F.P., KECK, D.B., and MAURER, R.D.: 'Radiation losses in glass optical waveguides' *in* 'Trunk telecommunications by guided waves', IEE Conf. Publ. 71, 1970
32 VINES, J.C., and ENDERSBY, J.C.: 'Twelve circuit carrier on deloaded audio cable' Proceeding of the conference on 'Transmission aspects of communication networks', IEE, London, Feb. 1964, pp. 52–55
33 DAVIS, C.G.: 'An experimental pulse code modulation system for short haul trunks', *Bell Syst. Tech. J.*, 1962, 41, pp. 1–24

Chapter 2

Digital-system impairments

2.1 Sources of impairment

It may seem strange to the reader that, in a work purporting to deal with transmission, he should find an early Chapter largely concerned with the peculiarities of analogue/digital conversion. Would it not have sufficed to concentrate on the transmission channel and its effects on the signals transmitted by it? We feel that such a limited approach would ignore the distinguishing features of digital systems, and that, in fact, it is impossible to design a digital transmission system without proper regard to its end-to-end performance. Before proceeding, it is appropriate to consider the object of communication, because 'performance' must be judged against some objective criterion. An excellent appraisal, due to Fano,[1] may be condensed to the statement that the object of communication is to make the information source available to the user, within some degree of acceptability. In turn, acceptability will be determined by the impairment suffered by the information, and therefore any communication system must be designed to work within specified acceptable impairment levels. The impairment allocation for the transmission channel, as such, will therefore be dependent on the other sources of impairment in the complete system.

We further justify our treatment by observing that most information sources are inherently analogue, and hence require either actual or implied conversion to digital form. Examples of such analogue sources are speech, music, picture and radar signals and source data for telemetry. One of the few examples not requiring direct digital conversion by electrical means is telegraphy although, in fact, the conversion is effected by the operator in encoding the written message by means of the teletype machine. This may be regarded as a unique communication system which allows the luxury of using a human operator as part of the analogue/digital conversion process.

Our first task is to identify the sources of impairment present in a digital transmission system conveying information from an analogue source. To do this, it is advisable to examine the constituent parts of a practical digital system, and then to reduce them to a model preserving only the essential features. Fig. 2.1*a* is a functional description of the system, but it is not sufficiently penetrating to be of use; this must be decomposed to show the constituent parts of an actual system, Fig. 2.1*b*. The analogue signal is first low-pass filtered to remove energy above the nominal pass band. The signal is then sampled at a rate sufficient for good reproduction, an aspect which we shall consider a little later. The samples are then held and presented to the analogue/digital coder, whose function it is to assign a coded binary number corresponding to the nearest permitted amplitude or quantising level. The binary digits are then coded into a form suitable for transmission over the channel, which will add random noise to the signal. This, as we shall see, is the mechanism responsible for errors in the binary signal at the output of the channel decoder. The output of the following digital/analogue convertor is a quantised version of the analogue signal which is then resampled and filtered to produce a received replica of the signal. It is important to discriminate at this stage between impairments caused by imperfect implementation and those inherent *de jure*. Let us

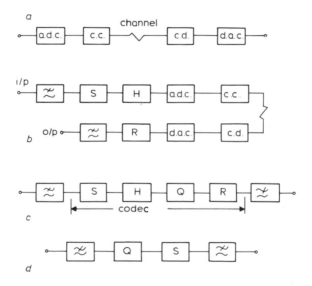

Fig. 2.1 Development of transmission-system model
    a Functional description
    b Actual system
    c Terminals linked by ideal channel
    d Basic model
    a.d.c. = analogue/digital convertor
    d.a.c. = digital/analogue convertor
    c.c.   = channel coder
    S   = sampler
    R   = resampler
    c.d.   = channel decoder
    H   = hold circuit
    Q   = quantiser

first eliminate the channel by labelling it perfect; this leads to Fig 2.1*c* in which we have isolated the complete digital coder and decoder into a single unit normally referred to as a 'codec', part of which, the quantiser Q, is the result of telescoping the a.d.c. and the d.a.c. Finally, if we assume that the sampler and resampler are perfectly synchronised then we can dispense with one, leading to the skeletal model of Fig. 2.1*d*.

The above reduction has finally condensed a rather complex looking chain of operations into the two basic processes of quantisation and sampling in the codec. It is these two operations, inherent in the transmission of an analogue signal over a digital system, which we shall consider in the following two Sections.

Two further mechanisms responsible for overall impairment may be easily identified from Fig. 2.1*b*. These may be attributed to the channel which we have assumed to be ideal in considering the terminal impairments. First, there will be an impairment due to random errors at the binary output of the channel decoder, caused by random noise present at various low-level signal points along the channel. The effect of these errors is translated by the digital/analogue convertor into erroneous random jumps in the quantised signal causing the final analogue signal to deviate significantly from the original transmitted signal. Although these deviations occur at a rate equal to the system error rate and may therefore be quite rare, nevertheless, their magnitude is large whenever the most significant binary digits of a code word are affected. In a digital system carrying time-division-multiplexed speech these errors give rise to received clicks, the impairment being then referred to as 'click noise'. We prefer to call this impairment error distortion, because in other systems, such as digitally transmitted f.d.m., the subjective effect is totally different.

A second impairment will arise when the receiver resampling operation is not ideally synchronised to that at the transmitter. In most practical systems, synchronisation is derived from the received signal itself and, as we shall see later, cannot be made perfect. The result is that, taking the timing of the signal samples at the transmitter as a reference, those at the receiver will slightly deviate, or jitter, in their timing from the reference. The effect of this on the final analogue signal may be loosely described as a phase modulation causing a certain amount of distortion.

In summary, we identify two major classes of possible impairment in digital systems transmitting analogue signals, attributable to either the terminal or the channel

  (i) terminal effects — quantising
                    sampling
  (ii) channel effects — random errors
                 jitter.

The terminal effects may be described as *de jure,* and the channel effects as *de facto.*

2.2 Quantisation effects

A quantiser of the simplest linear type has an output-input characteristic in the form of a uniform staircase; the output jumps by one riser or quantising step q when the input changes by one tread. It should be noted that the definition of this functional block has not involved time as a parameter. The output is thus a staircase approximation of the input and, depending on the fineness of the quantiser characteristic, will be a near replica of the input within an error term often referred to in the engineering literature as quantising noise. In fact, the description of quantising error as 'noise' is highly inappropriate because it has several features quite uncharacteristic of a true noise signal. First, the quantising error is causally related to, and dependent on, the input; furthermore, the quantising error at any point may always be predicted from the input. Secondly, it may be argued that the quantiser characteristic is simply a special case of a high-order nonlinearity and, as such, the error produced may be likened to high-order distortion products. We shall therefore speak of quantising distortion rather than quantising noise in our treatment.

2.2.1 Three regions of a quantiser

Let us consider an input signal, such as a sine wave, applied to a quantiser with M permitted output levels. As long as the signal lies within the quantiser permitted range of $-Mq/2$ to $+Mq/2$ then normal quantisation, as described above, will be obtained. If, however, the input signal exceeds this range the quantiser output will remain at the maximum permitted level of $\pm Mq/2$. (It may, in some designs, behave quite anomalously and, for instance, switch to the other extreme level.) This is synonymous with a very abrupt overload point and will be referred to as clipping. The control of signal levels with respect to the clipping level is thus of far more importance in digital systems than in analogue systems, in which the equivalent overload point is less well defined.

Finally, a third feature or region of quantiser performance is reached as the input is progressively reduced. As the number of quantising steps energised by the input excursion is reduced, the output deteriorates as a replica of the input, until the two show little mutual resemblance. In the limit, when the input excursion becomes less than one quantum step q two possible extreme conditions occur. If the mean level of the input sinusoid falls on the threshold halfway between adjacent quantiser levels then a dichotomised output will be obtained of peak-to-peak value q, thus causing an enhancement effect. Alternatively, if the input mean falls between adjacent quantiser thresholds then the output will remain constant at one of the permitted levels. These low-level input conditions define the idling region of the quantiser and are illustrated in Fig. 2.2d.

Hence, the quantiser performance may be resolved into three quite disparate features:
the quantising region, the clipping region and the idling region.

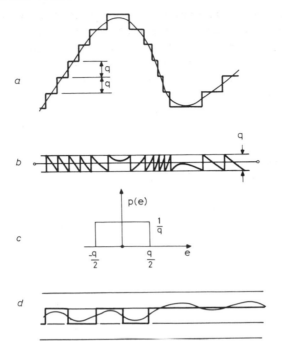

Fig. 2.2 *a* Quantisation of a waveform
   b Quantising error, *e*
   c Probability — density function of *e*
   d Quantiser idling conditions

2.2.2 Distribution and power of quantising distortion

Consider a quantiser excited by an input signal as shown in Fig. 2.2. It is noticeable that even with the coarse quantisation illustrated, the quantising error tends to be sawtooth shaped except in dwell regions of the input such as maxima and points of inflection. Clearly, if the quantisation were made finer this tendency would be even more pronounced. It is intuitively clear that the probability distribution of amplitudes of quantising error $p(e)$ will tend to be uniform; in other words, all amplitude values will be equally likely in the range $-q/2$ to $+q/2$. The probability-density function will thus be uniform as shown in Fig. 2.2c. A more thorough demonstration is given by Widrow.[2] By definition, the power of the quantising distortion D_q is the mean square value σ_e^2 of the quantising error e assuming a unity resistance load

$$D_q = \sigma_e^2 = \int_{-q/2}^{q/2} e^2 \, p(e) \, de = \frac{1}{q} \int_{-q/2}^{q/2} e^2 \, de = \frac{q^2}{12} \tag{2.1}$$

It should be noted that eqn. 2.1 holds for signals which are reasonably random and otherwise well behaved; this excludes signals with frequent dwell regions. At this point one might ask what constitutes a well behaved signal vis-à-vis a quantiser, and at what stage of fineness does a quantiser fail to perform as such. This question has prompted new work which we shall consider in due course. At this stage, it is possible to turn the question by defining the normal quantising region as one where the quantising distortion is $q^2/12$.

Several features of quantising distortion are noteworthy. As we have already implied, it is independent of the input statistics within the stated proviso. We stress that it is the *statistics* of the distortion (i.e. averaged quantities such as variance) which are independent of the *statistics* of the input; the actual input and the distortion are, of course, causally related and not independent, as stated earlier. A further feature is that the quantising-distortion power is constant and depends solely on the quantum step size q.

It is helpful to compute the maximum available signal/quantising distortion ratio (s.d.r.) which occurs when the quantiser is excited by a signal whose peak excursions coincide with the quantiser outermost levels V.

We have, if N is the number of binary code digits,

$$D_q = \frac{q^2}{12} = \frac{1}{12}\left(\frac{2V}{2^N}\right)^2 = \frac{V^2}{3 \times 4^N} \tag{2.2}$$

and, expressed in decibels relative to the peak quantiser level,

$$10 \log \frac{D_q}{V^2} = -(4\cdot77 + 6\cdot02N),\ \mathrm{dB} \tag{2.3}$$

Now, if the signal energy is S,

$$\mathrm{s.d.r.} = S/D_q = \frac{S}{V^2}\,3 \times 4^N \tag{2.4}$$

If the input signal is a sine wave it will fully load the quantiser when its amplitude is V, and $S = V^2/2$.
Under these conditions

s.d.r. = full load sine wave power/D_q
$$= (1\cdot76 + 6\cdot02\,N),\ \mathrm{dB} \tag{2.5}$$

2.2.3 Power of quantiser clipping distortion
If $+V$ and $-V$ are the outermost quantiser levels ($V = Mq/2$), and if the input signal p.d.f. is $p(x)$, then clipping will occur with a probability P_c given by

$$P_c = \mathrm{prob}\,(|X| > V) = 2 \int_V^\infty p(x)\,dx \tag{2.6}$$

We may define a distortion power due to quantiser clipping D_{qc} as the mean of the squared error between output and input in the region where $x > V$

$$D_{qc} = \underset{x > V}{Av} \{(V - x)^2\}$$

$$= 2 \int_V^\infty (V - x)^2 \, p(x) \, dx \tag{2.7}$$

In the above the integral is the averaged quantity while the range of integration defines the clipping region.

We cannot proceed without specifying the source signal p.d.f. If the input signal has noise-like characteristics then the normal or Gaussian p.d.f. is the usual model used. Certain signals which are essentially the sum of a large number of independent sources, are, by the statistical law of large numbers, closely described by the Gaussian distribution. Important examples of the class are f.d.m. multichannel signals,[3] and crosstalk in multipair telephone cable. A Gaussian p.d.f. is described by

$$p(x) = \frac{1}{\sigma\sqrt{2\pi}} \, e^{-x^2/2\sigma^2} \tag{2.8}$$

and

$$\text{prob} \,(x > V) = \frac{1}{\sigma\sqrt{2\pi}} \int_V^\infty e^{-x^2/2\sigma^2} \, dx \tag{2.9}$$

or, by a change of variable,

$$\text{prob} \,(x > V) = \frac{1}{\sqrt{2\pi}} \int_{\frac{V}{\sigma}}^\infty e^{-x^2/2} \, dx \tag{2.10}$$

$$= \tfrac{1}{2} \, \text{erfc} \left(\frac{V}{\sigma}\right) = \bar{\Phi} \left(\frac{V}{\sigma}\right), \text{ say} \tag{2.11}$$

In the above, σ is the root mean square (r.m.s.) value, and σ^2 is synonymous with the signal power in a unit load. Eqn. 2.11 denotes the above probability in a shorthand notation; we shall have occasion to use this function throughout the book. We purposely use the bar sign to distinguish the function from the standard error function which differs in the limits of integration and which normally results in more cumbersome expressions in our context.

Substituting the Gaussian p.d.f. into eqns. 2.6 and 2.7 we obtain

$$P_c = 2 \, \bar{\Phi} \left(\frac{V}{\sigma}\right) \tag{2.12}$$

$$D_{qc} = 2\,\bar{\Phi}\left(\frac{V}{\sigma}\right)(V^2 + \sigma^2) - \sqrt{\frac{2}{\pi}}\ V\,\sigma\,e^{-V^2/2\sigma^2} \tag{2.13}$$

2.2.4 Power of quantiser idling distortion

Under zero input conditions, or for an input whose peak-to-peak value is less than one quantum step q, the quantiser output will be anomalous. We can identify two limiting cases. When the input-signal mean coincides with a threshold level a dichotomised spurious output of peak-to-peak value q will be produced, whose energy is $q^2/4$. This condition will also pertain with no input drive, the spurious output being then generated by thermal noise effects.

When, however, the input resides between adjacent threshold levels without at any time crossing them, in principle, a more favourable 'quiet' idling condition is obtained, with a constant quantiser output. In fact, the ever-present thermal noise will cause random crossings of the thresholds with a probability exactly synonymous with the clipping probability discussed above; this idling probability, P_i say, may be obtained from eqn. 2.12, changing V to $q/2$. The idling distortion power will then be $P_i q^2/4$, and will clearly depend on the ratio of q to r.m.s. noise.

The actual idling performance of the quantiser will obviously lie somewhere between the above two extremes. In practice, system design may circumvent the problem; either the idling region may be avoided by ensuring that a certain minimum excitation or loading is always present, or steps taken to ensure near-zero drift from the quiet idling position.

2.2.5 Quantiser impairment curves

We are now in a position to assess the complete impairment characteristic of a quantiser. This may be expressed conventionally as the behaviour of the s.d.r. with input signal level. We assume that the input distribution is Gaussian (eqns. 2.8–2.11).

From eqn. 2.4 for quantising distortion, we have

$$\text{s.d.r.} = \frac{S}{V^2}\,3 \times 4^N = \frac{\sigma^2}{V^2}\,3 \times 4^N \tag{2.14}$$

or in dB

$$10\log\text{s.d.r.} = 20\log\frac{\sigma}{V} + 4{\cdot}77 + 6{\cdot}02N \tag{2.15}$$

When s.d.r. is plotted against σ/V, the ratio of r.m.s. input signal to peak quantiser level, a series of straight lines is obtained for different N, as shown in Fig. 2.3.

In the clipping region using eqn. 2.13 we obtain

$$\text{s.d.r.} = \frac{\sigma^2}{D_{qc}} = \frac{\sigma^2}{V^2}\left\{2\,\bar{\Phi}\left(\frac{V}{\sigma}\right)\left(1 + \frac{\sigma^2}{V^2}\right) - \sqrt{\frac{2}{\pi}}\frac{\sigma}{V}\exp\left(\frac{-V^2}{2\sigma^2}\right)\right\}^{-1} \tag{2.16}$$

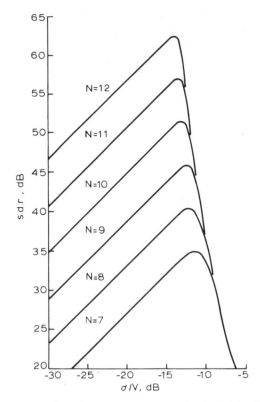

Fig. 2.3 Signal/distortion ratio (s.d.r.) for a linear quantiser loaded with a Gaussian input
σ = r.m.s. input V = chipping level

This is also a function of σ/V, and is again plotted in Fig. 2.3, forming a common boundary with which the quantising distortion lines merge. The region where clipping and quantising distortion have similar magnitudes obviously defines the minimum of overall distortion, which is obtained by power addition of the two types of distortion.

The shape of the quantiser impairment characteristic of Fig. 2.3 has important system implications. Clearly, the minimum defines the desirable input-signal loading condition; in short, the input r.m.s. value σ should be adjusted to be a particular ratio of V. Table 2.1 gives the value of σ/V for minimum distortion for a range of coding precision expressed in terms of binary-code digits.

A further implication must be that the maximum s.d.r. is achievable only with a signal whose power is controllable to a fine degree. This may not be the case in many practical situations. For instance, a digital system whose quantiser input is set to give minimum impairment at a given level will perform significantly worse if the input loading falls; this can happen when an f.d.m. signal goes from busy hour to idle condition. This particular example will be treated in Chapter 4.

Table 2.1 Minimum quantiser distortion against number
of coding bits for a Gaussian input

Coding precision, bits	Input (σ/V), dB	s.d.r.$_{max}$, dB
12	−14·0	62·7
11	−13·7	57·0
10	−13·1	51·5
9	−12·5	46·0
8	−12·0	40·5
7	−11·4	35·0

Input-signal statistics other than Gaussian result in different clipping performance, thus modifying the clipping boundary region and the maxima of Fig. 2.3.
An important signal having non-Gaussian statistics is the telephony speech signal.[4]
Space does not allow consideration of the various complex statistical models which
have been proposed; however, it may be helpful to note here that for a speech
model based on a gamma distribution,[5] the positions of the maxima of s.d.r. given
in Table 2.1 occur at values of σ/V approximately 5dB lower for $N = 7$, and up to
7dB lower for $N = 12$.

2.2.6 Companding
We have assumed throughout that the quantiser is linear; strictly, such a quantiser
gives minimum mean distortion for a signal with a uniform probability-density
function and it performs predictably only with signals whose statistics are reasonably constant (stationary). Unfortunately, the most basic signal, telephony speech,
is in this respect rather badly behaved. It has a large dynamic range, badly defined
average energy and a high probability of low-level intervals. It is obvious that to
ensure adequate rendering over the whole range of speech levels, linear quantisation
of high precision (11-bit coding) would be necessary, including a proportionately
large transmission bandwidth. The alternative approach, which is now the adopted
standard, is to increase the number of quantisation steps in the region around zero
level, and correspondingly decrease the number nearer the extremes of the input
range. This may be achieved by compression and expansion of signal levels before
and after the quantiser, a process known as companding.[4,7,8]
If x is the input, $F(x)$ and $F'(x)$ the compression characteristic and its first
derivative, then the expanded riser corresponding to a tread q, will be approximately $q/F'(x)$. If we decompose the companded quantiser, into a number of linear
quantisers, each contributing a quantising distortion power $q^2/\{12F'^2(x)\}$ then the
resulting total power will be the sum of such contributions, tending to an approximate limit given by

$$D_q \simeq \frac{q^2}{12} \int_X \frac{p(x)}{F'^2(x)} \, dx = \alpha \frac{q^2}{12} \tag{2.17}$$

The choice of $F(x)$ to extend the region over which the s.d.r. S/D_q is reasonably uniform is outside the scope of this book; References 7, 8 and 12 present a comprehensive treatment. It should be added that the detailed form of companding characteristic has now acquired international standardisation[9] in the form of the so-called A-law and μ-law, see Section 4.2. The quantity α in eqn. 2.17 is referred to as the companding improvement. The effect of companding on the quantiser impairment characteristic, Fig. 2.3, is to produce a wide region of σ/V over which s.d.r. is nearly constant, the maximum s.d.r. being, of course, less than the maximum of the peak s.d.r. for a linear quantiser, this representing the incurred penalty of the trade-off. The s.d.r. curve with standard companding will be found in Chapter 4.

Minimisation of the distortion power D_q of eqn. 2.17 is an alternative criterion for quantiser optimisation and is quite meaningful in cases where $p(x)$ is well defined, such as for Gaussian noise-like signals. Procedures for such optimisation are described in References 10 and 11.

2.2.7 Spectra of quantising distortion

Inspection of the quantiser waveform of Fig. 2.2 suggests that its spectrum must contain energy at frequencies many times the highest frequency component of the input signal. Bennett[6] confirms this for the case of a Gaussian input applied to a linear quantiser. For instance, with 8-bit coding and $V/\sigma = 12$ dB, the power spectral density of the quantising distortion is 3 dB down, relative to its maximum d.c. value, at a frequency equal to about 150 input signal bandwidths. The significance of this is that the quantising distortion energy $q^2/12$ is spread fairly evenly over an extremely wide band. However, it must be remembered that this spectrum pertains to the unsampled quantiser, and the implications of sampling will be seen to quite alter the situation.

2.2.8 Quantising theorem and rule

In our earlier discussion of quantisation we encountered and only parried the question of 'proper quantiser action' in regimes where the quantisation becomes rather coarse. A definition of the limits of useful quantiser performance, going beyond intuitive guesses, is obviously desirable; the question has prompted new work leading to the establishment of a quantising theorem,[2] which may be stated as follows:

The statistics of the input signal (mean, variance etc.) may be recovered from the quantised version, if the input-signal statistics obey a simple condition; the characteristic function (Fourier transform of the p.d.f.) must be zero outside the range $\pm 1/2q$. This is the exact analogue of the sampling theorem (Section 2.3) which states that the input signal can be recovered from its sampled version if the input signal bandwidth is zero outside the range $1/2T$. It also emerges that the p.d.f. of the quantising error is then independent of the input.

The quantising theorem offers rather limited practical insight because of the abstract nature of the characteristic function. Surprisingly, however, it may be condensed into a simple rule,[31] if the input has no favourite values, then the quantising theorem is satisfied, providing at least eight quantising steps are used.

2.2.9 Dither
The above strikingly simple rule puts in perspective recent practical work on the reduction of quantising effects by the use of independent 'dither' signals added at the quantiser input and subtracted at the output.[13,14] It may be shown that proper choice of the dither may, in spite of a coarse quantiser or a signal which dwells on favourite values, satisfy the quantising theorem.

2.3 Implications of sampling

Having considered the quantisation process we naturally progress to the sampling process as indicated in the decomposed system block diagram of Fig. 2.1d. In what follows, our intention is to derive the spectrum of the sampled signal which will indicate the relation, known as the sampling theorem, between sampling rate and spectrum of the original signal. Furthermore, we wish to examine the effect of sampling a signal in the presence of an additive noise or distortion component.

Consider a signal $g(t)$ sampled at regular intervals T with ideal, infinitessimally narrow pulses $\delta(t)$. For subsequent convenience we assume each pulse has an area T. The sampled signal $g_s(t)$ may be written as a series of such pulses multiplied by the original signal $g(t)$

$$g_s(t) = \sum_{-\infty}^{\infty} g(t)\, \delta(t - kT) \tag{2.18}$$

Now, the train of sampling pulses each of area T may be shown, via some Fourier series analysis,[26] to be

$$\sum_{-\infty}^{\infty} \delta(t - kT) = \sum_{-\infty}^{\infty} e^{jk2\pi t/T} \tag{2.19}$$

Substituting into eqn. 2.18 we have

$$g_s(t) = \sum_{-\infty}^{\infty} g(t)\, e^{jk2\pi t/T} \tag{2.20}$$

Taking the Fourier transform of $g_s(t)$ we obtain its spectrum $G_s(\omega)$;

$$G_s(\omega) = \frac{1}{2\pi} \int_{-\infty}^{\infty} \sum_{-\infty}^{\infty} g(t)\, e^{jk2\pi t/T}\, e^{-j\omega t}\, dt \qquad (2.21)$$

and inverting the order of the summation and integration

$$G_s(\omega) = \frac{1}{2\pi} \sum_{-\infty}^{\infty} G\left(\omega - \frac{2\pi k}{T}\right) \qquad (2.22)$$

where $G(\omega)$ is the spectrum of $g(t)$. This may be rewritten as

$$G_s(f) = \sum_{-\infty}^{\infty} G\left(f - \frac{k}{T}\right) \qquad (2.23)$$

The sampled spectrum is thus the superposition of an infinite series of spectra of the original signal centred at frequency intervals $1/T$, as shown in Fig. 2.4*a*. It is clear that recovery of the original signal $G(f)$ may be effected by means of a sharp low-pass filter with a cutoff at W hertz if $G(f)$ is band limited to below W hertz, since every member of the spectral series is identical. Further, a limiting condition will occur when $2W = 1/T$; beyond this value an overlap, known as aliasing, of the highest and lowest components of adjacent spectral members will take place, thus affecting the output of the low-pass filter, see Fig. 2.4. This limiting condition defines the sampling theorem.

2.3.1 Sampling theorem

Statement 1
Given a function whose energy is wholly contained within a band zero to W hertz; let the function be sampled at intervals of T seconds.

Theorem
The original function may be completely recovered if the sampling rate $1/T$ is at least $2W$ hertz. This may then be effected by means of a low-pass filter having rectangular frequency response with a cutoff of W hertz.
In the western literature the sampling theorem has been accredited to Nyquist, whose paper in 1928[15] described the principle. Its first explicit statement and use as a theorem, however, appears in Shannon.[16] Russian literature credits Kotelnikov with the first strict formulation in 1933; the reader should therefore be prepared to meet both Nyquist's and Kotelnikov's sampling theorem and we leave the choice to him.

In fact, the sampling theorem as stated above is excessively restricted in its applicability solely to band-limited signals. It has been recognised for some time[18,17] that complete recovery can also be effected in the case of gradual cutoff of the signal spectrum, provided the amplitude response is symmetric and the phase

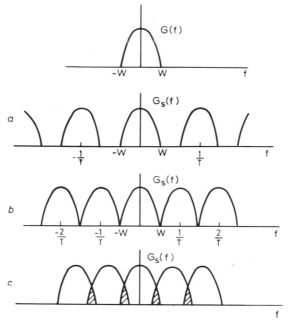

Fig. 2.4 Spectra of sampled signal $G_s(f)$

(a) $\dfrac{1}{T} > 2W$

(b) $\dfrac{1}{T} = 2W$, limiting condition defining sampling theorem

(c) $\dfrac{1}{T} = <2W$, aliasing condition

response linear in the cutoff region. This principle has been extended by Gibby and Smith[19] who have derived conditions relating the real and imaginary parts of the frequency response for gradual cutoff systems.

A separate, although related, extension has been suggested by Kharkievich,[20] stemming from the conceptual difficulty of strict band limitation of a random signal. Kharkievich argues that the sampling theorem should be regarded as exact for deterministic functions, and as approximate for functions for which strict band limiting raises formal difficulties.

Statement 2

Given a function (approximately) band limited to W hertz and existing for a duration of T seconds.

Theorem

The function may be completely described by $2WT$ samples. Such a function is said to have $2WT$ degrees of freedom, and the formulation is the basis of geometric signal-space concepts.[27] (see Chapter 13.) The approximate qualification, in brackets, is inserted to satisfy mathematical rigour; because, by the nature of the Fourier transform, no function can be simultaneously strictly band limited in both frequency and time. An alternative requirement is that the number of samples $2WT$ should be large. It is noteworthy that the latter formulation of the theorem does not prescribe the manner in which the sampling should be performed. In fact, it may be shown[21] that the samples may be taken irregularly, although reconstitution of the original signal then becomes complex.

2.3.2 Inband power in sampled system

Our next concern is to consider the conditions at the output of a sampled and filtered system, the input to which is the sum of a band-limited signal and a spectrally unlimited signal, as shown in Fig. 2.5. The wanted signal $s(t)$ has power S and a well defined spectrum with zero energy beyond W hertz. The unwanted signal $n(t)$ has an undefined spectrum but a known power N, and is independent of $s(t)$. Invoking the sampling theorem, if $s(t)$ is sampled at a rate $2W$ hertz, then a low-pass rectangular filter with cutoff at W hertz will reconstitute $s(t)$ exactly. However, the question arises of the effect of this process on the accompanying unwanted signal; we wish, therefore, to derive the output power, S_0 and N_0, of the two components. It is more convenient to work with power spectra, and eqn. 2.23 may be adapted to this form by replacing all the items by power terms, $P(f) = |G(f)|^2$, and assuming a unity resistance load. We have, at the output of the sampler

$$P_s(f) = \sum_{-\infty}^{\infty} P\left(f - \frac{k}{T}\right) \tag{2.24}$$

$P_s(f)$ and $P(f)$ being now power-density spectra. If it is assumed that $s(t)$ and $n(t)$ are independent, then summation on a power basis holds and we may write

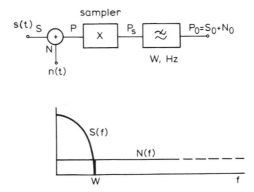

Fig. 2.5 Sampling in the presence of an unrestricted band signal

$$P(f) = S(f) + N(f) \qquad (2.25)$$

The power of the combined sampled and reconstituted signals P_0 is then

$$P_0 = \int_{-W}^{W} P_s(f)\, df$$

$$= \sum_{-\infty}^{\infty} \int_{-W}^{W} P\left(f - \frac{k}{T}\right) df$$

$$= \sum_{-\infty}^{\infty} \int_{-W+\frac{k}{T}}^{W+\frac{k}{T}} P(f)\, df \qquad (2.26)$$

If sampling is performed at the Nyquist rate, $1/T = 2W$, and

$$P_0 = \sum_{-\infty}^{\infty} \int_{W(2k-1)}^{W(2k+1)} P(f)\, df \qquad (2.27)$$

The above expression is a summation of integrated regions each of width $2W$ whose centres are $2W$ apart, and hence the whole of $P(f)$ is just fully covered without overlap. Thus

$$P_0 = \int_{-\infty}^{\infty} P(f)\, df$$

$$= \int_{-\infty}^{\infty} \{S(f) + N(f)\}\, df$$

$$= S + N \qquad (2.28)$$

The above analysis has a very significant outcome. Eqn. 2.28 states that the process of sampling (at $2W$ hertz) and reconstituting the signal by means of a low-pass filter of cutoff W hertz, leaves the power of both the band-limited *and the unlimited-band signal unaffected.* We restate more formally.

Theorem
In a sampled system, for a pass band defined as half the sampling rate, the total power, falling within the pass band is equal to the unsampled signal power, irrespective of the spectral nature of the input signal.

We should point out that the strict equality in eqn. 2.28 and in the above theorem applies as a result of our earlier normalisation of sampling-pulse area to T; in any other case, an appropriate scaling factor other than unity would be introduced, and the equality would then be replaced by a proportionality.

One important practical consideration is highlighted by the above; *it is essential that the input signal to be sampled and coded should be suitably prefiltered,* otherwise all unwanted signals accompanying the wanted input will be aliased at full power into the system pass band. This, in fact, is the *raison d'etre* for the input filter of Fig. 2.1*d.*

2.3.3 Sampling at a rate higher than 2*W*

In certain cases, there may be a requirement for sampling rates greater than the Nyquist rate of twice the system pass band. We wish to analyse the effect of the energy of both the wanted band-limited signal and the unwanted signal after reconstitution. The analysis leading to eqn. 2.26 obviously holds; we then let the sampling rate exceed the Nyquist rate by a positive ratio *r* greater than unity

$$1/T = 2Wr \tag{2.29}$$

Eqn. 2.27 is then modified to

$$P_0 = \sum_{-\infty}^{\infty} \int_{W(2kr-1)}^{W(2kr+1)} P(f)\, df \tag{2.30}$$

The situation may now be interpreted by reference to Fig. 2.6. It is seen that the regions of integration no longer adjoin, but that the wanted signal energy spectrum $S(f)$ is still wholly accommodated within the integrated areas. The unwanted signal energy, is, however, only a fraction of its fully adjoining value. Taking any one frequency interval, $1/T$ wide, this fraction is clearly $2W/2Wr$ or $1/r$. Hence, assuming $N(f)$, the unwanted power spectral density, is uniform

$$P_0 = S + N/r \tag{2.31}$$

We can therefore state that, if the sampling rate is increased to $2rW$, where W is the system pass band, the unwanted signal power falling within the pass band is a fraction $1/r$ of the unsampled input power.

The process of sampling beyond the Nyquist rate obviously results in an improvement of wanted/unwanted signal ratio. Specifically, sampling at double the

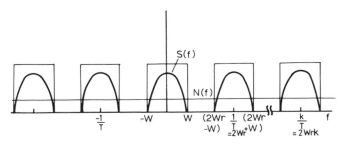

Fig. 2.6 Sampling at a rate higher than 2*W*. Rectangular areas signify regions of integration in eqn. 2.30

Nyquist rate, namely at $4W$ hertz, increases the ratio by 3 dB. Clearly, in most practical situations this is a costly trade-off.

A further complication occurs in cases where the input signal is bandpass rather than baseband. This application of the sampling theorem is covered in Chapter 4.

2.4 Quantisation with sampling

In the preceeding two Sections we have considered the functions of the quantiser and the sampling unit separately. We are now in a position to translate the results to the case, usually met in practice, of both units co-existing, as shown in Fig. 2.1d. We therefore have to consider a quantised signal presented as an input to a sampling unit. We have seen, from Section 2.2, that the quantised signal may be regarded as the wanted input signal plus a distortion component arising from the quantisation. Further, we indicated that the quantising-distortion spectrum is uniform to frequencies many times the system pass band W. This is, therefore, exactly the case treated in Section 2.3 of a band-limited signal with power S, and a spectrally unlimited signal with power N. It follows that we can apply the results, defined by eqn. 2.28 and by the associated theorem, directly to the case of a sampled quantiser. The result is summarised below.

2.4.1 Distortion from sampled quantiser
At the output of a sampled quantiser, for a pass band defined as half the sampling rate, the distortion power falling within the pass band is equal to the distortion power from the quantiser alone. (Remarks made in Section 2.3 concerning normalisation also apply here.)

The above has many implications as regards the results derived in Section 2.2. It means that *expressions for signal/distortion ratios remain valid without any change*, so long as sampling at the Nyquist rate is assumed. In particular, the quantiser-impairment curves shown in Fig. 2.3 can be associated either with an unsampled quantiser or with a quantiser sampled at the Nyquist rate. If sampling is carried out at a higher rate, then the s.d.r. should be simply adjusted as described at the end of the previous Section.

2.4.2 Spectra of quantising distortion in sampled system
In Section 2.2 we argued that the distortion spectrum of a quantiser alone extended to many times the system pass band W. The process of sampling such a spectrum causes the folding or aliasing of seemingly out-of-band unsampled energy back into the pass band, and is described by eqn. 2.24. As, in principle, the number of summed spectral components is infinite, then it may be expected that the resultant sampled spectrum will be uniform even if the unsampled spectrum departs from uniformity. This has, in fact, been demonstrated formally by several workers. The case of a white Gaussian-noise input signal was shown by Bennett[6] to result in a

uniform sampled spectrum within the system pass band. Robertson[22] computed results for input signals having quite highly peaked spectra and with quite small Gaussian components, for arbitrary quantiser laws. His results indicate that the quantising-distortion spectrum remains substantially uniform.

Fujii and Azegami[23] report measurements and predictions of sampled quantiser spectra for single- and multiple-sine-wave inputs. Their results underline the importance of dissociating the nature of quantiser distortion from random-noise effects, particularly with a deterministic input signal. A single sine wave causes a distortion spectrum consisting of discrete high-order intermodulation components, based on the input and sampling frequencies, ranging in level over some 20 dB; two sine waves generated independently cause a spectrum whose discrete components range over some 3 dB. The range for a set of ten independent sine waves is about 1 dB, and the intermodulation harmonics are so numerous as to approximate to a continuous spectrum. This remarkable tendency towards spectral uniformity only holds *if the input signal components are independent.*

If, for instance, the input components are all harmonically related then their intermodulation components are no longer well 'shuffled', being themselves harmonically related. The sampling process then produces a distortion spectrum consisting of high-order intermodulation products based on the fundamental input frequency and the sampling frequency. An important example of this arises in the digital conversion of a broadcast video signal (References 24 and 30) whose spectrum consists of harmonics which are all derived from a basic frequency.

2.5 Impairment due to jitter

The transmission of an analogue signal over a digital system entails a sampling operation at the transmitter and a resampling operation at the receiver, as indicated in Fig. 2.1. Ideally, the sampling and resampling periods would be exactly equal; in practice, a host of mechanisms contribute to make such a situation unlikely, if not impossible. We leave any discussion of these mechanisms to a later Chapter. Here we accept the existence of a nonideal situation in which the resampling instants exhibit small deviations with respect to the sampling instants. The condition is defined by describing the sampling signal $f_s(t)$ and the resampling signal $f_r(t)$ as follows:

$$f_s(t) = \sum \delta(t - nT) \tag{2.32}$$

$$f_r(t) = \sum \delta(t - nT + \epsilon_n) \tag{2.33}$$

where ϵ_n is the temporal deviation of each resampling instant, and is known as 'jitter'. ϵ_n may be regarded as the values of a function $\epsilon(t)$ taken at intervals T. The nature of the jitter $\epsilon(t)$ depends on the configuration of the transmission system; at this stage we must anticipate later results by stating that it is possible to modify $\epsilon(t)$

by suitable design. In particular, the spectrum of $\epsilon(t)$ may be progressively limited at the expense of increasing circuit investment. It is important, therefore, to assess the effect of jitter on the decoded analogue signal to be able to predict the extent of the required circuit hardware.

The actual effect of the jitter becomes noticeable at the output of the filter which follows the resampler, Fig. 2.1. The output filter cannot reconstitute the transmitted signal from irregular samples, and hence the output acquires a distortion component.

In the above, we quite readily mentioned a jitter spectrum, a quantity which, in isolated reflection, presents conceptual difficulties. Because jitter has the dimensions of time, a jitter-amplitude spectrum implies a given value in seconds existing at a given frequency. Worse still, if the jitter $\epsilon(t)$ is a random function for which only a mean-square density spectrum exists analytically,[26] normally associated with power, then we are faced with a quantity referred to as 'a jitter power density spectrum' whose dimensions are [seconds]^2/[seconds] [hertz]. Furthermore, power in this context is strictly a misnomer, since there is no associated work rate. We shall therefore use the term 'jitter mean square density spectrum'.

The derivation of the distortion due to jitter is quite lengthy, and is presented in a major work by Bennett.[25] Bennett analyses the case where the jitter $\epsilon(t)$ has a Gaussian distribution, an assumption which should be examined in the circumstances pertaining to any particular application. The main result of the cited Reference is an expression for the power-density spectrum $W_{sj}(f)$ of the jittered signal in terms of the spectrum $W_s(f)$ of the original signal. We find that

$$W_{sj}(f) \simeq W_s(f)\,(1 - 4\pi^2 f^2 \sigma_j^2) + 4\pi^2 f^2\, W_d(f) \tag{2.34}$$

$$W_d(f) = W_s(f) * W_j(f) + W_s(f) * W_j(f_s + f)$$

$$+ W_s(f) * W_j(f_s - f) \tag{2.35}$$

where $W_j(f)$ — jitter mean square density spectrum

σ_j — r.m.s. value of jitter

f_s — sampling rate

and * denotes a convolution,[26] i.e.

$$x(f) * y(f) = \int x(\lambda)\, y(f + \lambda)\, d\lambda$$

The first term of eqn. 2.34 is the original signal spectrum shaped by a frequency-dependent factor. The second term involves convolution or 'smearing' of the original spectrum with the jitter spectrum. As this process produces spectral terms not originally present, this term constitutes a distortion component. It should be noted that this component involves an f^2 weighting term, thus affecting, with increasing severity, signal components of increasing frequency. This feature is particularly relevant to the case of a multichannel f.d.m. signal. Here the worst sufferers will be channels located near the top of the f.d.m. band; allocation of jitter-impairment allowance must therefore refer to this worst case.

The spectral relationships described by eqns. 2.34 and 2.35 are illustrated in Fig. 2.7. We assume a uniform jitter spectral density $W_j(f) = \sigma_j^2/2f_j$ extending to a frequency f_j, and a uniform signal power spectral density S extending to the maximum frequency $f_s/2$ allowed by the sampling theorem. The three convolution terms of eqn. 2.35 are shown as heavy lines; their sum equals $W_d(f)$ which is the distortion component without the f^2 weighting. We consider the case of a channel of width B hertz at a frequency f_c. The distortion power D_j due to jitter in this channel is the integrated area given by

$$D_j = 4\pi^2 \int_{f_c - \frac{B}{2}}^{f_c + \frac{B}{2}} f^2 \, W_d(f) \, df \qquad (2.36)$$

if $B \ll f_c$, as is usually the case, the integration is confined to the small region centred on f_c, and

$$D_j = 4\pi^2 f_c^2 \int_{f_c - \frac{B}{2}}^{f_c + \frac{B}{2}} W_d(f) \, df \qquad (2.37)$$

With reference to Fig. 2.7

$$D_j = 4\pi^2 f_c^2 \, (\text{area } 1 + \text{area } 2)$$

$$= 4\pi^2 f_c^2 \left(2 + \frac{2f_c - f_s}{4f_j}\right) S \, \sigma_j^2 B \qquad (2.38)$$

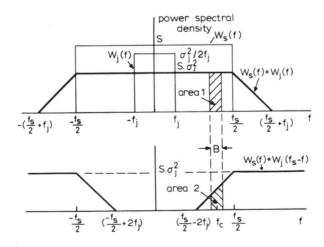

Fig. 2.7 Jitter spectral components

For a channel near the top of the signal band, for which $(2f_c - f_s) \ll 4f_j$, this reduces to

$$D_j = 8\pi^2 f_c^2 S \sigma_j^2 B \tag{2.39}$$

The signal/distortion ratio (s.d.r.) is therefore

$$\text{s.d.r.} = Sf_s/D_j = f_s/(8\pi^2 f_c^2 B \sigma_j^2) \tag{2.40}$$

The worst distortion D_{jmax} will occur in the top channel for which $f_c = f_s/2 - B/2$, which on substitution in eqn. 2.38 gives

$$D_{jmax} = \pi^2 (f_s - B)^2 \left(2 - \frac{B}{4f_j}\right) S \sigma_j^2 B$$

$$= 2\pi^2 f_s^2 S \sigma_j^2 B \tag{2.41}$$

if, as assumed earlier, $f_s \gg B$, and if, as is usual, $B \ll 4f_j$. The signal/distortion ratio in the top channel is, therefore

$$\text{s.d.r.}_{top} = 2SB/D_{jmax} = (\pi^2 f_s^2 \sigma_j^2)^{-1} \tag{2.42}$$

Because of the simple form of eqns. 2.40 and 2.42 it is only too easy to overlook the basic assumptions used in achieving such simplicity. We reiterate these as follows:

(i) Gaussian jitter distribution
(ii) uniform signal and jitter spectra
(iii) sampling rate = twice signal bandwidth
(iv) $B \ll f_c; B \ll 4f_j$.

2.6 Impairment due to decoded errors

As described earlier, the digital transmission system will always be subject to random errors caused by channel noise and other disturbances. In a system conveying a coded analogue signal these errors will result in inversion of the binary received digits which are subsequently used in decoding. These digit inversions will, in turn, insert jump-type errors into the decoded analogue signal thus adding yet another form of distortion. The distortion power due to decoded errors D_e may be computed on the basis of the mean square value of the decoded errors.[28] If N is the number of binary digits describing one analogue-signal sample, then if the ith digit is in error an error in decoded sample value $e = 2^{i-1}q$ will occur. If this event occurs with a probability p_i, then the mean square value (or power in a unit load) is given by

$$D_e = \sum_{i=1}^{N} e^2 p_i = \sum_{i=1}^{N} 4^{i-1} q^2 p_i \tag{2.43}$$

For low probabilities of error p_e in the channel only single errors per code word are significant, and p_i is equal to p_e. If the system is conveying a number of time-multiplexed analogue signals then the rate of error 'hits' in individual signals is proportionately reduced, and eqn. 2.41 should be suitably modified. Given that the probabilities p_i are equal to p_e we have

$$D_e = p_e q^2 \sum_{i=1}^{N} 4^{i-1} = p_e q^2 \frac{4^N - 1}{3} \tag{2.44}$$

The above form is obtained as the sum of the geometric progression. In computing the signal/distortion ratio (s.d.r.) we may differentiate between the two cases of practical importance; the case of a signal having Gaussian statistics for which the appropriate relative reference level is the full-load signal power, and, secondly, the case where the peak signal power V^2 is an accepted reference, as for video signals.

For a Gaussian signal with a loading factor $k = \sigma/V$ (see Section 2.2), the power $S = \sigma^2 = k^2 V^2$ and if $4^N \gg 1$, eqn. 2.44 simplifies to

$$D_e \simeq p_e q^2 \frac{4^N}{3}$$

and

$$\text{s.d.r.} = \frac{S}{D_e} = \frac{3k^2 V^2}{4^N p_e q^2} = \frac{3k^2}{4p_e} \tag{2.45}$$

Similarly, specified with respect to the peak level V, we have

$$\text{s.d.r.} \simeq 3/(4p_e) \tag{2.46}$$

A range of numerical values is presented in Table 2.2

2.7 Impairment due to loss of frame alignment

A sequence of digital symbols conveying information would be undecodable without certain identifying patterns. Such patterns inserted at relatively infrequent intervals provide the basic temporal frame of reference for decoding, and facilitate the process known as frame alignment. Frame-alignment pulses may be envisaged as the temporal dual of pilot tones in frequency-division multiplexing. Loss of frame alignment clearly renders the whole information stream undecodable, and is therefore much more serious than the loss of the same proportion of information bits. Frame alignment is normally achieved by a means of a specific combination of

Table 2.2   Distortion due to decoded errors

Error probability	s.d.r., dB	
	Relative peak level V	With Gaussian signal $V/\sigma = 12$ dB
10^{-4}	39·1	27·1
10^{-5}	49·1	37·1
10^{-6}	59·1	47·1
10^{-7}	69·1	57·1
10^{-8}	79·1	67·1

symbols; these are chosen such that the probability of their simulation by an information signal is as small as possible. However, this probability remains finite, as does the probability that random errors in transmission coincide with the frame-alignment sequence. For this reason, some sort of flywheel action is normally incorporated whereby a reframe hunt is initiated only after a predetermined number of failures at framing recognition.

At low error rates the probability of random errors causing 'hits' in two (or more) successive frame-alignment sequences becomes infinitessimally small, being approximately the square (or higher power) of the probability of an error in a single framing word. It is clear, therefore, that random errors generated by normal internal mechanisms, such as thermal noise, may be exonerated as contributors to loss of frame alignment, and other sources, correlated with external phenomena, have to be considered. These may be lightning discharges, man-made interference, power lines and any other sources causing relatively long bursts of severe interference affecting two or more consecutive frames. Little more can be said of a general nature about these burst effects, except that the resulting impairment must be covered by the allocation available for severe disturbances of an infrequent nature, see Chapter 3. In addition, the severity of the impairment depends on the reframe time and on the exact design of the multiplex system; topics that are covered in Chapter 5.

2.8 References

1   FANO, R.M.: 'Transmission of information' (Wiley, 1961)
2   WIDROW, B.: 'A study of rough amplitude quantisation by means of Nyquist sampling theory', *IRE Trans.*, 1956, CT-3, pp. 266–276
3   HOLBROOK, B.D., and DIXON, J.T.: 'Load rating theory for multi-channel amplifiers', *Bell Syst. Tech. J.*, 1939, **43**, pp. 624–644
4   PURTON, R.F.: 'A survey of telephone speech-signal statistics and their significance in the choice of a p.c.m. companding law', *Proc. IEE*, 1962, **109-B**, (1), pp. 60–66
5   RICHARDS, D.L.: 'Statistical properties of speech signals', *ibid.*, 1964, **111**, (5), pp. 941–949

6   BENNETT, W.R. 'Spectra of quantised signals', *Bell Syst. Tech. J.*, 1948, **27**, pp. 446–472
7   Bell Telephone Laboratories: 'Transmission systems for communications' (Western Electric Co., Winston-Salem, N. Carolina, 1970) 4th. edn.
8   CATTERMOLE, K.W. 'Principles of pulse code modulation' (Illife, 1969)
9   CCITT: Green Book, vol. III-2, 1973, Rec G 711
10   MAX, J.: 'Quantising for minimum distortion', *IEEE. Trans.*, 1960, **IT-6**, pp. 7–12
11   ALGAZI, V.R.: 'Useful approximations to optimum quantisation', *ibid.*, 1966, **COM-14**, pp 297–301
12   RICHARDS, D.L.: 'Transmission performance of telephone networks containing p.c.m. links', *Proc. IEE*, 1968, **115**, (9), pp. 1245–1258
13   ROBERTS, L.G.: 'Picture coding using pseudo-random noise', *IRE Trans.*, 1962, **IT-8**, pp. 145–154
14   THOMPSON, J.E.: 'A 36 Mbit/s television codec employing pseudo-random quantisation', *IEEE Trans.*, 1971, **COM-19**, pp. 872–879
15   NYQUIST, H.: 'Certain topics in telegraph transmission theory', *Trans. AIEE*, 1928, **47**, pp. 617–644
16   SHANNON, C.E.: 'A mathematical theory of communication', *Bell Syst. Tech. J.*, 1948, **27**, pp. 623–656
17   WHEELER, H.A.: 'The solution of unsymmetrical sideband problems', *Proc. IRE*, 1941, **29**, p. 446
18   BLACK, H.S.: 'Modulation theory' (Van Nostrand, 1953)
19   GIBBY, R.A., and SMITH, J.W.: 'Some extensions of Nyquist's telegraph transmission theory', *Bell Syst. Tech. J.*, 1965, **44**, pp. 1487–1510
20   KHARKIEVICH, A.A.: 'Kotelnikov's theorem – a review', *Radio Eng. & Electron. Phys.*, 1958, **13**, pp. 1–10
21   SUGIYAMA, H.: 'Band-limited signals and the sampling theorem', *Electron. & Commun. Jap.*, 1966, **49**, pp. 100–108
22   ROBERTSON, G.H.: 'Computer study of quantiser output spectra', *Bell Syst. Tech. J.*, 1969, **48**, pp. 2391–2403
23   FUJII, A., and AZEGAMI, K.: 'Quantising noise for a sine wave and a set of sine waves', *Rev. Electr. Commun. Lab.*, 1967, **15**, pp. 145–153
24   DORWARD, R.M.: 'Aspect of the quantisation noise associated with the digital coding of colour television signals', *Electron. Lett.*, 1970, **6** (1), pp. 5–7
25   BENNETT, W.R.: 'Statistics of regenerative digital transmission', *Bell Syst. Tech. J.*, 1958, **37**, pp. 1501–1542
26   LEE, Y.W.: 'Statistical theory of communication' (Wiley, 1960)
27   SHANNON, C.E.: 'Communication in the presence of noise', *Proc. IRE*, 1949, **37**, pp. 10–21
28   BEDROSIAN, E.: 'Weighted p.c.m.', *Trans. IRE*, 1958, **IT-4**, pp. 45–49
29   US National Bureau of Standards: 'Tables of the error function', 1954
30   VERHOEVEN, L.: 'More aspects of quantisation noise associated with digital coding of colour television signals', *Electron. Lett.*, 1973, **9**, (3), pp. 69–70
31   ROSS, D.T.: 'Sampling and quantising' *in* 'Notes on analog-digital conversion techniques', SUSSKIND (Ed.) (MIT Press, 1957)

Chapter 3

Principles of system planning

In the last Chapter we considered the types of impairments experienced by digital signals. We must next consider the problems of system planning, namely, how to specify the performance of individual pieces of equipment so that, when these are put together to form a complete transmission system, the overall transmission objectives will be met. This involves three subsidiary problems.

(i) What equipment configurations should be used for specifying the complete system?
(ii) How do the various contributions to the total transmission impairment accumulate along the length of the system?
(iii) How should the total impairment be allocated to the various sources of impairment, and to the separate pieces of equipment?

In the following Chapter it will be seen that transmission objectives are ultimately based on subjective judgements, and, as a result, can only be expressed quantitatively in statistical terms. System planning must also be based on statistical estimates, since no two transmission systems have identical configurations, and the conditions under which they are operating (signal loading, state of equipment etc.) are continually varying. However, it should be noted that, while in the case of transmission objectives we are able to assume that our statistics are ergodic, this is not the case in transmission planning.

A number of factors introduce regular time dependent changes. For example, in many parts of the world seasonal temperature changes will cause the loss of transmission lines to increase slightly during the summer, with a consequent shift in average noise levels. In a similar fashion, in some tropical regions microwave radio links will exhibit degraded performance during the rainy season. In the short term there will be regular daily variations in the signal loading, and in the long term there will be a continual slow drift in the network characteristics as the older parts are replaced by newer equipment with improved performance.

3.1 System configurations

The problem of defining suitable system configurations has been dealt with by introducing the concepts of the *hypothetical reference connection* (h.r.x.) and the *hypothetical reference circuit* (h.r.c.). However, before entering into a detailed discussion of these ideas we must caution the reader to treat our statements in the remainder of this Chapter as a provisional survey of the subject, that will require later verification. The reason for this is that the topic of hypothetical reference circuits for digital systems is currently under active consideration by the International Telegraph & Telephone Consultative Committee* (CCITT) and it is quite possible that some of our remarks will have been overtaken by events before this book is published. For accurate quantitative information the reader is urged to consult the most recent sources, in particular, the publications of the CCITT.

A specification for a hypothetical reference connection or circuit usually consists of four parts

(i) a specification of a circuit configuration
(ii) a specification of operating conditions in terms of traffic loading or test signals
(iii) a statement of the performance to be achieved under these conditions
(iv) some indication of the way in which the overall transmission impairments are to be allocated to the various parts of the h.r.x. (or h.r.c.).

There is frequently some confusion about the significance to be attached to an h.r.x. A hypothetical reference connection or circuit *is not* meant to represent a *typical* configuration. The intention is to define a configuration that will cover the majority of practical cases, while at the same time recognising that it would be unrealistic to attempt to design for the worst possible conditions that could arise in a long international link. Thus, we find in recommendation G 212 (Green Book, Vol. III—1 p. 118) the hypothetical reference circuit defined as 'a hypothetical circuit of defined length and with a specified number of terminal and intermediate equipments, this number being *sufficient but not excessive*'.

---

* The CCITT operates under the auspices of the International Telecommunications Union. Its main task is to ensure that the various national communication networks are compatible, so that long international connections may be set up by interconnecting national networks. It has been found that this necessitates detailed specification of the systems, and these specifications are commonly used in design studies. The CCITT hold a full plenary session every four years, and an account of these meetings is published as a series of books. (The current series are the Green Books, published in 1973.) These books contain full details of the technical recommendations made by the CCITT, together with a considerable amount of supporting material. Between plenary sessions various subcommittees meet. The most important of these, with respect to digital systems, is Special Study Group D. The papers of this committee, while not formally published, are generally available within the Telecommunications Industry, and give a great deal of background material on system design. A sister body, the CCIR deals with point-to-point and broadcast radio communications, and as such is concerned with digital transmission over radio circuits.

In a similar manner the test conditions are specified to cover fairly adverse situations, but not the worst possible that could arise. In this case, the idea of the 'busy hour' is often introduced. This takes account of the fact that many forms of transmission impairment depend on the loading on the system, and are likely to be most pronounced when the system is heavily loaded. The traffic will be continually fluctuating, but will tend to rise to its peak at certain times, usually during mid-morning and mid-afternoon. This is covered by assuming that the performance is measured as the average over one hour, at a time when this peak loading is present.

Then, if the equipment is designed to meet the specified performance objectives under these conditions, it is anticipated that real systems using this equipment will normally operate with transmission impairments within the specified limits. (See recommendation G 226 'Noise on a real link' for further comment on this point.) On the rare occasions when the target degradations are exceeded, the circuits will almost always be still usable as communication channels; albeit with some difficulty.

3.2 Accumulation of impairments

It is obviously important to know how the total transmission impairment builds up in proportion to the length and complexity of the system. As usual, we find that this question does not have any single and simple answer. To begin with, the various types of impairment are often, at least to some extent, interchangeable. The way in which this arises is fairly obvious for analogue signals. For example, in the case of telephony one can see that if a signal is transmitted with very little frequency and amplitude distortion a comparatively high transmission loss can be tolerated before the received speech becomes difficult to understand, whereas, in the presence of distortion a louder signal will be helpful. This is dealt with in CCITT recommendation G 113 (Green Book Vol. III—1 p. 19). It has also been shown that, provided a suitable measuring technique is employed, the various impairments incurred during analogue transmission of television can be combined.[1-3] It is less obvious that this sort of situation can exist in digital systems, but examples do arise, particularly in the case of digital transmission of analogue signals. For these signals any impairment that appears subjectively as a noise can usually be combined with other noise sources. Thus, in the digital transmission of f.d.m. telephony the contributions from digital errors and jitter may be combined by power addition, and the system designer is free to allocate the total noise impairment between these sources according to his requirements. In the case of purely digital transmissions, e.g. data, this situation cannot arise, but in this instance there is sometimes no clear distinction between a single digital error and a burst of closely associated errors, since both result in retransmission of a complete data block. Under these conditions, the designer then has some freedom in apportioning the various sources of digital errors, which may include errors caused by regeneration failures, errors due to error extension characteristics of the line code and errors arising in the framing and multiplexing system.

For each type of impairment we must consider the way in which the contributions from a number of sources distributed along the system will accumulate. Two limiting cases exist. (i) If there is no correlation between the contributions from the different sources they will add in the manner of random variables on a root mean square basis. Thermal noise will accumulate in this manner. (ii) If the contributions are completely correlated they will add linearly. Attenuation-frequency distortion terms are often of this type. In practice, two further cases arise. (iii) The sources are, in theory, correlated, but the correlation is disturbed by other factors. In this case, the law of addition will lie somewhere between the root mean square and the linear cases. 2nd-order intermodulation distortion in f.d.m. systems[4] tends to accumulate in this way. (iv) The measured impairment is a 2nd-order effect of some other underlying cause, to which it is related in a nonlinear manner. This leads to highly nonlinear build up of total impairment.

The error rate in a digital system sometimes behaves in this way, since the errors caused by regeneration failures are a nonlinear function of the signal/noise ratios existing in the system (see Chapter 9).

3.3 Reference circuits for analogue transmission

Before considering the allocation of transmission impairment to the different parts of a digital transmission system, it will be useful to study what has been done in the past in the case of analogue systems. This will help us to identify the differences between analogue and digital transmission techniques, and will therefore indicate where the differences in the approaches to system planning are likely to arise. There is also one other consideration of over-riding importance. As was discussed in Chapter 1, almost all of the world's existing communication facilities were developed with analogue transmission in mind, and their parameters have been chosen accordingly. Digital equipment is now being introduced into an existing analogue network, and even if digital transmission eventually completely replaces the present analogue systems there will be a very long period during which the two techniques must coexist. During this time a large number of transmission connections will be set up over mixed systems, part analogue and part digital. It is essential that the performance of these hybrid systems is satisfactory, and this can only be achieved if the system planning procedures used for the analogue and digital cases are themselves compatible. This has been recognised by the CCITT and the reader is refered to Section (*d*) of recommendation G 103 for an account of their current views.

3.3.1 Hypothetical reference connections for telephony
Since telephony forms the major component of communication traffic we will start
by reviewing the present approach to the planning of analogue telephony
connections.

Configuration. Almost all communication networks are based on a tree structure. In the case of telephony this takes the form shown in Fig. 3.1. On making a
long-distance call the subscriber is connected through his local line to his local
switching centre (l.s.c.), then over a junction line to a main switching centre
(m.s.c.), and then, by further links in the national trunk network, to a remote
m.s.c. or, for an international call, to an international switching centre in his
country (i.s.c.). In this latter case, he will then be connected over a series of
international and intercontinental links to an international switching centre in the
terminal country, and then through a series of links in the remote national network
to the required destination. The actual connection can therefore be depicted as a
simple sequential set of links between switching nodes. Although, from an operational point of view, the subscriber is provided with an independent transmission
path through the network, beyond the l.s.c. he is normally allocated one channel in
a multichannel transmission system, and is sharing common transmission equipment
with other users. This introduces two factors (i) the performance of the circuit may
be dependent on the presence of these other users, and (ii) at each node (switching
centre) it is necessary to extract the signals from the multichannel structure, which
involves signal processing at these points.

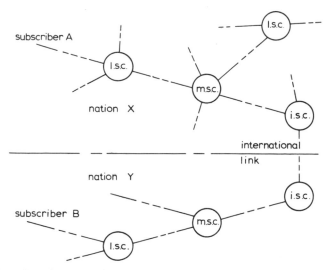

Fig. 3.1 Configuration of a switched telephone network
    l.s.c. = local switching centre
    m.s.c. = main switching centre
    i.s.c. = international switching centre

The CCITT have defined three hypothetical reference connections for telephony, covering the cases of the longest international connection envisaged, a typical moderate length international connection and a typical long international connection. The first of these is reproduced as Fig. 3.2. We may deduce that a long national connection could be represented by the end sections of this model. Details of these proposals are given in recommendation G 103.

A typical transmission system will form one of the links in these connections. Such transmission systems are defined in terms of hypothetical reference circuits, using similar approaches to those employed for the h.r.x. A series of these h.r.c.s are defined by the CCITT for different types of transmission equipment. As an example, the circuit for a 12 MHz coaxial cable system (recommendation G 332) is shown in Fig. 3.3. The total length of this circuit is 2500 km, this being divided into nine sections with modulation and demodulation operations of varying complexity between each section. This division leads to certain recommendations for the allocation of noise impairments that will be discussed later.

Operating conditions. The assessment of the h.r.c. must be carried out in a way that will reveal any deficiencies in the transmission of typical signals, and this in turn will determine the nature of the tests and the type of test signals used. Two types of degradation must be considered (i) those which are imposed directly on the signal; namely, distortions due to attenuation and phase characteristics, time effects (e.g. echo) and amplitude nonlinearity, and (ii) added impairments owing to the presence of other signals; namely, thermal noise, intermodulation noise and

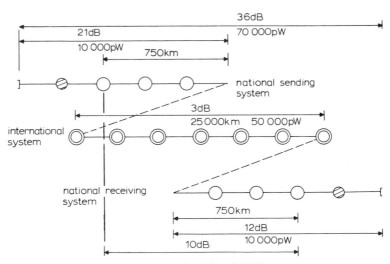

Fig. 3.2 Longest international connection envisaged by CCITT
⊢ subscriber
⊘ local exchange
○ main exchange
◎ international exchange

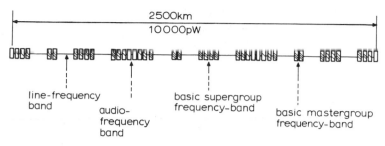

Fig. 3.3 Hypothetical reference circuit for a 12 MHz coaxial-pair system
[] 1st-level multiplex
[] 2nd-level multiplex
[] 3rd-level multiplex
[] 4th-level multiplex

crosstalk. For assessment purposes, telephony signals may be represented either by sinusoidal tones, or by a weighted noise spectrum. It is important to ensure that the amplitudes of the test signals are consistent with the amplitudes of real speech signals, and, because the actual signal levels vary from point to point through the connection, the practice of referring all levels back to a 'zero relative level point' is adopted. This point is usually taken to be equivalent to the terminals at the local-exchange end of the subscriber's line, and at this point, the mean power of a telephony signal is taken to be −15 dBm. This mean value takes into account the pauses in normal conversation. (See Green Book, Vol. 111, pp. 127–135, 575–578 for further details.) The peak level of the signal is rather more difficult to assess, but for practical purposes it may be taken to be equal to the peak-to-peak excursions of a sinusoidal tone of +2 dBm amplitude.[5]

Acceptable performance. The attenuation of a telephony connection is normally expressed in terms of the reference equivalent[6] which includes allowance for the performance of the transducers (the microphone and telephone earpiece) and is assessed by subjective comparisons using a standard reference system. In general, the overall reference equivalent should not exceed 36 dB. (recommendations G 103, G 111 and G 112.) Bandwidth limitations, which effect the clarity of speech, are subjectively equivalent to an increase in loss, and may be included as part of the reference equivalent (recommendation G 113). The noise objective for a long international circuit is −43 dBm0 (50 000 pW) leading to a minimum signal to noise ratio of 28 dB for telephony.

As previously discussed, it is assumed that not all the component links in a long connection will have the worst possible performance, and so the specification for a single hypothetical reference circuit is somewhat more severe than the average used for the complete connection. Typically, a single h.r.c. is allocated 10 000 pW noise. For line systems, this represents the limit reached during the busy hour, but in the case of radio systems it is recognised that more severe conditions can arise due to propagation effects, and that occasional excursions beyond the 10 000 pW limit are admissible. Thus, additional clauses in the specification allow a noise level exceeding 10 000 pW for 20% of the time during any month, 50 000 pW for 0·1% of the time during the month and 10^6 pW for 0·01% of the time (see recommendation G 222).

Allocation of impairments. We have so far used the terms 'noise' and 'distortion' without paying much attention to the difference between these two forms of impairment. Some authors differentiate these effects by identifying noise with extraneous interfering signals, and distortion with effects originating from the actual transmitted signal. We prefer a different distinction, namely, distortion is considered as an impairment that can, in principle, *be corrected;* while noise cannot be corrected by any subsequent processing of the signal. Thus, in analogue systems attenuation and phase versus frequency distortions can be, and are, removed by the use of equalisers, and, in principle, nonlinear distortion of the signals can be corrected by introducing a complementary nonlinearity. Thermal noise cannot be eliminated by any correcting operation. Intermodulation distortion represents an intermediate case. In a single transmission link, although the composite signal is still available it can, in principle, be eliminated by introducing complementary nonlinearities; but once the signal has passed through a node and becomes separated from the sources of the interference, nothing can be done, and the intermodulation then becomes a noise source.

The importance of these distinctions is that they affect the way in which impairments build up along a transmission system, and so determine the way in which the total impairment may be allocated to the subsections of the h.r.c. In the case of telephony, we must consider the accumulation of three types of impairments; (i) attenuation (ii) frequency distortion and (iii) noise.

Attenuation. This must add directly throughout the connection. It is desirable, to ensure network stability, that each individual link in the connection should have some loss; and the present practice is to allocate 0·5 dB to each of the international circuits. The transmitting end of the national sections of the connection is allocated a maximum loss of 21 dB, and the receiving end is allocated a maximum of 12 dB, yielding an overall reference equivalent of 36 dB. Within the national networks, most of the loss occurs in the local connections from the subscribers to the l.s.c.

Frequency distortion. In analogue systems using frequency-division-multiplex transmission most of the frequency distortion occurs in the first stage of modulation. The longest international circuit may contain up to 12 pairs of channel modulators (recommendation G 103). The overall attenuation frequency distortion will then be within the limits shown in Fig. 3.4; but note that this is not the linear sum of the contributions from the individual modulator-demodulator pairs, the average values being greater than 1/12th of the overall limit (see G 232). The actual assumed law of accumulation is not stated but approximates to an 0·65 power law.

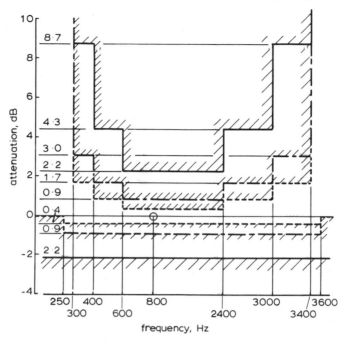

Fig. 3.4 Attenuation-frequency distortion for a single terminal and for a worldwide reference connection
---------- limit for the average loss of 12 pairs of equipment
— · — limit for any single pair of transmitting and receiving equipments
———— limit for a worldwide connection of 12 circuits

Noise. Noise for each individual link in the complete connection will accumulate by direct-power addition. Within a single link noise is divided between contributions from the modulation equipment, taken to be 2500 pW, and contributions from thermal noise, intermodulation and crosstalk arising in the transmission path. These latter components are assumed to accumulate uniformly along the line, and so are equivalent to a contribution of 3 pW/km. This assumption of uniform distribution is in fact not strictly correct, since it is known that some forms of intermodulation noise add coherently.[4] However, it is left to the system designer to deal with this by making appropriate adjustments in his design calculations.

3.3.2 Hypothetical reference connections for television
Although less important than telephony, the case of television deserves attention, since it introduces a number of factors that are neglected in telephony transmission. Because television is normally transmitted over permanent or semipermanent circuits, rather than over a switched network, the connections tend to be simpler and so the CCITT assume that the hypothetical reference connection for television consists of only three links (recommendation J 61). However, some national authorities, for example the British Post Office, have used rather more complex

connections for national planning.[7] The UK hypothetical reference chain proposed by the British Post Office is shown in Fig. 3.5. The test signals used for assessing television connections are relatively complex, a typical example being a combination of a sine squared pulse and bar, appearing as a modulated carrier signal in the case of colour television. (see annex 2, recommendation J 62). The assessment of the response to these tests signals must also take into account a number of different features; for example, rise times, overshoot, droop on the bar and overall magnitudes. In one technique (recommendation J 62) these are all related back to a single performance parameter K which gives an overall rating for the transmission quality. A number of other types of impairments have to be considered separately. For instance, thermal noise and periodic noise components must receive separate attention, because these have differing subjective effects. The various impairments will accumulate along the connection at differing rates. The expressions covering these rates are discussed in the References cited above.

3.4 System planning for digital transmission

Details of the planning for various services will be discussed in the next two Chapters, and so for the present we will confine our remarks to the general principles. There seem to be no reasons for abandoning the basic approach to system planning used for analogue systems, and so the ideas of the hypothetical reference connection and the hypothetical reference circuit will continue to be used. These configurations are largely determined by geography, and so the structure of digital reference connections and circuits will look very similar to that of their corresponding analogue counterparts, right down to the number of sections in an h.r.c. and the average length of these sections. Since connections consisting of a mixture of analogue and digital sections will be used it is necessary that the total apportionment of transmission impairment is also, to a large extent, maintained in the present manner down to the individual sections in the h.r.c. At this level the designer of a digital systems will then be free to alter the way in which the total impairment is divided between the various contributory sources. In many cases, all the previous types of impairment found in analogue systems will still be present, since the digital systems still include analogue operations in parts of the transmission path. In addition, new types of impairment peculiar to digital transmission will appear.

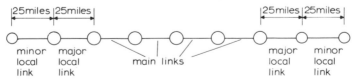

Fig. 3.5 UK hypothetical reference chain for television

Signals will, however, only appear in analogue forms in small portions of the complete system, and so the contributions from analogue forms of transmission impairments will be relatively small. On the major part of the transmission path signals will appear in digital form. There is then no counterpart of the accumulation of loss, noise and distortion that occurs in an analogue transmission path, this being replaced by contributions from error rate and jitter.

As will be seen later, it is usually possible to maintain a low digital-error rate, at least for most of the time. Digital errors form a noise contribution that cannot be directly corrected; the only remedy is to employ a redundant error-correcting code, which is the digital equivalent of providing noise immunity by wide-band modulation schemes, such as frequency modulation. Jitter, on the other hand, may be considered to be the digital equivalent of intermodulation distortion, since it builds up along the transmission path in a similar manner, and, like distortion, is a reversible effect. The removal of jitter is however a rather easier process than the correction of nonlinearity in an analogue system.

Wherever analogue/digital conversion takes place quantising distortion will appear, and this will frequently be a major source of transmission impairment. At most of the nodes between transmission links some form of digital processing will take place, such as code conversion, digital multiplexing and justification. These operations frequently give rise to error propagation effects, in which each single digital error which has appeared during transmission can be converted into a number of errors in the processed signal. Additional jitter, in particular low-frequency components that are difficult to eliminate, can be introduced at these points by the justification operations. The final result of all this is that it is desirable to reverse the practice used for analogue systems, and allocate the major part of the transmission impairments to contributions arising at the system nodes.

The time dependence of transmission impairment in digital systems can be very different from that in analogue systems. To start with, once the signals have been converted into a digital form, there is no direct relation between the characteristics of the digital signal and the amplitude or power of the original analogue contribution. Thus, the performance of digital systems is largely independent of signal loading (there are in fact some load-dependent effects in coders, but these are usually small) and so the busy hour looses its significance. On the other hand, as will be seen in a later Chapter, the change in error rate with changes in transmission-path loss can be very rapid, typically an order of magnitude change in error rate for a 1 dB change in attenuation, and as a result there is a virtual threshold in digital systems, below which they are almost perfect, and above which they are unusable. This is important in the case of radio systems in which fading occurs. The limits set for analogue systems are based on the assumption that there is no abrupt change from a usable to an unusable system, and must therefore be reconsidered. Present thinking in the CCITT is presented in contribution 153 to Special Study Group D (Sept. 1972). This suggests that the error rate for a 2500 km circuit should not exceed 1×10^{-6} for more than 20% of the time in any month, and should not exceed 1×10^{-3} for more than 0·01% of the time in any month. (These proposals are very tentative.)

A somewhat analogous effect arises in the consideration of the build up of error rate with system length. Since the error rate is very strongly dependent on the transmission-path loss it is found that the total system error rate is dominated by the contribution from the worst section of the route, usually the longest section. This means that error rate does not build up in proportion to the absolute length of the system, but is dependent on the statistical distribution of repeater section lengths. This point will be discussed again in Chapter 7.

3.5 References

1   ALLNATT, J.W.: 'Subjective assessment of impairments in television pictures', *Appl. Ergon.*, Dec. 1973, **2**, p. 231
2   ALLNATT, J.W., and LEWIS, N.W.: 'Subjective and objective impairments in television pictures', International Broadcasting Convention, 1968
3   SISCOSS, C.A.: 'Numerical values for subjective picture quality in television', *J. Soc. Motion Pict. & Televis. Eng.*, Dec. 1971, **80**, p. 958
4   BROCKBANK, R.A., and WASS, C.A.A.: 'Nonlinear distortion in transmission systems', *J. IEE*, 1945, **92**, Pt. III, pp. 45–56
5   RICHARDS, D.L.: 'Transmission performance of telephone networks containing p.c.m. links', *Proc. IEE*, 1968, **115**, (9), pp. 1245–1258
6   RICHARDS, D.L.: 'Telecommunication by speech' (Butterworth, 1973)
7   MACDIARMID, I.F.: 'Performance requirements of links for colour television', Post Office Research Department report 20, June 1968

Chapter 4

Services for transmission

4.1 General plan

The performance objectives for a telecommunications system depend primarily upon the requirements set by the various services which it is designed to convey. In this Chapter, we shall consider the services most commonly met in practice and indicate how their demands may be translated into a specification of digital transmission performance. As the widespread penetration of digital techniques is still some years away, much of the discussion and quantitative examples will have to be tentative and should not be read as established fact; standardisation of the various questions involved is the subject of current CCITT and CCIR study and inevitably some of our conclusions will be overtaken by events. However, the recommendations presented by the above international organisations are the collation of a mass of diverse expert contributions and are normally succinct and lacking any elaboration; our main aim here is to provide the explanatory background which forms the basis of future recommendations. Chapters 2 and 3 provide the theoretical and system planning groundwork, and we shall refer to their material.

Before considering the various services which require transmission, it is instructive to enumerate the logical steps involved in deducing the performance requirements for the transmission equipment for any one service; this basically formulates a method of attack which is applicable in most cases.

(i) formulate a hypothetical reference connection (h.r.c.)
(ii) specify impairment objectives for h.r.c.
(iii) allocate impairment objectives to subsystems
(iv) specify subsystem performance requirements.

Factors involved in formulating the h.r.c. have been largely covered in the previous Chapter, the dominant one being geographical. This simply means that some services in association with some transmission systems may require very different path lengths, ranging from quite short sections (100 km, say) to the longest international specification (25 000 km). The h.r.c. must also specify a breakdown of the complete digital path into a number of digital sections and the type of equipment involved in terms of information rates. For example, in the case of the longest h.r.c. of Fig. 3.2, it is likely that the 25 000 km international and 750 km national links will use high-speed ($>$ 100 Mbit/s), high-capacity ($>$ 1000 channels) digital transmission systems, the remainder being made up by primary systems at 1·5 or 2 Mbit/s (24 or 30 channels), and possibly secondary systems at 6·3 or 8·4 Mbit/s.

The specification of impairment objectives, step (ii), is obviously specific to the service under consideration, as described in Chapter 3. In this, subjective effects play a dominant role in services which convey human communication, such as speech and visual systems. The number of parameters requiring specification is dependent on the transmission technology and is different for the cases of digital and analogue systems. In the case of digital systems, experience will tell whether the impairments discussed in Chapter 2, namely, quantisation, error rate and jitter, form a sufficient description.

Having established an h.r.c. and its overall impairment objectives, step (iii) involves the allocation or subdivision of these impairments among the subsystems defined in the h.r.c. At this stage foreknowledge of the technical and economic feasibility of meeting a given specification is required for each subsystem. For instance, in analogue f.d.m. practice it is known that the line system contributes noise proportional to system length, and this clearly affects the allocation of noise between modulating and line equipments. The objective here must be to strike an economic balance avoiding disproportionate expense by overspecifying the performance of any one subsystem.

Finally, step (iv) translates the subsystem impairment objectives into subsystem performance requirements and, as implied above, is tacitly anticipated in step (iii). This translation is performed using the relations developed in Chapter 2 between s.n.r. and coding accuracy, error rate and jitter.

The four steps described will be used in deriving, where possible, quantitative performance requirements for the various services. The main services which require consideration may be tabulated as follows, in what we feel is their order of importance in the predictable future:

(i) speech
(ii) data
(iii) f.d.m. assemblies
(iv) television
(v) visual telephone
(vi) sound programme.

This order is likely to change; in particular, f.d.m. assemblies will be progressively demoted as their share of the total transmission facilities decreases. We shall apply the principles developed in Chapters 2 and 3 to each of these services in the following Sections.

4.2 Speech

Here we consider transmission of speech over connections containing digital sections. Telephony is, and will almost certainly remain, the most important service in telecommunications and, to quote from Reference 1, 'the non-telephony services are always engineered bearing in mind that the design criteria of the transmission systems were chosen with the requirements of long-distance telephony principally in view'. As a starting point, there is wide agreement[1] that the plan for telephony involving digital systems should be based on the three hypothetical reference connections[2a] described in Chapter 3, as regards length. Furthermore, during the gradual transition phase from analogue to digital transmission and from space to digital trunk switching, there will inevitably exist complicated combinations of digital and analogue facilities; adequate performance will have to be ensured during this awkward phase.[2a]

The next step is to establish performance objectives for speech which has been subjected to coding and decoding. These must be based on subjective evaluation within the framework of the telephony network. Richards[3] suggests that the minimum s.d.r. for quantised speech is some 22 dB for middle-range volumes (−5 to −25 dBm at the input to a listening system having a receive reference equivalent of 0 dB) and some 16 dB for lower-speech volumes. The latter lower requirement is a result of the quantising-distortion components being then below the threshold of audibility. Conventional telephony speech level falls within the above middle range, and hence we can take 22 dB as the minimum s.d.r. requirement. Referring now to the h.r.c. of Fig. 3.2, and assuming the extreme case of digital links between every exchange and space switching within the exchange, it is seen that the total (maximum) number of coding-decoding operations can be 14. Assuming no other noise sources, the required s.d.r. per codec operation is then $(22 + 10 \log_{10} 14) = 33.2$ dB. This would be within the capabilities of a 7-bit linear coder (Fig. 2.2), but as we have already discussed in Chapter 2, the wide deviation of speech level from its nominal mean can only be accommodated by means of companding.

4.2.1 *A*-law and μ-law companding
While we cannot afford to dwell on the choice of suitable companding laws,[11] the importance of companding warrants a brief factual diversion. After much international effort, two companding laws have been accepted as desirable standards by the CCITT,[2b,c] the N. American μ-law and the European *A*-law. Both are segmented piecewise-linear approximations to the original continuous laws whose compression characteristics $F(x)$, for a range of ±1, are described by

$$F_A(x) = \text{sgn}\,(x)\,\frac{A|x|}{1 + \log_e A}, \text{ for } 0 \leqslant |x| \leqslant 1/A$$

$$= \text{sgn}(x) \frac{1 + \log_e A |x|}{1 + \log_e A}, \text{ for } 1/A \leqslant |x| \leqslant 1 \tag{4.1}$$

and

$$F_\mu(x) = \text{sgn}(x) \frac{\log_e(1 + \mu|x|)}{\log_e(1 + \mu)}, \text{ for } -1 \leqslant x \leqslant 1 \tag{4.2}$$

The segmented version of the A-law uses 13 segments with $A = 87 \cdot 6$; it is defined for positive values in Table 4.1. It is in fact convenient to visualise an additional segment 0 which is colinear with segment 1. The segmented version of the μ-law uses 15 segments with $\mu = 255$; a definition for positive values is given in Table 4.2. There are clearly strong similarities between the laws; in the 8-bit code word which consists of a polarity digit P followed by three segment-defining digits XYZ followed by four digits $ABCD$ which specify the intrasegment value on a linear scale of 16 values. For ease of realisation, segment ranges, segment slopes and within-segment quantising intervals are binarily related in successive segments (except for segments 0 and 1 in the A-law). For low-level signals, the companding advantage, namely the ratio of smallest quanta with and without companding, is 24 dB (4096/256) for the A-law and 30 dB (8192/256) for the μ-law.

Returning now to our previous discussion, Fig. 4.1 reproduces the recommended CCITT specification together with theoretical curves for 7- and 8-bit coding using the A-law. It is seen that the 7-bit theoretical maximum just fails to meet the s.d.r. requirement, whereas the 8-bit recommended lower limit provides some

Table 4.1   Definition of segmented p.c.m. A-law

Segment no., S	Coder input range	Output code	Quantum interval
0	$0 - V/128$	$P\,000\,ABCD$	2
1	$V/128 - V/64$	$P\,001\,ABCD$	2
2	$V/64\ - V/32$	$P\,010\,ABCD$	4
3	$V/32\ - V/16$	$P\,011\,ABCD$	8
4	$V/16\ - V/8$	$P\,100\,ABCD$	16
5	$V/8\ \ - V/4$	$P\,101\,ABCD$	32
6	$V/4\ \ - V/2$	$P\,110\,ABCD$	64
7	$V/2\ \ - V$	$P\,111\,ABCD$	128

$V = 4096$ Binary code word: $P\,XYZ\,ABCD$, where P = polarity digit, XYZ = segment digits, $ABCD$ = linear digits
$S = 4X + 2Y + Z, L = 8A + 4B + 2C + D$
Decoded output $= \alpha(\beta 32 + 2L + 1)2^{S-\beta}$ where
$\qquad \alpha = 1$ for $P = 1, \alpha = -1$ for $P = 0$
$\qquad \beta = 0$ for $S = 0, \beta = 1$ for $S \neq 0$

Table 4.2   Definition of segmented p.c.m. μ-law

Segment no., S	Coder Input range	Output code	Quantum interval
1	−31 − 31, $V/128$	$P\,111\,ABCD$	2
2	31 − 95, $V/128$	$P\,110\,ABCD$	4
3	95 − 223, $V/64$	$P\,101\,ABCD$	8
4	223 − 479, $V/32$	$P\,100\,ABCD$	16
5	479 − 991, $V/16$	$P\,011\,ABCD$	32
6	991 − 2015, $V/8$	$P\,010\,ABCD$	64
7	2015 − 4063, $V/4$	$P\,001\,ABCD$	128
8	4063 − 8159, $V/2$	$P\,000\,ABCD$	256

$V = 8192$ Binary code word: $P\,XYZ\,ABCD$, where P = polarity digit, XYZ = segment digits, $ABCD$ = linear digits
$S - 1 = 4\bar{X} + 2\bar{Y} + \bar{Z}, L = 8\bar{A} + 4\bar{B} + 2\bar{C} + \bar{D}$
Decoded output $= \alpha\left\{2^{S}L + 33\,(2^{S-1} - 1)\right\}$ where
$\alpha = 1$ for $P = 1$, $\alpha = -1$ for $P = 0$

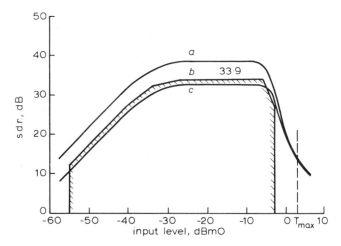

Fig.4.1 Quantising distortion curves for A-law coding with noise-like input
  a 8-bit theoretical
  b Limits of CCITT recommendation G 712
  c 7-bit theoretical

accommodation for codec imperfections. It should be remembered that the situation considered represents the worst combination of space switching and digital transmission for the longest h.r.c. and might be regarded as over pessimistic; however, it is not the whole story.

A further requirement arises from the need for graceful replacement of analogue transmission facilities by digital systems; in particular, the displacement of any number of analogue links by digital equivalents should not be allowed to degrade overall noise performance. Recommendation G 103 specifies limits on the noise contributions of individual interexchange links, and we must therefore check that the 8-bit p.c.m. system described above can form a viable subunit of the h.r.c. It is important to note here that the recommendation was formulated for analogue systems in which the noise contribution is largely constant irrespective of the presence or level of the signal. As pointed out in Chapter 2, this situation does not hold where a quantiser is involved, and the recommendation must be used with caution. Basically it applies to the no-signal (idle channel) and very low-level regimes in which the listener tends to assess the quality of a connection on the basis of a noise level unmasked by a signal. We proceed to calculate the idle channel distortion power D_{qi}; this is given by $q^2/4$ (see Section 2.2) where q is the smallest quantum step around zero input level. Using the expression of Table 4.1, the maximum zero-to-peak excursion V for a decoded A-law companded signal is $2016\,q$ and, denoting the sinusoidal power level necessary to fully load the quantiser by T_{max}, we have

$$T_{max} = V^2/2 = \tfrac{1}{2}\,2016^2\,q^2 = (4032^2/2)D_{qi}$$

or, in dBm

$$D_{qi} = T_{max} - 69\cdot09 \text{ dBm.}$$

This assumes that the distortion power is present in a uniform band equal to half the sampling rate (see Sections 2.3 and 2.4), whereas conventionally telephony noise is specified in a band of 300 Hz–3·4 kHz. Psophometric shaping reduces the noise power by a further 2·5 dB[2d] resulting in a downward correction of $(10 \log_{10} \dfrac{3\cdot1}{4\cdot0} - 2\cdot5)$ dB $= -3\cdot6$ dB. Hence, the psophometrically shaped idling-channel distortion power for A-law becomes

$$D_{qi} = T_{max} - 72\cdot69, \text{ dBmp} \tag{4.3}$$

T_{max} is specified for A-law as 3·14 dBm0 leading to a theoretical $D_{qi} = -69\cdot55$ dBm0p or 220 pW0p. The limit specified in recommendation G 712 is in fact -65 dBm0p or 320 pW0p. This level is significantly smaller than the noise contributions recommended for individual links of the h.r.c. of Fig. 3.2 (recommendation G 103); the most stringent recommendation refers to the link connecting the local exchange to the primary centre, being 500 pW0p with a maximum limit of 2000 pW0p.

We may conclude therefore, that, using idle-channel distortion as a criterion, the displacement of existing links in the longest h.r.c. by digital systems using 8-bit A-law coding should progressively improve speech-transmission performance. The values quoted apply within fractions of a decibel to the μ-law as well.

In addition to quantising effects, errors in transmission cause speech degradation as described in Chapter 2. It is currently accepted that an overall error rate of about 1 in 10^5 is just discernible on low-level speech,[1] and it is therefore reasonable to design the maximal length h.r.c. for an error performance somewhat better than this. Let us assume that the h.r.c. is composed of high-speed digital links in its international and national trunk sections, a total of 26 500 km, the remaining circuit consisting of primary p.c.m. short-haul links. Then a design error rate of 10^{-10}/km in the former accounts for 26·5 errors in 10^7, still leaving a margin for errors in the short-haul links and digital switching centres. A reasonable total error rate might be between 40 and 70 in 10^7 (reference 1), which is still within the 10^{-5} limit.

Having considered both quantising and error effects, it is clear that the worst degradation would prevail in the hypothetical case of a connection made up exclusively from digital links between space-switching exchanges involving quantising at each exchange, to a total of 14 links and 14 codecs. Conversely, the best connection will consist exclusively of digital links (14) between digital switches (15) and only one codec; this would represent the final evolution of the digital telephony network; Chapter 16 elaborates this point. Estimates of jitter limits for speech are derived in Chapter 10; these are fortunately high enough, being over 1 μs r.m.s., as to present no problems in practice.

4.3 Data

Data are expected to account for an increasing fraction of telecommunication-network traffic, although the dominance of telephony criteria for transmission objectives is not in doubt. Hence, any data services designed to share the telephony network must accept its specified characteristics and, where necessary, adopt special measures to achieve adequate performance. Current data services involve modulation and demodulation of the source binary information using a variety of digital-modulation schemes described in Chapter 13, over single analogue channels, predominantly at voiceband, and exceptionally occupying whole f.d.m. blocks. The equipments used, known as modems,[4] operate on a range of input information rates, typical being 100, 200, 600, 2400, 4800 and 9600 bit/s over voiceband facilities, and 48 kbit/s and above over f.d.m. assemblies. A prerequisite for data services is error control, and this is commonly realised by transmitting the data in specified block lengths of a few thousand bits, the transmitter then waiting for an acknowledgment of valid reception from the receiving modem or for a request for a repeat. Validity is checked by some form of error detection, the simplest being word repetition or alternatively a parity check. If connections are set up over the switched telephone network, rather than over leased lines, then adaptive equalisation becomes necessary at the higher-speed data rates.

Data services using digital transmission facilities are in the advanced planning stage. Again, such services will be required to comply with digital transmission systems already established for telephony. Thus, the digital p.c.m. telephony channel at 64 kbit/s may accommodate any reasonably compatible combinations of data rates, such as 600, 2400, 9600 and 48 000 bit/s, in a UK proposal.[5] In this scheme, the 48 kbit/s data rate is made up to 64 kbit/s in two stages; first, two bits are added to head each 8-bit information word thus bringing the rate to 60 kbit/s, and finally filling bits are added on the basis of 1 bit in 16 resulting in the required 64 kbit/s rate. The two additional header bits are a 'status' bit, indicating whether the following word consists of data or signalling information, and an alignment bit referring to the 10-bit envelope.

Although the design of the early p.c.m. systems was aimed at the telephony application, data characteristics have been instrumental in highlighting some of their shortcomings and so expediting basic improvements. A particular case in point was the use of bipolar line coding in association with self-timed repeaters; satisfactory operation of this system (see Chapter 14) depends on the restriction of the number of consecutive spaces (typically to ten), a condition easily realised in telephony but not in data. This unsatisfactory situation is obviated by the use of filled bipolar codes (see Chapter 11) recommended for the new generation of p.c.m. systems.

4.4 F.D.M. assemblies

Frequency-division-multiplex (f.d.m.) systems have for some time been the mainstay of telephony transmission, and are precisely standardised. As digital transmission links are progressively introduced, occasions will arise where an f.d.m. assembly will be confronted with a digital transmission system. As f.d.m. modulation equipment is expensive it would not be economical to demultiplex the assembly into individual channels and recode these into p.c.m., and a more convenient procedure is to encode the f.d.m. assembly en bloc by means of special codecs and apply the resulting digital output into one of the available digital multiplex ports described in Chapter 5. In fact, current technology limits such 'transmultiplexing' to certain f.d.m. assemblies only and those of particular interest are listed in Table 4.3; they are specified by the CCITT,[2e] but the last two do not apply to North American practice. The calculation of digital objectives for these services will serve as a good example of the application of the principles of Chapter 2.

We start in deciding on a suitable h.r.c. by examining the current recommendations for analogue links. In this the situation is seemingly complicated by the existence of several such recommended connections, for instance, over symmetric pair cable using groups and supergroups,[2f] or over coaxial cable as shown in Fig. 3.3. It is, however, readily deduced that the latter calls for more stringent noise objectives because of the larger number of sections and translations permitted, and

Table 4.3 F.D.M. assemblies

Assembly	Frequency limits, kHz	Bandwidth, kHz	Number of channels	Load level, dBm0	Noise target,[1] pW
Group	60 – 108	48	12	3·3	60 – 100
Supergroup	312 – 552	240	60	6·1	60 – 100
Mastergroup	812 – 2044	1232	300	9·8	40 – 60
16-supergroup	60 – 4028	3968	960	14·8	40 – 60

Note: 1 refers to modulation equipment

may therefore be taken as a realistic model. The 2500 km connection is allowed 10 000 pW of noise and may be split into nine sections. The noise allocation is divided between line and terminal equipment in the ratio 7500 pW (3 pW/km) and 2500 pW, respectively, and additional targets for modulator-demodulator pairs are also recommended, see Table 4.3. Replacement of a typical f.d.m. section by a digital section is illustrated in Fig. 4.2. An essential feature to note is that whereas the line is the major noise contributor in the f.d.m. system, it is the terminal codec

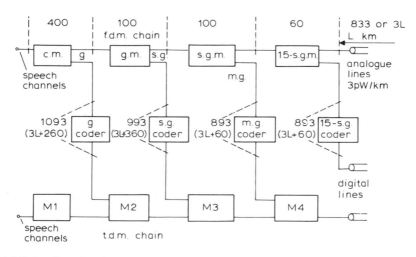

Fig.4.2 Noise allocations in digital conversion of f.d.m.
c.m.   = channel modulator
g.m.   = group modulator
s.g.m. = supergroup modulator
g.   = group
m.g.   = master group
s.g.   = supergroup
Figures refer to noise allocations, pW

which is the culprit in the digital case, and we shall see that some 100 pW will cover the contribution of the 278 km digital line link. Referring to Fig. 4.2 we see that if, for instance, a group coder is provided to enter the 2nd-order digital multiplex *M2* then its noise allocation is the sum of the allocations of the f.d.m. units it displaces; namely, 833 pW of line, and $(60 + 100 + 100)$ pW of 15- supergroup, supergroup and group modulators, totalling 1093 pW. Of this we shall allow 100 pW for the digital line. In this way, codec noise objectives for the various f.d.m. assemblies can be deduced. It should be noted, however, that since allocation is largely a result of the displacement of the analogue line link of L km and its length-dependent contribution $3L$ pW the above figures must be reduced if the actual section lengths are significantly shorter than 278 km (as, for instance, in the UK). This is an example of a penalty incurred by digital systems as a result of manipulation of parameters which would have been harmless for analogue systems. We must now consider the technicalities of converting these objectives to codec specifications.

4.4.1 F.D.M. loading of p.c.m. codecs

A fully loaded f.d.m. assembly, being the sum of a large number of independent random signals, is known to approach Gaussian properties. Hence, the quantiser impairment curves of Fig. 2.3, and the data of Table 2.1 are directly applicable; clearly the conventional f.d.m. load should be made to coincide with the maximum of the s.d.r. curve. (In practice, however, the conventional f.d.m. load level may vary over a range of some ±3 dB, resulting in a possible reduction of s.d.r. by some 3 dB.) Gaussian statistics do not apply to assemblies with a small number of channels such as the 12-channel group. Here it is found that, as for speech (see Section 2.2), the maxima of s.d.r. occur at input levels some 5 dB lower than for a Gaussian signal, with a corresponding reduction of the maximum value by approximately 5 dB.

Companding applied to Gaussian inputs makes a small improvement to the s.d.r. which is not considered worthwhile. An interesting consequence is that the small-signal and idling performance of the linear codecs is, in principle, potentially suspect when the f.d.m. bank is abnormally lightly loaded. However, the situation is saved, even in the absence of any speech, by the contribution of the everpresent pilot tones which load the codec away from the idle region.

We now deduce the quantising noise contribution in one f.d.m. channel of an *M*-channel assembly from the s.d.r. data of Fig. 2.3 and Table 2.1, assuming that the conventional f.d.m. load L_{con} is set at the codec input level corresponding to the maximum s.d.r. We have, in dBm

$$\text{noise for coded f.d.m. assembly} = L_{con} - \text{s.d.r.}_{max} = N_1$$

$$\text{noise per 4 kHz band} = N_1 - 10 \log_{10} M = N_2$$

$$\text{noise in 3·1 kHz channel} = N_2 - 10 \log_{10} \frac{4 \cdot 0}{3 \cdot 1} = N_3$$

noise after psophometric
weighting per telephony channel $= N_3 - 2{\cdot}5 = D_{qp}$

and

$$D_{qp} = L_{con} - \text{s.d.r.}_{max} - 10 \log_{10} M - 3{\cdot}6 \qquad (4.4)$$

L_{con} is given in Table 4.3, s.d.r.$_{max}$ in Table 2.1.

In a similar way, the noise from decoded errors in one f.d.m. channel after psophometric weighting D_{ep} may be deduced from eqn. 2.45 to be, in dBm

$$D_{ep} = L_{con} - 20 \log_{10} k + 10 \log_{10} P_e - 10 \log_{10} M - 2{\cdot}35 \qquad (4.5)$$

The loading factor k may be taken from Table 2.1. Alternatively, if high accuracy is not required, then eqn. 4.4 may be used in conjunction with the s.d.r. values given in Table 2.2 assuming $k = -12$ dB.

Finally, the noise due to jitter in the most seriously affected top channel of the f.d.m. assembly after psophometric weighting D_{jp} may be deduced from eqn. 2.42 to be, in dBm

$$D_{jp} = L_{con} - 10 \log_{10} M - \text{s.d.r.}_{top} - 3{\cdot}6$$

$$= L_{con} - 10 \log_{10} M + 20 \log_{10} \pi f_s \, \sigma_j - 3{\cdot}6 \qquad (4.6)$$

where f_s is the codec sampling rate and σ_j the r.m.s. jitter.

Eqns. 4.4 to 4.6 enable the translation of the noise objectives derived from the h.r.c. into specifications of coding precision, error rate and jitter.

4.4.2 Example: f.d.m. supergroup

We now apply the above to calculate the distortion specifications for a supergroup codec. From Fig. 4.2, the target maximum noise allowance based on the h.r.c. of Fig. 3.2 is 993 pW. In anticipation we allow 100 pW for error and jitter impairment, leaving 893 pW for coding proper. We shall eventually have to allow for a 3 dB increase in noise due to f.d.m. loading variation and 1 dB, say, for codec imperfections, limiting the actual codec contribution to 357 pW. Applying eqn. 4.4 and using Table 4.3 we therefore require

$$-64{\cdot}51 \text{ dBmp} \leqslant 6{\cdot}1 - \text{s.d.r.}_{max} - 17{\cdot}78 - 3{\cdot}6$$

$$\text{s.d.r.}_{max} \geqslant 49{\cdot}23 \text{ dB}$$

Consulting Fig. 2.3 we see that 10-bit coding gives s.d.r.$_{max}$ = 52 dB which offers a further margin for a certain amount of thermal circuit noise. Since the h.r.c. target-noise objective is highly dependent on section length, we present values of D_{qp}, an independent quantity, in Table 4.4 which may be used as a basis for calculation in other circumstances.

Using eqn. 4.5 it may be checked that an error rate of 10^{-10}/km contributes only 17 pW of distortion, thus leaving the lion's share of the 100 pW, left earlier in reserve, for jitter distortion. If this value is in fact used, then, from eqn. 4.6 the

Table 4.4   F.D.M. digital impairments

F.D.M. assembly	h.r.c. target, 278 km section	Quantising distortion D_{qp}, for N bits				Error distortion D_{ep}, $p_e = 10^{-10}$/km
		$N = 9$	$N = 10$	$N = 11$	$N = 12$	
Group	1093	–	1740	490	130	180
Supergroup	993	740	210	60	–	17
Mastergroup	893	350	100	–	–	10
16-supergroup	893	350	100	–	–	10

Note: all values in pW0p

required product $f_s \sigma_j$ is found to be 0·001. Anticipating the choice of $f_s = 576$ kHz, the required r.m.s. jitter is about 1 ns (see Table 4.5).

It should be apparent from Table 4.4 that a choice of a given set of quantising, error and jitter specifications is a compromise based on the technical difficulties of achieving each constituent specification, and only future experience will shape the optimum solution.

4.4.3 Sampling of f.d.m. assemblies

Most f.d.m. signals have bandpass rather than baseband spectra, so that the sampling theorem of Chapter 2 requires interpretation. In principle, of course, any signal may be translated to baseband by modulation allowing sampling at twice the basebandwidth, but, since sampling itself is basically a modulation process, judicious choice of the sampling frequency can in many cases neatly combine the two operations. We illustrate this in Fig. 4.3, showing some nonaliasing sampling conditions. It is seen that these are satisfied if, in addition to the sampling theorem requirements, the signal spectrum, f_1 to f_2, is wholly contained within the band $n f_s/2$ and $(n + 1)f_s/2$, $n = 0, 1, 2, \ldots$ i.e.

$$f_2 \leqslant (n + 1) f_s/2$$

and

$$f_1 \geqslant n f_s/2 \tag{4.7}$$

If these conditions are met and if, further, the signal spectrum is placed symmetrically between adjacent half-sampling-frequency harmonics, as shown in Fig. 4.3, then equal separation between sidebands is obtained; this condition is defined by

$$(2n + 1) f_s = 2(f_1 + f_2) \tag{4.8}$$

The sideband separation then becomes

$$\frac{f_s}{2} - (f_2 - f_1) \tag{4.9}$$

Table 4.5 Possible parameters for digital services

Service	Coding resolution, bits	Sampling rate, kHz	Information rate kbit/s	Information rate displaced speech channels[1]	Error rate overall	Error rate per km of h.r.c.[7]	Jitter r.m.s., ns
speech	8 C	8	64	1	10^{-5}	10^{-10}	1400
f.d.m. group	12 L	112	1344	21		10^{-10}	3·1[2]
f.d.m. supergroup	10 L	576	5760	90		10^{-10}	1·0[2]
f.d.m. mastergroup	10 L	2600[3]	26000[3]	406		10^{-10}	0·32[2]
f.d.m. 16-supergroup	10 L	8432[4]	84320[4]	1318		10^{-10}	0·10[2]
television[5,6]	8/9 L	13300[3]	120000[3]	1875	10^{-5}	10^{-9}	0·2
visual telephone			6336/8448	96/120			
sound programme	13 L	32	448	7	10^{-5}	10^{-9}	200

Notes: C = Companded, L = Linear
1 - Disregarding Multiplex structure   2 - For jitter distortion of 100 pWp0, no jitter reduction   3 - Approximate   4 - Nonunique solution
5 - Uncompressed   6 - With error concealment   7 - Between primary centres

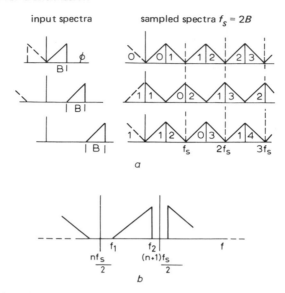

input spectra   sampled spectra $f_s = 2B$

a

b

Fig.4.3 Sampling of band-pass signals
   a Nonaliasing sampling conditions (numbers refer to harmonics of f_s)
   b Sideband relations

Since the spacing between f.d.m. channels is 4 kHz and all carrier translation frequencies are integers of this, it is desirable to choose f_s to be an integer of 4 kHz, also ensuring the absence of incoherent intermodulation components resulting from the sampling and pilot carrier frequencies. A particularly convenient situation is where one of the available carrier frequencies can be chosen as the sampling frequency.

We can apply the above to the choice of sampling frequencies for the various f.d.m. assemblies. For the f.d.m. group (60–108 kHz), application of eqn. 4.7 shows that for $n = 0$ f_s must exceed 216 kHz, clearly in excess of the 96 kHz required by the sampling theorem. However, $n = 1$ offers reasonable conditions, namely

$$108 \text{ kHz} \leqslant f_s \leqslant 120 \text{ kHz}$$

For equal sideband separation, applying eqn. 4.8 leads to $f_s = 2/3 \ 168 = 112$ kHz and thus an automatic choice. From eqn. 4.9, the sideband separation is then 8 kHz.

The same procedure leads to an f.d.m. supergroup sampling frequency of 576 kHz with a sideband separation of 48 kHz.

The 16-supergroup band of 60–4028 kHz necessitates a choice of $n = 0$ in eqn. 4.7 with a minimum $f_s = 8056$ kHz. In fact, the nearest carrier frequency of 8432 kHz is a possible choice.

The mastergroup (812–2044 kHz) is interesting inasmuch as it precludes the direct sampling approach, occupying more than one octave; application of eqn. 4.7 yields a permissible condition for $n = 0$ only, implying f_s in excess of 4088 kHz compared to the theoretical requirement of 2464 kHz, an unacceptably inefficient solution. To avoid this the spectrum must be first translated to a band for which the choice of $n = 1$ or $n = 2$ offers sampling at a rate acceptably higher than 2464 kHz, say 2600 kHz. The exact choice of f_s and of n depends on detailed consideration of f.d.m. translation practice.

Having determined the sampling rate for each assembly, the required r.m.s. jitter may be found following the method of the earlier supergroup example; values are presented in Table 4.5. The severity of the jitter specifications is quite apparent and it is doubtful whether they can be met without jitter reducing equipment, see Chapter 10.

4.5 Television

We restrict the discussion to the so-called System I, with 625 lines and a video bandwidth of 5·5 MHz, which is used in Europe,[28] although the method and quantitative conclusions would be very similar for the North American System M. The h.r.c. for television is significantly simpler than for telephony, being a dedicated rather than a switched, network. CCITT recommendation J 62 suggests an h.r.c. of 2500 km composed of three equal sections for an international link. It does not, however, specify the national part of the connection; geographically compact countries such as the UK cater for a total of eight sections, see Chapter 3. As each section ends in a video interconnecting point the possibility of eight coding-decoding operations arises.

Codec requirements are primarily determined by noise and visible interference criteria, although many other transmission parameters have to be specified. Reference 2, recommendation J 62 specifies a luminance channel peak-to-peak signal/r.m.s. noise ratio of 52 dB with noise weighted with a network providing 6·5 dB reduction over white noise. Let X_q be the peak-to-peak signal/r.m.s. quantising distortion produced by the codec. This becomes $(X_q + 6·5)$ after weighting. Anticipating a sampling rate of 13 MHz this represents a noise reduction of 6·5/5·5 or about 1 dB. Codec imperfections of 1 dB, say, exactly offset the latter gain. Chrominance signal excursions can exceed the peak (white) signal level, necessitating a further reduction of some 2 dB to $(X_q + 4·5)$. For the complete connection involving eight codecs, the s.n.r. becomes $(X_q - 4·5)$ and must not be less than the necessary 52 dB, i.e. $X_q \geqslant 56·5$ dB. Referring to eqns. 2.2 and 2.3, our definition of X_q implies

$$X_q = 20 \log \frac{2V}{\sqrt{D_q}} = 10·79 - 6·02 \, N \qquad (4.10)$$

For $N = 8$, $X_q = 59$ dB, which meets the target for the 8-section connection. However, there is some doubt on subjective grounds whether 8-bit coding is adequate in a chain of codecs without more sophisticated processing such as the addition of dither.[7]

As we have already discussed in Chapter 2, the choice of sampling frequency for a composite television signal is critical because of the highly structured harmonic spectrum in which all components have determinate relationships, being derived from one master clock frequency. As a result, the sampled quantising-distortion spectrum will consist of discrete intermodulation components which must be placed relative to the video components so as to produce minimum picture impairment by careful choice of sampling frequency. To this end, three selection strategies are possible (i) choice of f_s equal to a harmonic of the colour subcarrier (ii) choice of f_s equal to a harmonic of the line frequency, and (iii) choice of f_s unrelated to signal components, for minimum visible impairment. Both (i) and (ii) preclude the presence of subjectively disturbing beat effects, since the quantising distortion components are then locked to the signal structure. If (iii) is adopted then a harmonic of 4 kHz would alleviate potential crosstalk effects with f.d.m. systems.

If strategy (i) is adopted then the obvious choice for f_s is the third harmonic of the colour subcarrier, or 13·300856 MHz, which is comfortably, although not extravagantly, higher than the sampling theorem requirement of 11 MHz. Taking this figure as a guide we have an information rate for television of 106·4 Mbit/s.

At this stage we should point out that current work on television-signal compression may well lead to significant savings in information rate. The use of line and frame predictive algorithms, differential p.c.m. and sampling below the Nyquist rate may well result in broadcast quality television at 40–60 Mbit/s, (e.g. Reference 8), at the cost of more 'brittle' error performance.

In terms of random error-rate requirements, subjective tests for directly coded video signals[6] indicate that an error rate of 10^{-7} is imperceptible and 10^{-6} barely so. Use of an error-concealment strategy by, say, parity checking and previous sample retention on detection of an error, lowers the requirement for imperceptible impairment to an error rate of 10^{-4} for random errors and 10^{-5} for error bursts.[6] The conclusion here is, therefore, that if the line error rate of 10^{-10}/km is accepted as a requirement for other services, the television case will be covered without additional error protection; however, a modest increase in information rate to allow for error protection would obviously render the service almost insensitive to the design error rates envisaged in practice.

Finally, digital television has certain inherent advantages; studio operations such as mixing and fading can be digitally controlled, and, more importantly, the control of group delay in long transmission links ceases to be a problem.

4.6 Visual telephone

Specifications for a visual-telephone service are currently the subject of much international study and in a state of flux owing to the increasingly powerful signal processing technology which is becoming available. A basebandwidth of 1 MHz is currently favoured, and this implies a sampling rate of around 2 MHz assuming aliasing is not permitted. Further assuming that more than one bit per sample encoding is required, we can conclude that the service cannot be accommodated within the 1st-order multiplex rate of 1·5 or 2 Mbit/s (see chapter 3). The question, therefore, which remains is how many visual telephone channels can be accommodated within the 2nd-order multiplex of 6·3 or 8·4 Mbit/s. A current conservative estimate would be one, allowing either 3 or 4 bits/per sample. However, the more sophisticated coding and processing already mentioned in connection with television may well increase the number of channels or alternatively allow a colour service within the 2nd-order multiplex.

Quite obviously, the susceptibility to errors and jitter will be specific to the processing involved, and we can only reiterate the philosophy that the requirements on error rate must be compatible with those envisaged for the 'senior' service, namely speech. In general, it should be added that, although there is little doubt that a viable picture-communication service will emerge in the long term, it may show little resemblance to the currently considered service and might well be a fast data-based system, without being strictly real-time, having a much lower information rate.

4.7 Sound programme

The sound-programme service is another specialised utility which has to make use of circuits designed for speech. The recommended h.r.c.[2h] has a length of 2500 km with two intermediate audio interconnection points, although again smaller countries such as the UK favour a somewhat larger number. Recommendation J 21[2] calls for a circuit s.n.r. with an unweighted noise of 47 dB; this refers to full-load sine wave/r.m.s. noise ratio. This value is simply the best that can be reasonably expected from the number of telephone circuits displaced in an f.d.m. assembly by the programme signal. The bandwidth of the signal for high-quality monophonic sound is specified as 15 kHz, allowing a comfortable fit for two programme circuits in an existing f.d.m. group. To achieve a subjectively acceptable performance, current analogue transmission practice[21] calls for the use of pre-emphasis/de-emphasis networks and compandors. The latter provide a gain at low levels of 17 dB implying an s.n.r. with unweighted noise of 47 + 17 = 64 dB. A further 2 dB must be allowed for level uncertainty, resulting in an overall s.n.r. of 66 dB.

If now the h.r.c. includes digital links then, assuming three intermediate audio points implies a worst combination of four codecs in tandem, and hence a signal/ quantising-distortion ratio of better than 72 dB per codec. Applying eqn. 2.5, the number of coding bits N must be not less than 12; N = 12 leaves some 2 dB to spare for all other impairments and hence N = 13 is a safer value. This condition is reinforced by the possibility of some codecs in the connection being in their idling region (see Chapter 2) during the quietest programme passages; in such a case, the normal $q^2/12$ distortion may increase to $q^2/4$, and although the presence of random effects makes this unlikely, 13 coding bits would be preferable.

The above derivation of codec specifications has been based purely on current analogue-system practice and it is clearly necessary to re-examine them by direct subjective assessment of the effects of digital impairments on programme quality. Reference 9 indicates that for the quietest passages an s.n.r. of 75 dB is required for imperceptible subjective impairment, implying 81 dB per codec or 14-bit coding. However, application of dither to randomise idling effects (see Chapter 2) reduces the requirement to 13 bits. If an error-concealment strategy is further incorporated (as in television) by the addition of one bit for parity checking, an error rate of 10^{-5} is rendered imperceptible. A tentative target for jitter performance is some 200 ns r.m.s. (Reference 9). The above digital requirements assume linear coding, and it is possible that development of a companding law for programme signals may reduce the number of digits.

The choice of sampling frequency here is fairly easy; f_s = 32 kHz is comfortably above the minimum of 30 kHz, while being an integer of both 4 kHz and 8 kHz, the f.d.m. spacing and p.c.m. speech-sampling frequencies, respectively.

It is interesting to note the inherent advantages offered here by digital techniques; the possibility of digital mixing, fading and recording of programme material. It seems, however, that this is achievable at a cost of increased transmission capacity, this being 14 × 32 = 448 kbit/s, or exactly seven 64 kbit/s speech channels, compared to four analogue speech channels. For transmission a number of digital programme channels would be combined to fit into the first or 2nd-order multiplex ports.

4.8 Summary

Collecting together the above considerations we can summarise our conclusions in Table 4.5. We again stress that the data presented is tentative and that we expect authoritative recommendations to take shape in the near future under the auspices of the CCITT and CCIR.[10] Several explanatory remarks are required. The information rate given in terms of displaced speech channels draws attention to the basic inefficiency of digitising certain services; for example, the 12-channel f.d.m. group displaces 21 digital channels. This situation is in fact even worse when attention is directed to the possible entry ports into the digital multiplex hierarchy; in the case of the group assembly, the nearest p.c.m. multiplex would be the 24-channel format, if available, implying a wastage of one digital channel for every f.d.m. channel.

Jitter requirements, particularly for t.d.m. assemblies are theoretical values and will obviously be relaxed when experience with jitter reducers has been established, see Chapter 10.

The quoted error rate per kilometre of the h.r.c. is rather more stringent than a simple division of overall error rate by h.r.c. length would imply. The margin left allows a somewhat worse error performance for the end connections, (see Section 4.2).

Some combinations of services and multiplex hierarchies (see Chapter 5) fit rather badly on the basis of information rates; for example, the f.d.m. group and supergroup do not 'fill' the 30 and 120 channel p.c.m. multiplex levels adequately. In such cases, there may be circumstances where a subsidiary multiplex combining a number of input signals for entry to the next-order multiplex might be warranted; in the case of the f.d.m. group this might involve a ×5 multiplex bringing the information rate up to 105 channels, which allows sufficient capacity for additional digital housekeeping and entry into the 120-channel 2nd-order multiplex level.

The manner in which the services considered may be incorporated within the digital systems now under consideration is indicated in Chapter 16.

4.9 References

1   CCITT: 'Planning of digital systems', Special Study Group D, contribution 103, June 1974
2   CCITT: Green Book, Geneva, 1973 vol. III-1 covers recommendations G 100—455 inclusive vol. III-2 covers remaining G series and J series
    a G 103, *b* G 711, *c* G 712, *d* G 223, *e* G 211, *f* G 322, *g* J 62, *h* J 11, *i* J 31
3   RICHARDS, D.L.: 'Transmission performance of telephone networks containing p.c.m. links', *Proc. IEE*, 1968, **115**, (9), pp. 1245—1258
4   DAVEY, J.R.: 'Modems', *Proc. IEEE*, 1972, **60**, pp. 1284—1292
5   WILLIAMS, M.B.: 'Developments in data communications', *Post Off. Electr. Eng. J.*, 1971, **64**, pp. 70—80
6   DEVEREUX, V.G., and MEARES, D.J.: 'Pulse code modulation of video signals: subjective effects of random digit errors', BBC Research Dept. Report 1972/14
7   THOMPSON, J.E.: 'A 36 Mbit/s television codec employing pseudo-random quantisation', *IEEE Trans.* 1971, **COM-19**, pp. 872—879
8   GOLDING, L.: 'Comsat technical Memo., CL-8—67, available on microfiche PB 178993, USGRDR
9   SHORTER, D.E.L., and CHEW, J.R.: 'Application of pulse-code modulation to sound-signal distribution in a broadcasting network', *Proc. IEE*, 1972, **119**, pp. 1442—1448
10   MUNDAY, S.: 'Error-rate objectives for a multi-service digital network', *in* 'Telecommunication transmission', *IEE Conf. Publ.* 1975, **131**, pp. 15—21
11   CATTERMOLE, K.W.: 'Principles of pulse code modulation' (Illiffe, 1969)

Chapter 5

Multiplexing

5.1 Requirements and definitions

In previous Chapters we have examined the characteristics of signal sources, and we have found that each type of source tends to be associated with a characteristic digit rate. For example, telephony grade speech requires 64 kbit/s, while a broadcast quality colour television signal requires about 60 Mbit/s. In later Chapters we will find that there are also characteristic digit rates associated with the various transmission media; at one extreme a rate of about 2 Mbit/s is appropriate for a balanced pair transmission line, while at the other, the optimum digital channel on a waveguide system provides a capacity in the region of 500 Mbit/s. To make efficient use of these transmission media the information rate of the signal source must be matched to the capacity of the transmission path. This is achieved by multiplexing, which is the process of combining a number of the primary sources to form a composite signal. In this sense, multiplexing may be regarded as a signal matching process, and may be considered to complement the waveform matching used to optimise the signal/noise performance of the transmission path, which will be described in later Chapters.

There is an important subsidiary benefit of multiplexing. It leads naturally to arrangements in which various primary signals are processed into almost identical formats, so that a single design of transmission equipment may then be used to carry a number of different types of traffic. As an extension of this approach, it is found to be convenient to adopt a hierarchical multiplexing structure, in which the signals carried by low-capacity transmission systems form the basic units from which the inputs for the higher-capacity systems are assembled. This tends to define an overall structure for the complete transmission network, while maintaining the flexibility essential to cover operational requirements and future developments.

5.2 Basic ideas

An obvious first requirement is that the multiplexing process must be reversible, that is, it must be possible to separate the signals again at the receiving end of the transmission system. This requirement is met if the components of the multiplexed signal are orthogonal. Mathematically, the condition for a set of signals $f_1(x), f_2(x), f_3(x) \ldots f_n(x)$ to be orthogonal is

$$\int_0^s f_n(x) f_m(x) \, dx = 0 \qquad\qquad \text{for } m \neq n \text{ and } s \text{ suitably chosen} \qquad (5.1)$$

(See Pipes and Harvill, Reference 1, Chapter 12.) One such set of orthogonal functions is the set of circular functions, for which we have

$$\int_0^{2\pi(m+n)} \sin(2\pi n t) \sin(2\pi m t) \, dt = 0 \qquad\qquad \text{for } m \neq n \qquad (5.2)$$

Alternatively, if we define a set of functions such that $f_m(t)$ is nonzero only for values of t where all other functions $f_n(t)$ in the set have zero value, then these also form an orthogonal set. These two sets form the bases of the classical schemes of frequency-division multiplexing (f.d.m.) and time-division multiplexing (t.d.m.). In frequency-division multiplexing, each primary signal is shifted in frequency by modulation, to form a set of sinusoidal components which do not overlap the sets formed by any other signals, as shown in Fig. 5.1a. In time-division multiplexing each signal is quantised in time (see the discussion of the sampling theorem in Chapter 2) the sampling instants being chosen to interleave with those employed for the other signals, as shown in Fig. 5.1b.

In a sense, these two schemes may be considered to represent partitions in a time-frequency space. In f.d.m., the signals are processed to occupy separate segments of the frequency dimension, while sharing common time co-ordinates. T.D.M. signals are the converse; they occur at separated times, but since the samples have infinite frequency spectra they simultaneously occupy the entire frequency dimension. The reader may have noticed that this division between time and frequency was not implicit in the original definition of orthogonality, which leads to the question of the possible existence of orthogonal sets of signals which coexist in both time and frequency. In recent years, it has been shown that such sets do exist, the best known being the set of Walsh functions.[2] Multiplexing schemes based on these have been devised,[3,4] but since these are rather specialised, we do not propose to present any further account of them here. For the remainder of this Chapter we shall concentrate on time-division schemes.

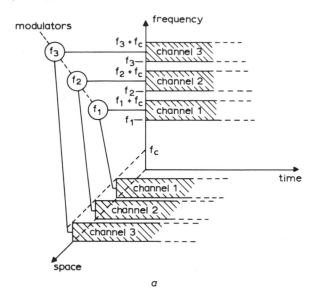

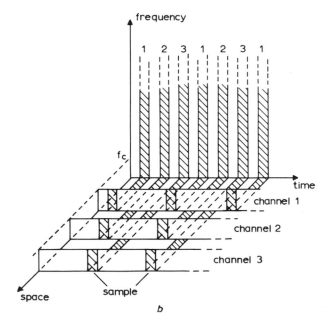

Fig.5.1 Frequency and time-division multiplexing
a Frequency division
b Time division

5.3 Frame structures

In a t.d.m. system the time slots in the outgoing signal will be allocated to the incoming channels in some predetermined manner. In theory this allocation can be done in an almost unlimited number of ways, but, in practice, the schemes employed usually fall into one of four catagories. The simplest cases arise when we are combining a number of similar signal sources, each of which is arriving as a continuous digit stream at a common fixed-digit rate. The first possibility is then to combine these streams digit by digit, as shown in Fig. 5.2*a*. This arrangement is known as 'digit interleaving'. The multiplexer can be considered to be functionally equivalent to a single rotating switch, and it can be seen that, apart from a small delay that may be needed in one or two channels to avoid sampling at the instant when the input is changing, no storage will be required in this scheme. A second alternative is to accept groups of digits from each input in turn, as shown in Fig. 5.2*b*. The arrangements are obviously rather more complicated in this case, since the multiplex switch has, in effect, to halt at each input while the entire group of digits is transferred, and it is clear that if the inputs are arriving continuously on each channel some local storage will be needed to accumulate the incoming signals while waiting for the next transfer. Despite this added complexity this scheme is

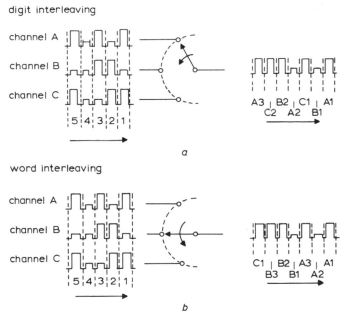

Fig.5.2 Multiplex structures
   a Digit interleaving
   b Word interleaving

fairly common. It becomes attractive when the incoming signals have some internal structure, for example, when the signals consist of groups of digits, each group being a code word, and for operational reasons it is advantageous to preserve the integrity of these groups throughout the transmission system. This scheme is therefore known as 'word interleaving'.

These two schemes can be extended to cover cases in which the incoming channels are not all at the same digit rate. As an example, consider the association of an incoming channel having a digit rate $3f_t$ with three other channels having digit rates f_t. This can be accomplished by using a frame of six time slots, with alternate slots allocated to the high-speed channel, as shown in Fig. 5.3. In a word-interleaved system a similar result can be attained by combining words of different lengths. It can be seen that the minimum length of the multiplex frame must be a multiple of the lowest common multiple of the incoming channel digit rates, and hence this sort of arrangement is only practical when there is some fairly simple relationship between these rates. The case of completely asynchronous tributaries will be considered in Section 5.6.

We have so far only considered inputs which are arriving continuously, but we know that in the majority of communication systems the individual channels are only active for a small part of the time. This leads to multiplexing schemes in which the number of inputs connected to the multiplexer exceeds the number of output channels available, with only those that are actually carrying traffic being included in the multiplexed output at any given time. This obviously involves much more complicated switching operations, and also requires rather careful system planning. In any random traffic situation we cannot guarantee that the number of transmission channels demanded will not exceed the number available, but, by taking account of the statistics of the signal sources, it is possible to ensure that the probability of this occuring becomes acceptably low. Multiplex structures of this type have been developed for satellite systems, and are known as 'time-division multiple-access' (t.d.m.a.) systems.[5,6]

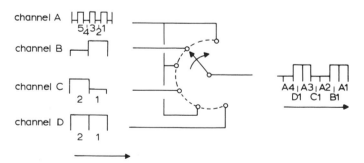

Fig.5.3 Multiplexing of tributaries having different digit rates

In t.d.m.a. systems employed for telephony the design parameters are chosen so that any overload condition only lasts for a fraction of a second, which leads to acceptable performance for speech communication. However, for many types of data, and for telegraphy, transmission delays are relatively unimportant, and it is then possible to use systems in which complete messages or data packets are accumulated at the terminals of the system and transmitted in sequence. These systems are referred to as 'message-switching' or 'packet-switching' systems.

At the receiving terminal the incoming digit stream must be divided into its component signals, which are then directed to the appropriate output channels. This implies that the receiving terminal must be able to correctly identify each incoming time slot. In fixed systems operating continuously this could, in theory, be organised during the initial setting up of the system which should then continue to operate in synchronism *ad infinitum,* but in the practical world frame alignment will sooner or later inevitably be lost through some transmission impairment or equipment malfunction, and it is essential to provide some means of automatically detecting this loss and regaining frame alignment. Similar requirements arise in systems which only operate intermittently, or which have variable transmission paths (i.e. mobile systems). This is usually accomplished by inserting a fixed digit pattern called a 'frame-alignment word' into the transmitted digit stream at regular intervals. The frame-alignment word (f.a.w.) may occupy a number of adjacent time slots (bunched framing), or may be allocated a number of isolated time slots distributed through the signal (distributed framing). The segment of the transmitted signal formed by a frame-alignment word and its associated message time slots is defined as a 'frame'.

It should be noted that it is not always necessary to include a frame-alignment word. If the transmitted signal contains any systematic feature that can be identified with the frame structure this can be used by the receiving terminal to establish alignment. For example, in a word-interleaved system if it is known that certain digit patterns are not included in the legal vocabulary these patterns can only arise through combinations of parts of adjacent words, and so their detection may be used to locate the boundaries between words. We shall encounter examples of this type when we consider line codes in Chapter 11. Another very interesting technique has been described by Gray and Pan.[7] They have shown that for a television signal converted to p.c.m. through the use of Gray code, the probability of a mark in the second digit in each code word is significantly different from that in any other location, and so this feature may be used as a framing identifier.

The message-switching schemes described earlier represent another variant of the framing structure. Each message may be considered to form a frame, but in this case the frames are of variable length. Each message is preceeded by a number of service digits. The first of these form a fixed digit pattern, the equivalent of the frame-alignment word, and are followed by further digits which provide information relevant to the message, for example, its destination, length and priority.

5.4 Frame-alignment systems

At this point it becomes appropriate to examine the methods used to detect loss of frame alignment and to regain it. For the present, we shall concentrate on bunched-alignment systems. It will be shown later that the analysis of bunched-framing schemes can be applied to the distributed case with only minor alterations. We define a frame to consist of m time slots, the first n of which contain the frame-alignment word, the remaining $m-n$ being allocated to c message channels, $c_1, c_2, c_3, \ldots c_c$. We will assume that the information being transmitted over these channels appears as random digit streams, so that the probability of a mark in any channel time slot is 0·5. The probability of a transmission error in any time slot is P_t. The transmission rate is r digits/s, hence the duration of each time slot is

$$t_s = 1/r \tag{5.3}$$

and the duration of each frame is

$$t_f = mt_s \tag{5.4}$$

The f.a.w. will be assumed to occupy a fraction α of the total transmission capacity of the systems, so that

$$\alpha = n/m \tag{5.5}$$

In the majority of cases the design objective will be to obtain satisfactory performance for the minimum value of α.

5.4.1 State-transition diagram

We now introduce the idea of describing the operation of the framing system in terms of a set of states and the probabilities of transitions between these states. This is an important technique that will be used again later. We start by making the rather obvious remark that the receiving demultiplexer can be in one of two conditions, it can either be in alignment with the incoming signals, or it can be out of alignment. As discussed in the last Section, the terminal includes facilities for detecting the f.a.w. (or its equivalent) and uses these to determine its current state. If a framing word is detected in the expected position, the terminal will assume that alignment exists, and will operate in the 'lock' mode. This usually involves distributing the next $m-n$ digits to the channel outputs, and then checking the following n digits to confirm the presence of the next f.a.w. If this test fails the terminal may assume that alignment has been lost, and initiate the procedure employed to regain alignment. A large number of techniques can be used for this, but the most common involve slipping one time slot and then examining the next n digits to see if these match the f.a.w. If not, the system slips another time slot, repeats the examination, and continues in this way unitl a framing word is recognised. While this is taking place the system is said to be in the 'search' condition. In the simplest framing schemes, the terminal returns directly to the lock condition on identifying a framing word.

It is clear that these operations are open to errors. For example, a transmission error may corrupt an f.a.w. so that it is not correctly identified, or during a search operation a random combination of signal digits may simulate the required framing pattern. Thus the condition that the receiving terminal believes to exist may not coincide with the actual situation and four possible states may then arise:

S_1 the system is in alignment and in the lock mode
S_2 the system is in alignment, but in the search mode
S_3 the system is out of alignment, but in the lock mode
S_4 the system is out of alignment and in the search mode.

The time spent in each of these states will depend on the probabilities of transitions between one state and another. For example, there will be a probability P_{31} of a transition from S_1 to S_3 and a probability P_{41} of a transition to S_4, which together represent the probability of a genuine loss of alignment. Similarly, the transition P_{21} represents the probability of a spurious detection of misalignment due to a failure to recognise the f.a.w. Considering all possible transitions, we arrive at the diagram shown in Fig. 5.4.

This diagram can be used in two ways. If we define the transitions to be the probabilities of a change between each separate test for the framing word (that is, the probabilities of transition during each frame period), and define the probability of the system being in state S_i at the kth test as $P_k(S_i)$ we can derive a set of equations describing the changes at each test of the f.a.w.

$$\begin{bmatrix} P_{k+1}(S_1) \\ P_{k+1}(S_2) \\ P_{k+1}(S_3) \\ P_{k+1}(S_4) \end{bmatrix} = \begin{bmatrix} P_{11} \dots P_{14} \\ \vdots \qquad \vdots \\ \vdots \qquad \vdots \\ P_{41} \dots P_{44} \end{bmatrix} \begin{bmatrix} P_k(S_1) \\ P_k(S_2) \\ P_k(S_3) \\ P_k(S_4) \end{bmatrix} \qquad (5.6)$$

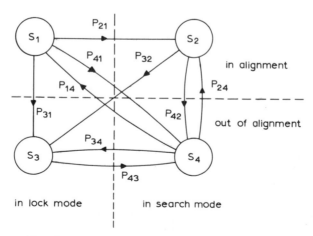

Fig. 5.4 State-transition diagram for a simple frame-alignment system

We can then define the initial state of the system, for example, that is initially in lock and in alignment, by

$$[P_1(S)] = \begin{bmatrix} 1 \\ 0 \\ 0 \\ 0 \end{bmatrix} \tag{5.7}$$

and by repeated application of eqn. 5.6. study the future operation. For the nth test

$$[P_n(S)] = [P]^n \, [P_1(S)] \tag{5.8}$$

In general, the system will tend to a stable condition after a large number of iterations. We also note that since all possible conditions have been covered by our state diagram

$$\sum_i P_\infty(S_i) P_{ki} = P_\infty(S_k)$$

$$\sum_i P_\infty(S_i) = 1 \tag{5.8a}$$

where $P_\infty(S_i)$ is the ultimate stable probability of state i. We may therefore set up an extended relationship describing this stable condition

$$\begin{bmatrix} P_{11}-1 & P_{12} \ldots \ldots P_{1n} \\ P_{21} & P_{22}-1 & \vdots \\ \vdots & & \ddots & \vdots \\ P_{n1} \ldots \ldots \ldots \cdot P_{nn}-1 \\ 1 & 1 & 1 \ldots . 1 & 1 \end{bmatrix} \begin{bmatrix} P_\infty(S_i) \\ \vdots \\ \vdots \\ P_\infty(S_n) \end{bmatrix} = \begin{bmatrix} 0 \\ 0 \\ \vdots \\ \vdots \\ 1 \end{bmatrix} \tag{5.9}$$

It is clear that since there are now more equations than unknowns, they cannot form an independent set. In general, we may eliminate one equation and then solve the reduced set. It should be noted that it does not necessarily follow that even this reduced set of equations will be linerly independent, in which case multiple solutions can arise. This would be undesirable in practical systems, and so the designer will attempt to avoid such conditions. The reader is referred to Braae[8] for an account of the mathematical background to this situation.

A little thought will show that the transition probabilities are not completely independent. For example, to achieve rapid realignment we would like to maximise P_{14} and minimise P_{34} which implies the use of a long frame-alignment word that is not easily simulated by random data. However, a long word is more easily corrupted by transmission errors, so this process results in an undesirable increase in P_{21} and P_{24}. In practical systems this leads to the adoption of rather more complex realignment strategies, a typical example being shown in Fig. 5.5. In this the probabilities of spurious transitions out of the lock mode have been reduced by using

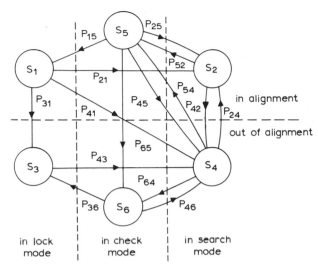

Fig. 5.5 State-transition diagram for a 3-stage frame-alignment system

rather more tolerant tests for the presence of the f.a.w. These involve accepting some corruption in any single test, provided that corruptions do not occur repeatedly. Errors must then be detected in μ successive tests, before a transition to the search mode occurs, which is acceptable under conditions where genuine loss of alignment is rare. However, this rather lengthy checking procedure is not acceptable during the search for realignment, and so a modified test, defined as the 'check' mode is introduced. In this, when a framing word is located during a search it is provisionally assumed to be correct, but any failure to identify it during the next β tests results in immediate resumption of the search. This introduces two additional states.

S_5 the system is in alignment and in the check mode
S_6 the system is out of alignment but in the check mode.

Detailed analysis of schemes of this type are given by Taylor[9] and Williard.[10] These lead to extended versions of eqns. 5.6 and 5.9 which may be solved for the transient behaviour, and the long-term average performance of the system. However, as we have seen in earlier Chapters, the subjective effect of an impairment may be strongly dependent on its duration, and hence information on the mean time spent out of alignment will not give a complete assessment of the resulting transmission degradation. We also have to consider the durations of the individual events and the statistics of these durations. These will be examined in the following Sections.

5.4.2 Detection of misalignment
We will first consider transitions from the lock to search modes. Two typical schemes will be examined; in the first each f.a.w. is checked, and if μ successive

tests fail to confirm the presence of the word a search is started; in the second an up-down counter is used. This is initially set to zero, each failure to identify the f.a.w., then increments the count by x, each successful identification decrements the count by y, and alignment is deemed to be lost if the total count exceeds z.

We will examine the performance of the first scheme when a genuine loss of synchronism has occurred. On average, half a frame will elapse before the next test, and a minimum of μ tests must be made before loss of alignment is confirmed. The minimum detection time is therefore

$$t_{min} = t_f(\mu - 0\cdot5) \qquad (5.10)$$

where t_f is the duration of one frame.

There will be a probability P_τ that τ tests have to be made before μ successive identifications of loss of alignment take place, resulting in a detection time

$$t_\tau = t_f(\tau - 0\cdot5) \qquad (5.11)$$

The average or expected number of tests required is ϵ yielding an average detection time of

$$t_\epsilon = t_f(\epsilon - 0\cdot5) \qquad (5.12)$$

We must next determine the probability of detection of misalignment in each separate test. This will depend on the length of the alignment word and the severity of the test criterion. If we accept the f.a.w. to be present when the correct digit pattern appears in all but e of the n time slots occupied by the word, the probability of simulation by random data becomes the sum of the probability that perfect simulation will occur, the probability that simulation will occur in all but one time slot, and so on. This can be shown to be (see for example Fry[11])

$$P_e^n = \frac{1}{2^n} \sum_{x=0}^{e} \frac{n!}{x!\,(n-x)!} \qquad (5.13)$$

The probability of detection of loss of alignment is then

$$Q_e^n = 1 - P_e^n \qquad (5.14)$$

Values of Q_e^n are given in Table 5.1.

We now have to calculate the probability that τ tests will have to be made before μ successive identifications of misalignment take place. While this requirement is easy to state, its evaluation is surprisingly difficult, and involves fairly sophisticated mathematical techniques. Since these techniques will be applicable to a number of other problems in digital transmission, it is appropriate at this point to include a brief summary of the theory involved. Readers interested in a more extensive treatment should consult Feller's book.[12]

5.4.3 Generating functions

Suppose a series of tests are carried out (in our case, tests for loss of alignment) and it is found that the probability of an event x (e.g. detection of misalignment) at the ith test is P_i. This sequence of probabilities can be expressed as a power series

$$P(s) = p_0 + p_1 s + p_2 s^2 + \ldots \tag{5.15}$$

In this s is merely a dummy variable since we are only interested in the coefficient of each term. The significance of this mathematical trick is that it is frequently possible to identify the series as the expansion of some known function. For example

$$P(s) = p(1 + s + s^2 + s^3 + \ldots)$$
$$= p/(1 - s) \tag{5.16}$$

This makes it possible to describe the probabilities in a concise closed form. The function employed is called the 'generating function'. Study of generating functions shows that a number of important properties of the probability distributions can be derived directly from them. We can, for example, set up a second series giving the probability that the event occurs after the jth test, namely

$$Q(s) = q_0 + q_1 s + q_2 s^2 + \ldots$$

It is found that

$$Q(s) = (1 - P(s))/(1 - s) \tag{5.17}$$

Differentiating $P(s)$ in eqn. 5.15 and then putting $s = 1$ gives the expectation or mean value of the number of tests needed to produce the event, i.e.

$$E(x) = P'(1) = Q(1) \tag{5.18}$$

Similarly, the variance of the number of tests is

$$V(x) = P''(1) + P'(1) - P'^2(1)$$
$$= 2Q'(1) + Q(1) - Q^2(1) \tag{5.19}$$

It can be shown that if u_i is the probability that the event occurs at the ith trial and f_i is the probability that this is the first occurrence of the event, then the corresponding generating function is

$$F(s) = 1 - 1/U(s) \tag{5.20}$$

5.4.4 Detection times

The probability that an event has occurred μ times in succession at the jth trial is described by the generating function

$$U(s) = \frac{1 - s + qp^\mu s^{\mu+1}}{(1 - s)(1 - p^\mu s^\mu)} \tag{5.21}$$

where p is the probability of occurrence in any single trial and $q = 1 - p$.[12] By applying eqn. 5.20 it follows that the probability that this is the first occurrence of μ successive events is described by

$$F(s) = \frac{p^\mu s^\mu (1 - ps)}{1 - s + qp^\mu s^{\mu+1}}$$ (5.22)

Substitution of the value of Q_e^n derived in eqn. 5.14 in place of p then leads to a formal solution to the realignment problem. Applying eqn. 5.18 shows that the expected number of tests to detect loss of alignment will be

$$\epsilon = \frac{1 - (Q_e^n)^\mu}{(1 - Q_e^n)(Q_e^n)^\mu}$$ (5.23)

which may be substituted in eqn. 5.12.

However, if we are interested in the probability of detection in a specified number of tests, we require a term-by-term expansion of eqn. 5.22 and a few attempts will convince the reader that this is not a simple matter. One solution is to expand $F(s)$ as a series of partial fractions.

If

$$F(s) = \frac{A(s)}{B(s)}$$

and

$$B(s) = (b_1 - s)(b_2 - s)(b_3 - s) \ldots$$

Then

$$F(s) = \frac{a_1}{b_1 - s} + \frac{a_2}{b_2 - s} + \ldots$$ (5.24)

Then it can be shown that

$$f_i = \frac{a_1}{b_1^{i+1}} + \frac{a_2}{b_2^{i+1}} + \frac{a_3}{b_3^{i+1}} + \ldots$$ (5.25)

5.4.5 Detection probabilities for a counter system
Like the check-and-repeat system, the counter system is easy to describe but it is difficult to analyse, and as far as we are aware no formal solution for this scheme has been published.* The difficulties arise mainly from the boundary conditions, namely, that the counter does not count below zero, and on reaching the maximum count z terminates operations. However, numerical solutions can be obtained by a variant of the state-sequence method described earlier.

*We have had access to some unpublished work by Mr. J.M. Griffiths of the British Post Office, and the development here follows his analysis.

The state vector (S_i) is now considered to describe the probability of the count being i at a given test. If the probability of detecting misalignment at this test is Q_e^n there will be a probability of Q_e^n of transition to state S_{i+x} and a probability $1 - Q_e^n$ of transition to S_{i-y}. The lower boundary condition is introduced by specifying that if $i - y$ is negative the zero state is entered. Similarly, at the upper boundary any $x + i$ greater than z is put equal to z. Because operation ceases at this point the probability of a count down of $z-y$ is zero, and if P_{zz} is made unity we will evaluate the accumulated probability of detecting misalignment. The complete expression then becomes

$$
\begin{bmatrix}
P_{\mu+1}(S_0) \\
P_{\mu+1}(S_1) \\
\\
P_{\mu+1}(S_x) \\
\\
P_{\mu+1}(S_z)
\end{bmatrix}
=
\begin{bmatrix}
1-Q_e^n & 1-Q_e^n & \cdots & \cdots & 1-Q_e^n & \cdots & 0 \\
0 & \cdots & 0 & \cdots & \cdots & 1-Q_e^n & \vdots \\
\vdots & & \vdots & & & & 0 \\
Q_e^n & \cdots & 0 & & & & \vdots \\
\vdots & & Q_e^n & \ddots & & & \vdots \\
\vdots & & & \ddots & \vdots & & \vdots \\
0 & \cdots & \cdots & Q_e^n & Q_e^n & \cdots & 1
\end{bmatrix}
\begin{bmatrix}
P_\mu(S_0) \\
P_\mu(S_1) \\
\\
\\
\\
P_\mu(S_z)
\end{bmatrix}
\quad (5.26)
$$

As before the initial state is

$$
P_1(s) =
\begin{bmatrix}
1 \\
0 \\
0 \\
\vdots \\
0
\end{bmatrix}
$$

and after k tests

$$
[P_k(s)] = [Q]^n [P_1(s)] \qquad (5.27)
$$

5.4.6 Practical values and optimum frame lengths

It will be useful at this point to introduce some practical examples to give an idea of the magnitudes of the parameters involved. We will take two typical systems. System 1 initiates reframing after three successive framing words have been in error. System 2 uses a counter, counting up by 4 whenever a framing word is not detected, and down by 1 whenever a successful detection takes place. Reframing is initiated if the count reaches 10. In both systems we will assume that the f.a.w. is accepted only if it contains no errors. Thus, the probability of simulation of an n-digit word by random data is $\frac{1}{2}^n$ and $Q_0^n = 1 - \frac{1}{2}^n$. The number of tests needed to initiate reframing are then given by Table 5.2. If we assume that a framing content of 1% is employed, we can translate these results into the equivalent number of time slots needed to detect misalignment, as a function of the length of the f.a.w. This is given in Table 5.3. Note that since we do not know when the original misalignment occurred, we assume that on average it takes place half way through a frame, so 0·5 is subtracted from the values in Table 5.2 before calculating

Table 5.1   Probability of misalignment detection in a single test for random data

Word length, n	Number of errors per word, e					
	0	1	2	3	4	5
1	0·5	—				
2	0·75	0·25	—			
3	0·875	0·50	0·125	—		
4	0·9375	0·688	0·3125	0·0625	—	
5	0·9688	0·8125	0·50	0·1875	0·0313	—
6	0·9844	0·8906	0·6563	0·3437	0·1094	0·0157
7	0·9922	0·9375	0·7734	0·50	0·2266	0·0625
8	0·9961	0·9648	0·8555	0·6367	0·3633	0·1455
9	0·9980	0·9805	0·9102	0·7461	0·50	0·2539
10	0·9990	0·9893	0·9453	0·8281	0·6231	0·3770
11	0·9995	0·9942	0·9693	0·8867	0·7256	0·50
12	0·9998	0·9968	0·9807	0·9270	0·8062	0·6128

Table 5.2   Performance of typical frame-misalignment detection schemes

Detection probability in each test Q_e^n	Number of tests to detect misalignment					
	System 1			System 2		
	mean	90% probability	99% probability	mean	90% probability	99% probability
0·5	14·0	>20	>20	6·79	12	20
0·75	5·48	10	18	4·07	6	9
0·875	3·94	6	11	3·44	4	6
0·9375	3·42	5	8	3·20	4	5
0·9688	3·20	3	6	3·10	3	4
0·9844	3·10	3	6	3·05	3	4

System 1 Check for 3 successive words in error
System 2 Count up 4 for each error, down 1 for each correct word, reframe at count of 10

the detection times. Mean times are also included. In the case of system 1 these are calculated from eqn. 5.23. In the case of system 2 numerical evaluation is used.

In every case there is an optimum length of f.a.w. which will minimise the time needed for detection of misalignment. This optimum varies both with the system used, and with the form of the target criteria. Short framing words are best for minimising the mean detection times, but longer words are preferable if we are working on a measure of the 90% to 99% probabilities of detection.

Table 5.3 Time to detect misalignment with 1% framing content and no errors in any frame-alignment word

Length of frame word, n	Time in units of t_s					
	System 1			System 2		
	mean	90%	99%	mean	90%	99%
1	1350	—	—	629	1150	1950
2	996	1900	3500	714	1100	1700
3	1032	1650	3150	882	1050	1650
4	1168	1800	3000	1080	1400	1800
5	1350	1250	3750	1300	1250	1750
6	1560	1500	3300	1530	1500	2100

See Table 5.2 for system details

5.4.7 Spurious exit from the lock mode

In the presence of transmission errors there is some chance, albeit small, that the frame-alignment word will be corrupted in a number of successive tests, and that realignment operations will thereby be initiated. The estimation of this probability of spurious reframing employs very similar expressions to those developed previously. The probability of failure in each separate test is given by the probability that more than e errors occur in a single word of n digits. This is

$$R_e^n = \sum_{x=e+1}^{n} C_{n-x}^n (P_e)^x (1 - P_e)^{n-x}$$

$$= \sum_{x=e+1}^{n} C_x^n (P_e)^x (1 - P_e)^{n-x} \tag{5.28}$$

where

$$C_x^n = \frac{n!}{x! (n - x)!}$$

In this case, for the low probabilities of error involved, there is no point in carrying out a step-by-step iterative solution, so we will only consider the mean reframing probabilities. For the check and repeat scheme, eqn. 5.23 can be applied directly giving the mean number of frames between spurious realignments as

$$\epsilon_s = \frac{1 - (R_e^n)^\mu}{(1 - R_e^n) (R_e^n)^\mu}$$

$$\approx \frac{1}{(R_e^n)^\mu} \tag{5.29}$$

To make an estimate for the counter scheme we assume that the time actually taken to achieve realignment is insignificant, so that assuming that this always occurs within one frame period introduces no errors. Expression 5.26 is then modified to include realignment by putting

$$P_{zz} = 0$$

$$P_{1z} = 1$$

Under these conditions it follows that

$$\sum P_\mu(s_i) = 1$$

and an expression similar to eqn. 5.9 is obtained. This can be solved in the same manner as before.

For the numerical examples given in Section 5.4.6 if we now assume a transmission error rate of 1×10^{-6} we find that for an n-digit framing word the spurious reframing rate for system 1 is $n^3 \times 10^{-18}$ and for system 2 it is $6n^3 \times 10^{-18}$. The advantages of using these systems can now be seen since if only a single test had been considered adequate, the ratio of the spurious realignment rate to the digital error rate would only have been equal to the framing content α. As each realignment operation takes at least one frame, we would probably have doubled the effective error rate, and subjectively, due to the bunching of framing errors, the effects would be even more serious.

5.4.8 Realignment

Since the risk of simulation of the correct frame-alignment pattern by random data can become fairly high during the realignment process it is normal practice to include some checking procedure. In this, the first digit pattern that matches the f.a.w. is provisionally accepted, but is subjected to further checks in following frames before the normal lock mode is finally re-entered. A variety of search and check techniques may be used, the choice being determined by the trade-off between equipment complexity and performance that the designer is prepared to accept.

As an example of a fast complex scheme, suppose that on detection of misalignment, the next m digits (corresponding to a complete frame) are examined, and all locations where a match to the frame alignment word occurs are recorded in a store. The match nearest to the previous location of the f.a.w. is provisionally accepted as the required location. The next m digits are then checked, and all matching locations that have failed to reappear are then deleted from the store. When, after β tests, only one location survives this is taken to be verified as the correct f.a.w. and lock is re-established. A typical sequence of operations is depicted in Fig. 5.6.

From eqn. 5.14 the probability of simulation by any n digits is Q_e^n, and hence the total probability of spurious simulation in m digits is $(m-1) Q_e^n$. (The reader

word position

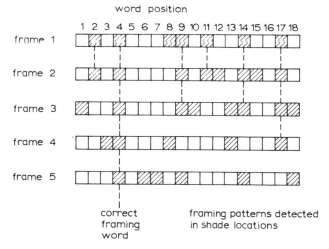

Fig. 5.6 Operation of a parallel frame-alignment system

should refer to the remarks in Section 5.4.9 for some qualification of this expression.) After β checks, the probability of a spurious simulation surviving is

$$S_p(\beta) = (m - 1) (Q_e^n)^\beta$$

For the special case of a framing content α, no errors permitted in the framing word, and random data, this becomes

$$S_p(\beta) = (n/\alpha - 1)/2^{\beta n} \tag{5.30}$$

Values from the 99% probability of successful realignment, derived from this expression, are given in Table 5.4. These may be added to the times given for detection of misalignment in Table 5.3, but it should be noted that, since provision has been made for carrying out a number of simultaneous checks, there is no reason why the process of searching for the f.a.w. should not be started as soon as the first failure to recognise it in its expected location takes place. The switch to control by any newly located f.a.w. is delayed until loss of alignment has been confirmed, by which time there is a high probability that the search will have been completed. Thus, allowing for a mean delay of half a frame before the first detection, the total time for the recovery of alignment can be estimated.

Schemes of this complexity are not often used. In most cases, only one potential framing word may be checked at a time, and so the operations of search and check alternate in sequence until alignment has been re-established. A detailed study of a system of this type has been carried out by Haberle.[13] To investigate the sequence of events in this case let us assume that by the time loss of alignment has been detected the receiving terminal has slipped to a point xm times slots before the next f.a.w. The pattern of n digits currently held in the detector is then examined, with probabilities P_e^n that a spurious simulation of the alignment word will occur, and

Table 5.4   Parallel frame-alignment system; number of frames for a 99% probability of realignment

Frame content, α %	10	5	3	2	1	0·5	0·3	0·2	0·1
Frame word length, n									
1	10	11	12	13	14	15	16	16	17
2	6	6	7	7	8	8	9	9	9
3	4	5	5	5	5	6	6	6	7
4	3	4	4	4	4	5	5	5	5
5	3	3	3	3	4	4	4	4	4
6	3	3	3	3	3	3	3	4	4
7	2	2	3	3	3	3	3	3	3
8	2	2	2	2	3	3	3	3	3
9	2	2	2	2	2	2	3	3	3
10	2	2	2	2	2	2	2	2	3
11	2	2	2	2	2	2	2	2	2
12	2	2	2	2	2	2	2	2	2

Q_e^n that there will be no such simulation. In this latter case, the terminal immediately slips one time slot and re-examines the digit pattern. Clearly, before the true framing word is reached there will, on average, be xmP_e^n false simulations, and for each of these the terminal will wait one complete framing period, of duration mt_s and then check the corresponding n time slots, again with a probability P_e^n of another false simulation. Referring back to eqn. 5.23 and putting $\beta = 1$ we find that the mean number of trials to detect each spurious alignment is $1/Q_e^n$. The total time spent is then $xmt_s \, P_e^n/Q_e^n$ to which musi be added the time xmt_s taken to slip xm time slots, yielding a total mean time to regain realignment of

$$t_r = \left(\frac{mP_e^n}{Q_e^n} + 1 \right) xmt_s \qquad (5.31)$$

If we again take the simplest case of perfect detection of an n digit word in random data, and replace n by αm, eqn. 5.31 becomes

$$t_r = \left(\frac{m}{2^{\alpha m} - 1} + 1 \right) xmt_s \qquad (5.32)$$

This is plotted in Fig. 5.7. The results prove to be highly interesting, since they show that for each value of the framing content α there are well defined optimum

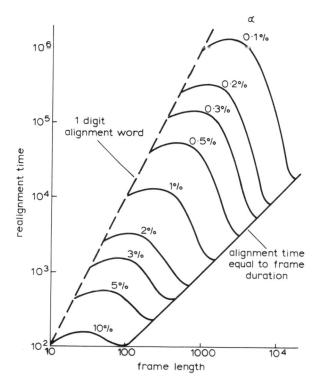

Fig. 5.7 Performance of sequential realignment systems
[Reproduced from *Electrical Communication*, 1969] [13]

lengths for the frame and the frame-alignment word. Furthermore, the optimum length of the framing word does not vary very much with the framing content, as shown in Table 5.5. Referring back to Table 5.3 shows that the optimum lengths in this case are rather greater than those needed for misalignment detection, and it can be seen that since the realignment time is more strongly dependent on the length of the f.a.w. this latter condition will dominate the overall performance.

5.4.9 Choice of the frame-alignment word
In Section 5.4.2 we derived expressions for the probability of spurious simulation of the frame-alignment word by the contents of other time slots. This derivation was based on the assumption that these other slots contained random data, and that each test for the f.a.w. was totally independent of all previous tests. This second assumption is clearly not true for the sequential search techniques described in the last Section, because, if the first few tests fail to register a match to the f.a.w., in following tests a number of the time slots examined will have already been included in previous examinations. Two situations then arise, in which the system performance becomes dependent on the actual choice of the frame-alignment word pattern.

Table 5.5   Optimum frame structures

Framing content, α %	Frame length, m	Frame-alignment word length, n	Mean time to reframe (units of t_s)
10	80	8	105·1
7	129	9	161·5
5	200	10	239·1
3	330	11	383·2
2	600	12	687·9
1	1300	13	1506
0·7	2000	14	2283
0·5	3000	15	3275
0·3	5000	15	5763
0·2	8000	16	8977
0·1	17000	17	19204

The first occurs when the system is only slightly out of alignment, so that the test aperture includes part of the correct framing word. As an example, let us consider the limit case when the system is only one time slot out of alignment. Then, as shown in Fig. 5.8., the pattern examined consists of $n - 1$ of the elements of the true alignment word all displaced by one position, and one time slot containing a random element. Now, if the f.a.w. had been chosen so that all positions, except possibly the last, were identical (for example, the words 11111111 or 11111110 were used) then the overlap positions would still offer a perfect match, and the probability of a spurious simulation would depend only on the single random time slot, and would be ½ rather than $\frac{1}{2}^n$ as previously derived. It can be seen that similar situations will arise for other degrees of overlap, and we conclude that an unfortunate choice of pattern for the alignment word can lead to a greatly degraded performance in the overlap region.

On the other hand, we could have chosen a pattern for which the probability is zero. For example, the pattern 10000001 cannot be matched by any combination of the displaced digits and random digits up to a total shift of $n - 1$ time slots. Thus, we can also find patterns whose performance in the overlap region is very much better than their performance for completely random data. In fact, the example given above does not completely justify this statement, since, as previously discussed, we may use a matching criterion which will accept up to e errors, and it is clear that the single displaced mark of our example would be treated as one of the allowable errors in this situation. The first comprehensive study of the choice of frame-alignment patterns was carried out by Barker.[14] He used the autocorrelation

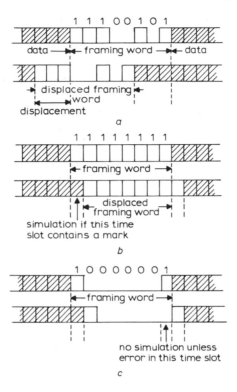

Fig. 5.8 Frame-alignment words in the overlap condition
  a General case
  b Poorly chosen framing word
  c Good choice of framing word

function as a measure of the pattern's performance, and proposed three 'ideal' patterns of lengths 3, 7 and 11. These are

110
1110010
11100010010

These patterns are found to exhibit a high tolerance to transmission errors. Since then, a number of investigations have been carried out, resulting in proposals for patterns, which, while not 'ideal' in Barker's sense, do yield very good performance. One such set have been derived by Williard[15] and Styles.[16] They have shown that for a given error rate, the sum total probability of simulation of the pattern in all overlap conditions is less than the probability of a single spurious simulation by random data. Thus, as far as the search operation is concerned, the total overlap condition can be approximated by a single test of random digits, so eqn. 5.31 becomes

$$t_r \approx \left(\frac{(m - n + 1) P_e^n}{Q_e^n} + 1 \right) xmt_s \tag{5.33}$$

A list of these patterns, which also appear in American telemetering standards[17] are reproduced in Table 5.6. It should be noted that the complementary and reverse

Table 5.6   Recommended frame-alignment codes

Word length	Pattern	Relative simulation probability, R_t
7	0001101	0·8342
8	00011101	0·8951
9	000011101	0·9127
10	0000111011	0·6972
11	00011101101	0·8967
12	000001101011	0·5804
13	0000011010111	0·5397
14	00000101100111	0·5532
15	000010100110111	0·4508
16	0000100111010111	0·4202
17	00000101011001111	0·3892
18	000000101011001111	0·3483
19	0000010100110011111	0·3226
20	00000100011110110111	0·3303
21	000000110100101110111	0·3252
22	0000000101011011001111	0·2936
23	00000001011001110101111	0·2899
24	000001001100111101011111	0·2847
25	0000001000111011010011111	0·2827
26	00000010001101011001011111	0·2730
27	000000011001100101101011111	0·2664
28	0000000110011010011110101111	0·2659
29	00000000101100110011110101111	0·2662
30	000000001011001100111101011111	0·2570
31	0000000000101100110011110101111	0·2578
32	00000000000101100110011110101111	0·2616
33	000000000000101100110011110101111	0·2685

R_t is the sum total probability of simulation for all overlap positions relative to the simulation by random data in a single test, for an error rate of 1×10^{-3}
[Reproduced from 'Optimum code patterns for p.c.m. synchronisation', by M.W. Williard][15]

patterns are equally satisfactory. A further set of patterns has been given by Codrington.[18]

A comparable situation can arise outside the overlap region in a rather more subtle manner. To demonstrate this let us imagine that we are using the 2-digit framing word 10, and that following loss of alignment a search for the f.a.w. has started. The situation is then depicted by Fig. 5.9a. In the first tests there will be one chance in four that the required word will be simulated by the data. If this occurs we enter the check mode and slip one frame period. If not, the next test follows after a slip of one digit, but, in this case, the probability of the first digit in the pair being a mark will have been modified by the outcome of the previous test, and this in turn alters the probability of a spurious simulation in the second test. The process continues into all subsequent tests, and, as shown in Fig. 5.9b choice of a different framing pattern yields a different set of probabilities. Thus, the probability of simulation during the search mode is variable and pattern dependent. This

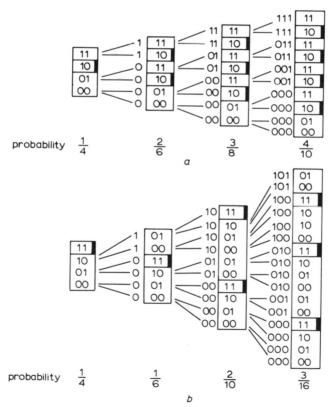

Fig. 5.9 Spurious simulation of frame-alignment words in successive tests
   a Word 10
   b Word 11
[Reproduced from *Electrical Communication*, 1969][13]

situation has been investigated in detail by Haberle, who has shown that, fortunately, if we choose framing-word lengths close to the optima given in Table 5.5 the changes produced by this effect are almost insignificant.

5.4.10 Distributed frame-alignment words

Up to the present we have assumed that bunched frame-alignment patterns have been employed. It is found that the general principles used in analysing these bunched systems can be applied with only minor alterations to schemes in which the time slots containing the alignment pattern are distributed throughout the frame. The process of detection of loss of alignment will be unchanged, and the analysis developed in Sections 5.4.2—5.4.7 can be used directly. The main differences arise during the reframing operations. While any set of time slots within the frame could be used to hold a distributed frame-alignment word, the equipment design is considerably simplified if regular spacings are employed, and so almost all practical systems locate the framing word in a set of uniformly spaced positions set r time slots apart. The frame may then be considered to be divided into a set of r tracks, only one of which contains framing information. The search then proceeds in two stages, each track being examined in turn and discarded if frame alignment is not established within a predefined period. Normally, this period will be slightly greater than the frame duration. A modified version of the analysis of Section 5.4.8 can then be applied.

As before, let us assume that the frame consists of m time slots, and that the frame-alignment word consists of n digits. If there are r tracks, we will, on average, have to examine $(r - 1)/2$ tracks before locating the track containing the frame-alignment word. Each of these tracks has m/r time slots, so, following Section 5.4.8, during the examination of each track an average of mP_e^n/r spurious simulations of the f.a.w. will occur. Testing these will occupy a total time of $mt_s P_e^n/rQ_e^n$ to which must be added a further contribution mt_s to cover the slips between successive tests. Ideally, having tested m time slots without regaining alignment we could then immediately transfer attention to the next track, but normally, a slight overlap is introduced, so that in fact γm tests are carried out. The total time to examine the $(r - 1)/2$ tracks is then

$$t_1 = \left(\frac{m}{r} \frac{P_e^n}{Q_e^n} + 1 \right) \frac{(r - 1) m\gamma}{2} t_s \tag{5.34}$$

Having arrived at the correct track a further xm positions in this track must be examined before the correct alignment word is reached, so that the total time to regain alignment becomes

$$t_r = \left(\frac{m}{r} \frac{P_e^n}{Q_e^n} + 1 \right) \left(\frac{r - 1}{2} \gamma + x \right) mt_s \tag{5.35}$$

The considerations governing the choice of frame alignment pattern remain the same and Section 5.4.9 may again be used in the case of distributed-alignment words.

5.5 Multiplex jitter and crosstalk

We have already pointed out that in many cases the incoming signals must be held in buffer stores until the correct transmission time slots become available. It is clear that if no further action is taken this will introduce some irregularity into the timing of the signal that will ultimately appear as timing jitter at the output of the receiving terminal demultiplexer. To eliminate this jitter further buffer storage must be provided at the terminal, and a regular local clock signal must be regenerated from the incoming frame timing. We must also recognise that due to unavoidable waveform distortion some 'spill over' of signals into adjacent time slots is likely to occur. In the case of pulse-amplitude modulated signals this is a very significant source of crosstalk. We propose in this book to consider only quantised signals, and hence will not include this topic. The reader is referred to Cattermole[30] for a discussion of p.a.m. crosstalk. An interesting variant of this problem has also been considered by Flood.[20]

In the case of quantised signals the effects of intersymbol interference in the multiplex systems are similar to those arising on the transmission path, and the reader is referred to the discussion of this problem in Chapter 9.

5.6 Asynchronous inputs and justification

We have so far assumed a one-to-one correspondence between the message time slots in the multiplexed signals and the corresponding time slots in the tributary inputs, which implies synchronisation between all the sources and the multiplexer system. In general, this will not be possible, and the best we can hope for is that the various input signals will be at the same nominal digit rate and will only differ slightly from one another owing to the tolerance and stability of their individual clock sources. The multiplex system will also have its own independent clock. The signals in this situation are referred to as being plesiochronous. Let us consider one input and the corresponding channel in the multiplexer, and assume that the multiplexer is running slightly faster than the incoming source. The situation will then be as shown in Fig. 5.10 and it can be seen that periodically there will be a surplus time slot in the transmitted signal that will either contain a repetition of one of the incoming digits, or may contain some random signal. In either case, the output at the far end of the system will contain digital errors and will suffer a change in mean digit rate.

In some cases, this situation might be quite acceptable, for example, in the case of delta-modulated speech signals occasional errors of this type will have little effect, but it would obviously be an improvement if we could identify the time slots containing these error signals, transmit information on them to the receiving terminal, and arrange for their deletion from the received signal. Schemes of this

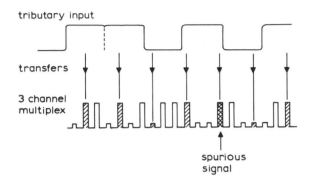

tributary input

transfers

3 channel
multiplex

spurious
signal

Fig. 5.10 Asynchronous multiplexing

type are now in common use under the name of 'justification systems'.*

Positive justification. In practice, the process becomes a little more involved, and several variants of the basic idea have been developed. We will first describe positive justification which, conceptually, is the simplest of the known schemes. In this, the digit rate B_m of the final multiplexed signal is made sufficiently high to accommodate all the tributaries at their maximum digit rate $B_t(max)$. Now, although it is quite possible to justify arbitrary time slots, this leads to considerable complications, and hence almost all practical systems restrict justification operations to certain preassigned time slots. It can be seen that once this assignment has been made this will also define the allowable range of digit rates for the tributary inputs. If each tributary is allowed J justifiable time slots per second it follows that

$$B_t(max) - B_t(min) = J \tag{5.36}$$

In a positive justification scheme time slots in the outgoing multiplexed signal will become available at a rate exceeding that of the incoming data, so that, as shown in Fig. 5.11, the tributary signal will tend to lag. At some stage the system will decide that this lag has become great enough to require justification; at this point a message will be sent to the receiving terminal to inform it of this decision, and the next assigned time slot will be justified. As shown in Fig. 5.11, the receiving terminal will then delete the next justifiable time slot from the signal.

There are a number of highly important features of this process. First, as the reader will appreciate, the security of the data channel used to transmit the justification-control information is of great importance. Any error in this channel will immediately cause either the deletion or the spurious insertion of a time slot in the tributary channel, thereby introducing an error and a loss of alignment. We have

---

* The term justification originates in the printing industry where it is used to describe the process of adjusting the spaces between printed words so that all the lines of print have the same length. In America the rather more descriptive term 'pulse stuffing' is used.

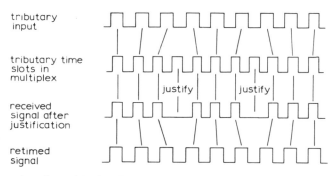

tributary
input

tributary time
slots in
multiplex

justify   justify

received
signal after
justification

retimed
signal

Fig. 5.11 Operation of a positive justification system

already seen in preceding Sections that the realignment operations involve a major transmission interruption, and so this single error results in a large transmission impairment. While ideally, one single bit would suffice to transmit the justification information, the use of an error-correcting code then is advisable. Most systems so far introduced have used the simple scheme of transmitting each stuffing-control bit in triplicate, and carrying out a majority decision on the received signal. Thus, if the system error rate is P_t the error rate for the justification-control channel becomes

$$P_j = P_t^3 + 3P_t^2 \qquad (5.37)$$

It also follows that since each justified digit and its control signal occupy four time slots, and allowing a framing content α the total digit rate required for a multiplex of c channels will be

$$B_m = c(B_t(min) + 4J)/(1 - \alpha) \qquad (5.38)$$

Secondly, the act of deletion at the receiving terminal results in an irregular digit stream, that is, we have introduced timing jitter. Critical examination reveals·that two mechanisms are involved. We have the jitter caused by the act of justification, called 'justification jitter', and we also have an additional component owing to the fact that justification does not take place on demand, but is delayed until one of the preassigned time slots appears. This is defined as 'waiting-time jitter'. The situation is shown in Fig. 5.12, and it can be seen that the waiting time introduces a rather irregular low-frequency component into the jitter. We will defer a full analysis of this effect to Chapter 10, and for the present note that it is a particularly awkward impairment to deal with.

Negative justification. As the name implies, this is effectively the complement of positive justification. The time slots in the multiplexed signal now appear at a slightly slower rate than those of the tributaries, so that it is necessary to periodically insert an extra time slot. This is achieved by transmitting both the notification of the act of insertion, and the content of the added time slot over the control channel. The control arrangements are therefore somewhat more complicated than

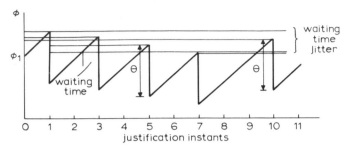

Fig. 5.12 Generation of jitter by justification operations
ϕ = phase delay of tributary relative to multiplex channel
ϕ_1 = phase delay at which justification is initiated
θ = phase delay introduced by justification operation

those of a positive justification system, while the transmission characteristics are virtually identical.

Positive/negative justification. If the nominal digit rate of the channels in the multiplex are made equal to those of the tributaries, the actual difference may be either positive or negative, and so a combination of the two arrangements described above is then required. The control channel must now transmit two, and possibly three signals (positive justify, negative justify, or 'do nothing'), and in the case of the negative justification must also transmit the content of the inserted time slot.

Despite its complexity this scheme does have attractive features. Its main advantage is that it can very easily be adapted to handle synchronous inputs, or mixed systems, in which some of the incoming tributaries are synchronised to the multiplex digit rate. This is of potential importance if large synchronous switching networks are developed. A second advantage is that, with some extensions,[24] the positive/negative scheme leads to reductions in the overall jitter magnitudes.

5.7 Multiplex hierarchies

As mentioned at the beginning of this Chapter, there are advantages in adopting a hierarchical multiplexing structure, and the definition of such a scheme for digital transmission in the international telecommunication network is currently receiving a great deal of attention. Investigations of this problem have been published by Geissler[21,22] and Decina[23] and Fig. 5.13 reproduced from Decina's paper, gives some indication of the scope of these studies. As indicated by this Figure, the choice of the multiplex structure must take into account the characteristics of the signals to be processed, technical aspects of the multiplexing process and the characteristics of the transmission systems which will handle the resulting signals.

One major question is the number of tributaries to be combined at each stage in the structure. This is discussed at some length in Geissler's report. In previous

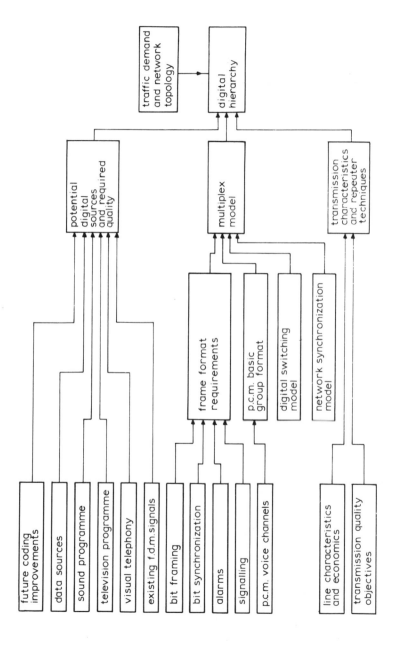

Fig. 5.13 Design factors of a digital hierarchy
[Reproduced from *IEEE Transactions* (Communications Technology), 1972][23]

analogue f.d.m. systems, the number of tributaries was different at each stage, typically lying between five and twelve. Geissler reaches the conclusion that in digital systems a uniform structure in which four tributaries are combined at each level represents the best choice. It now seems probable that this recommendation will be adopted in the system finally defined by the CCITT.

There are also some new technical problems introduced by the existence of the hierarchy. Notably, disturbances such as loss of frame alignment at a high level in the hierarchy, will tend to be propagated down through the lower levels. As yet there is little practical experience of such effects, and so we are only able to make tentative suggestions. The general belief is that realignment operations at the higher levels should take place more rapidly than at lower levels, which implies that frame lengths should remain more or less constant at all levels.

5.8 Practical multiplex structures

Most of the techniques described in previous Sections have, in one form or another, appeared in practical systems, and so to conclude this Chapter we will briefly review some of these embodiments. As a first example, we will take the 24-channel p.c.m. system installed by the British Post Office between 1965–74.[25-27] In this system the actual multiplexing operation takes place immediately after sampling, and the multiplexed samples are then processed by a common coder. Thus, a word-interleaved digital signal is formed. Eight digits are allocated to each word, with the coded speech signal given by digits 2–8. Digit 1 is used for both channel signalling and frame alignment. This is achieved by combining four of the basic frames (192 digits length) into a multiframe of 768 digits. In the first and third frames digit 1 of each word is used for channel signalling, in the second frame its use is undefined, and in the fourth frame, digit 1 of words 9–24 contains the frame-alignment pattern 1101010101010101.

A distributed frame-alignment word has therefore been used. It will be noted that the framing pattern is not one of the 'optimum' patterns previously listed. This pattern, however, does have the virtue that it combines reasonable framing characteristics with an easy technique for implementing reframing. This may be achieved[26] by first searching the digit stream for two marks spaced by eight time slots, and then testing for the repeated 01 pattern in the corresponding following digit time slots. Smith quotes a measured mean time to regain alignment of 2·28 ms. This is in good agreement with the value calculated from eqn. 5.34 and comfortably inside the specified limit of 7 ms set by the Post Office. A comparable American system is described by Davis.[28] This employs a single-digit frame-alignment word, added to each 192 digit frame, which may therefore be treated as a bunched frame-alignment scheme. The mean realignment time will then be about 2·4 ms.[29] The new standard European 32 time-slot 30-channel system employs a more complex bunched frame-alignment scheme (See CCITT recommendations and Table 5.7 for details).

Table 5.7 **Frame structure for primary p.c.m. multiplex equipment operating at 2048 kbit/s**

Number of bits per channel time slot	eight, numbered 1—8
Number of channel time slots per frame	thirty-two, numbered 0—31
Channel time-slot assignment	
Time slot 0	frame-alignment signal
odd frames	X0011011
even frames	X1YZZZZZ
Time slots 1—15	telephony channels 1—15
Time slot 16	common channel or channel associated signalling
Time slots 17—31	telephony channels 16—30

X = digit reserved for international use
Y = alarm signal
Z = reserved for national use

As a second example we will examine the 2nd-order multiplex system formed by combining four 32 time-slot systems (CCITT recommendation G 742). Details of the frame structure are given in Table 5.8. It is seen that the frame is divided into four sections, each section consisting of a group of service digits, followed by digit interleaved signal channels. The first section contains a bunched frame-alignment word, having a structure close to the optimum pattern given in Table 5.6, followed by two service bits. The first is used as an alarm for fault conditions in the multiplex, containing a '0' for no alarm and '1' for the alarm condition. On detecting an alarm the receiving terminal will disconnect the tributary outputs and feed these some predetermined pattern. This will minimise the risk of the disturbance being propagated through the network. The use of the second service digit is not defined.

Positive justification is used, the control digits being in groups of four at the beginning of each section. The control for each channel is distributed over the three sections, the signal '111' signifying justification, and '000' no justification. A majority decision is used. Justification is carried out in the first four time slots in the last section.

Loss of frame alignment is assumed when four consecutive frame-alignment signals have been incorrectly received in their expected positions. The frame-alignment circuit, having then detected the appearance of a single correct frame-alignment signal should immediately renew the search if the signal fails to appear in

Table 5.8   Frame structure for 2nd-order digital multiplex operating at 8448 kbit/s

Tributary bit rate (kbit/s)	2048
Number of tributaries	4

	bit numbering scheme
Frame section 1	
Frame-alignment signal 1111010000	1–10
Bit for international alarm	11
Bit for national use	12
Bits from tributaries	13–212
Frame section 2	
Justification control bits C_{j1}	1–4
Bits from tributaries	5–212
Frame section 3	
Justification control bits C_{j2}	1–4
Bits from tributaries	5–212
Frame section 4	
Justification control bits C_{j3}	1–4
Bits from tributaries available for justification	5–8
Bits from tributaries	9–212
Frame length	848 bits
Bits per tributary	206
Maximum justification rate per tributary	10 kbit/s
Maximum justification rate/nominal justification rate	2·36

C_{ji} is the ith justification control bit of the jth tributary

one of the following two frames. If three successive signals appear, it assumes that alignment has been correctly restored.

As yet, the European multiplex hierarchy is undefined beyond this second level, although it appears that a digit rate of 139·264 mbit/s will be used at the fourth level, corresponding to 16 2nd-level tributaries. In North America, development is proceeding along different lines. A regular hierarchical structure has not been adopted, and the present proposals (See contribution 159 to Special Study Group D) are shown in Fig. 5.14. The frame structures proposed for the American hierarchy are given in Table 5.9. In the 2nd-order multiplex 48 information digits are inserted between each H, or service digit. The information digits are in the sequence $I_1, I_2, I_3, I_4, I_1, \ldots$ where I_j is the digit from the jth tributary. The H

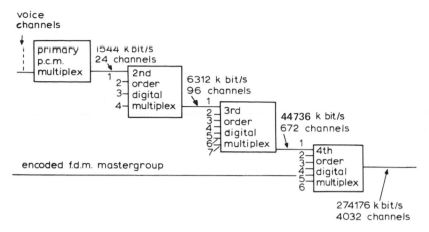

Fig. 5.14 Proposed N. American digital multiplex hierarchy

Table 5.9 Proposed frame formats for North American multiplex hierarchy

	2nd order	3rd order	4th order
Output gross digit rate, kbit/s	6312	44736	274176
Tolerance on output gross digit rate, p.p.m.	± 30	±20	±10
Output line code*	B6ZS	B3ZS	
Tributary nominal gross digit rate, kbit/s	1544	6312	44736
No of tributaries	4	7	6
Tributary line code*	Bipolar	B6ZS	B3ZS
Frame structure†	481-H	841-H	
Multiframe length, bits	1176	4760	
Bits per tributary per multiframe	288	672	
Maximum justification rate per tributary, Hz	5367	9398	
Nominal justification rate per tributary, Hz	1796	3670	
Maximum/nominal	2·99	2·56	

* See Chapter 11 for details of line codes
† See text

digit in every 49th time slot follows the sequence F_1, M_j, C_j, F_0, C_j, C_j cyclically repeated for $j = 1-4$. F_1, F_0 forms the frame-alignment signal 10 and M_1, M_2, M_3, M_4 is the multiframe alignment signal 0111. C_j, C_j, C_j is the triplicated justification-control signal, being 111 for justification and 000 for no justification. The bit allocated for justification in tributary j is the first bit of this tributary, following the F_1 digit after the C_j, C_j, C_j signal. The 3rd-order multiplex structure is very similar, except here seven tributaries are interleaved employing a sequence of 84 information time slots to every service time slot. The H-bit sequence is now F_1, M_j, F_1, C_j, F_0, C_j, F_0, C_j so that the frame alignment pattern is 1100 and the multiframe-alignment pattern is 1110010.

5.9 References

1   PIPES, L.A., and HARVIL, L.R.: 'Applied mathematics for engineers and physicists' (McGraw-Hill, 1970) 3rd edn.
2   HARMUTH, H.F.: 'Transmission of information by orthogonal functions' (Springer Verlag, New York, 1969)
3   BARRETT, R., and GORDON, J.A.: 'Correlation recovered adaptive majority multiplexing', *Proc. IEE*, 1971, **118**, (3/4), pp. 417–422
4   BARRETT, R., and GORDON, J.A.: 'Group multiplexing by concatenation of non-linear code division systems', 1972 Walsh Function Symposium, Washington
5   SCHMIDT, W.D., GABBARD, O.G., CACCIAMANI, E.R., MAILLET, W.G., and WU, W.W.: 'MAT-1 Intelsat's experimental 700 channel TDMA/DA system', *in* 'Digital satellite communications', *IEE Conf. Publ.*, 59, 1969, pp. 428–440
6   HALL, G.C.: 'Single channel per carrier, pulse code modulation, demand assignment equipment – Spade – for satellite communications', *Post Off. Electr. Eng. J.*, 1974, **67**, pp. 42–48
7   GRAY, J.R., and PAN, J.W.: 'Using digit statistics to word frame p.c.m. signals', *Bell Syst. Tech. J.*, 1964, **43**, p. 2985
8   BRAAE, R.: 'Matrix algebra for electrical engineers' (Pitman, 1963)
9   TAYLOR, V.L.: 'Optimum p.c.m. synchronising', Proceedings of the National Telemetering Conference (USA), 1965, pp. 46–49
10   WILLIARD, M.W.: 'Mean time to acquire p.c.m. synchronisation', Proceedings of the National Symposium on 'Space electronics and telemetry', Miami Beach, Oct. 1972
11   FRY, T.C.: 'Probability and its engineering uses' (Van Nostrand, 1928)
12   FELLER, W.: 'An introduction to probability theory and its applications' (Wiley, 1968) 3rd edn., vol. 1
13   HABERLE, H.: 'Frame synchronising p.c.m. systems', *Electr. Commun.*, 1969, **44**, pp. 280–287
14   BARKER, R.H.: 'Group synchronising of binary digital systems', *in* 'Communication theory (Ed.) JACKSON, W. (Academic Press, New York, 1953) pp. 273–287
15   WILLIARD, M.W.: 'Optimum code patterns for p.c.m. synchronisation', Proceedings of the National Telemetering Conference, Washington, May 1962
16   STYLES, F.J., and MAURY, J.L.: (*a*) 'Development of optimum frame synchronisation codes for GSFC PCM telemetry standards', Proceedings of the National Telemetering Conference (USA), 1964, Section 3-1, pp. 1–10 (*b*) 'Performance evaluation of frame synchronisation codes for p.c.m. telemetry', US Government report N70-15657, Goddard Space Flight Centre, 1967
17   'Pulse code modulation telemetry standard, appendix A; p.c.m. frame synchronisation codes', Aerospace Data Systems Standards, Pt. I, Section 1, GSFC X-560-63-2, Jan. 1963
18   CODRINGTON, R.S., and MAGNIN, J.P.: 'Legendre p.c.m. synchronisation codes', Proceedings of the symposium on 'Space electronics and telemetry', Miami Beach, 1962

20 FLOOD, J.E., URQUHART-PULLEN, D.I.: 'Time-compression-multiplex transmission', *Proc. IEE*, 1964, **111**, (4), pp. 647–668

21 GEISSLER, H.: 'Planning a p.c.m. system hierarchy', NTZ report 8, VDE-Verlag GmbH, Berlin, 1971

22 GEISSLER, H.: 'Planning a p.c.m. system hierarchy' (condensed version of Reference 21) NTZ, 1971, **24**, p. 217

23 DECINA, M.: 'Planning a digital system hierarchy', *IEEE Trans.*, 1972, **COM-20**, pp. 60–64

24 CCITT Green Book vol. III, Annex to recommendation G 742

25 BOLTON, L.J.: 'Design features and applications of the British Post Office 24 channel pulse code modulation system', *Post Off. Elect. Eng. J.*, 1968, **61**, pp. 95–101

26 SMITH, E., and GABRIEL, M.: 'A 24 channel pulse code modulation junction carrier system', *Electr. Commun.*, **43**, pp. 123–129

27 STEVENS, A.D.: 'A pulse code modulation system for junction telephone circuits', *Point to Point Telecomm.*, Feb. 1966, **10**, pp. 6–25

28 DAVIS, C.G.: 'An experimental pulse code modulation system for short haul trunks', *Bell Syst. Tech. J.*, 1962, **41**, pp. 1–24

29 SWARTZ, L.: 'Statistical distributions of frame resynchronisation times in D1 and D2 type p.c.m. terminals', Proceedings of the IEEE 1972 International Communications Conference, Philadelphia, pp. 21–23/21–29

30 CATTERMOLE, K.W.: 'Principles of pulse code modulation' (Iliffe, 1969)

Chapter 6
Transmission-line theory

The design of any transmission system must be largely determined by the characteristics of the transmission medium employed, and by far the most common
medium in use is the transmission line. It is also, theoretically, the most complex.
As discussed in Chapter 1, the development of transmission-line theory commenced
soon after the appearance of the earliest telegraph systems, and over the past
century a considerable body of material on this topic has been accumulated. In this
Chapter we will attempt to extract and summarise the parts of this work that are
relevant to digital transmission systems, and in the next Chapter we will discuss
practical aspects of transmission lines. Although our primary interest is in digital
transmission, the reader will find that a large part of this Chapter is devoted to
frequency-domain analysis. There are three reasons for this (i) much of the theory
is more easily presented, and its physical basis is more readily understood, in terms
of frequency responses (ii) in later Chapters we will find that it is often convenient
to carry out other parts of the design work in terms of frequency-domain characteristics (iii) in the case of digital systems employing modulated carrier techniques
interest is centred on the characteristics of the transmission path in the vicinity of
the carrier frequency.

Basic theory. A simple transmission line can be represented by a pair of conductors having resistance R, inductance L, leakance G and capacitance C per unit
length, as shown in Fig. 6.1. Then, if at a point x the current flowing in the
conductors is i, and the potential difference between them is v,

$$\frac{\partial v}{\partial x} = -\left(Ri + L\frac{\partial i}{\partial t} \right) \tag{6.1}$$

$$\frac{\partial i}{\partial x} = -\left(Gv + C\frac{\partial v}{\partial t} \right) \tag{6.2}$$

These form the basic transmission-line equations.

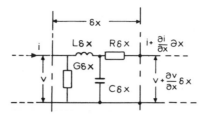

Fig. 6.1 Elementary conditions in a transmission line

6.1 Primary characteristics in the frequency domain

It can easily be shown that, for sinusoidal voltages and currents $Ve^{j\omega t}$ and $Ie^{j\omega t}$ the solutions are

$$V = Ae^{-px} + Be^{px} \tag{6.3}$$

$$I = \frac{1}{Z_0}(Ae^{-px} - Be^{px}) \tag{6.4}$$

where

$$p = \{(R + j\omega L)(G + j\omega C)\}^{\frac{1}{2}} \tag{6.5}$$

$$Z_0 = \left(\frac{R + j\omega L}{G + j\omega C}\right)^{\frac{1}{2}} \tag{6.6}$$

The negative exponential term then represents a wave travelling along the line from the source towards the load, and the positive exponential term represents a wave travelling in the reverse direction. The values A and B are determined by the boundary conditions, that is, the terminations at the two ends of the line, and by defining these, and then rearranging eqns. 6.3 and 6.4, a large number of alternative forms of the solutions may be obtained. The reader is referred to the standard texts (Willis Jackson,[1] Grivet,[2] Goldman,[3] and ITT[4]) for further details.

If the line is assumed to be inserted between a source E_g, of internal impedance Z_g, and a load Z_l, as shown in Fig. 6.2, then various expressions relating the input and output conditions may be obtained. Of particular importance are the reflection and transmission coefficients. These are

$$\text{voltage reflection coefficient} = \frac{Be^{pl}}{Ae^{-pl}} = \frac{Z_l - Z_0}{Z_l + Z_0} \tag{6.7}$$

$$\text{current reflection coefficient} = -\frac{Be^{pl}}{Ae^{pl}} = \frac{Z_0 - Z_l}{Z_0 + Z_l} \tag{6.8}$$

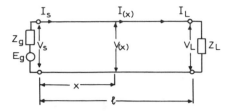

Fig. 6.2 Transmission line with terminations

$$\text{voltage transmission coefficient} = \frac{V_I}{Ae^{pl}} = \frac{2Z_I}{Z_I + Z_0} \tag{6.9}$$

$$\text{current transmission coefficient} = \frac{I_I Z_0}{Ae^{pl}} = \frac{2Z_0}{Z_I + Z_0} \tag{6.10}$$

By replacing Z_I by Z_s similar expressions may be derived for the source.

Loss-free line. If the resistance R and the conductance G are both negligible

$$Z_0 = (L/C)^{\frac{1}{2}} \tag{6.11}$$

$$p = (LC)^{\frac{1}{2}} = 1/v \tag{6.12}$$

where v is the propagation velocity on the line. Since v is related to v_C, the propagation velocity in free space, by

$$v = \frac{v_C}{(\epsilon_r \mu_r)^{\frac{1}{2}}} \tag{6.13}$$

where ϵ_r and μ_r are the permitivity and permeability of the dielectric relative to free space, it is then possible to derive simple numerical expressions for the primary constants of the line. These are

$$L = (\epsilon_r \mu_r)^{\frac{1}{2}} \frac{Z_0}{v_C} = 3 \cdot 33 \, (\epsilon_r \mu_r)^{\frac{1}{2}} Z_0 \qquad \text{nH/m} \tag{6.14}$$

$$C = (\epsilon_r \mu_r)^{\frac{1}{2}} \frac{1}{v_C Z_0} = \frac{3330 \, (\epsilon_r \mu_r)^{\frac{1}{2}}}{Z_0} \qquad \text{pF/m} \tag{6.15}$$

$$T = 3 \cdot 33 \, (\epsilon_r \mu_r)^{\frac{1}{2}} \qquad \text{ns/m} \tag{6.16}$$

where T is the propagation delay. Under these conditions Z_0 is determined by the line geometry and the dielectric characteristics. Expressions for the characteristic impedance for a few important cases are given in Table 6.1. Information on dielectric materials is given in Table 6.2. More extensive data will be found in many reference books. (See ITT,[4] and von Hippel[6]).

As might be expected, practical lines deviate from the idealised characteristics

Table 6.1 Characteristic impedance of transmission lines

Line type	Impedance
a 2-wire line, remote from ground	$Z_0 = 120 \cosh^{-1}\left(\dfrac{D}{d}\right)$
b 2-wire line, near ground	$Z_0 = 30 \log_e \left\{ \dfrac{4h}{d} \sqrt{1 + \left(\dfrac{2h}{D}\right)^2} \right\}$
c single wire, near ground	$Z_0 = 60 \cosh^{-1}\left(\dfrac{2h}{d}\right)$
d coaxial	$Z_0 = 60 \log_e\left(\dfrac{D}{d}\right)$
e parallel strip	$Z_0 = \dfrac{2368}{1 + 2\pi\,\dfrac{b}{w} + \log_e\left(1 + 2\pi\,\dfrac{b}{w}\right)}$

described in Section 6.1, and in the following Sections a number of the more important limitations will be considered.

6.2 Conductor impedance and skin effect

In simple transmission-line theory the conductor impedance is denoted by R, and the student might therefore assume that it is a real constant quantity. It is, in fact, a complex frequency-dependent impedance. Expressions for this impedance may be obtained by applying Maxwell's equations to the conditions within the conductor

Table 6.2   Characteristics of dielectric materials

Material	Dielectric constant				Dissipation factor $\times 10^4$			
	1 kHz	1 MHz	100 MHz	1 GHz	1 kHz	1 MHz	100 MHz	1 GHz
Aluminium oxide	8·8	8·8	8·8	8·8	6	3	3	6
Glass (borosilicate)	4·8	4·8	4·8	4·8	60	30	30	60
Quartz	3·78	3·78	3·78	3·78	2·6	0·1	0·3	1·0
Polyethylene	2·26	2·26	2·26	2·26	2	2	2	3
Polystyrene	2·56	2·56	2·55	2·55	0·5	0·7	1·0	3·5
Polytetra- fluoroethylene	2·1	2·1	2·1	2·1	3	2	2	1·5
Polyvinyl chloride	3·1	2·9	2·8	2·75	140	180	160	100
Nylon	3·8	3·4	3·2	3·0	180	270	180	130
Rubber (vulcanised)	2·9	2·8	2·4	2·3	30	450	180	50
Neoprene	6·6	6·3	4·5	4·0	110	380	900	400
Dry paper (as used in cables)	3·3	3·0	2·8	2·75	80	400	650	600

These are typical values, and, in practice, values will vary according to the precise composition and methods of fabrication employed. For accurate values manufacturer's data should be consulted.

(for details of the analysis see Ramo, Whinnery and Van Duzer,[7] Chapter 5). The simplest expressions are obtained if the diameters of the conductors are assumed to be large, so that they may be treated as plane conducting slabs of infinite thickness. A measure of the impedance of a unit area of the conductor, as measured at its surface, can then be obtained. This is

$$Z_s = \frac{1}{\delta\sigma}(1+j) = \left(\frac{\omega\mu}{2\sigma}\right)^{\frac{1}{2}}(1+j) \tag{6.17}$$

where

σ = conductivity of the conductor material

μ = permeability of the conductor material

$$\delta = \left(\frac{2}{\omega\mu\sigma}\right)^{\frac{1}{2}} \tag{6.18}$$

Z_s is defined as the surface impedance. It has equal real and imaginary components, and is proportional to the square root of the frequency. Each component is equal in magnitude to the d.c. resistance of a slab of conductor of thickness δ. δ is defined as the skin depth, since it is a measure of the depth of penetration of the current into the conductor. (The current has fallen to $1/e$ (36·8%) of its surface value at a depth δ.) R_s, the real part of Z_s, is defined as the surface resistance. Values of δ and R_s for common conductor materials are given in Table 6.3, which shows that δ becomes very small for frequencies in excess of a few kilohertz. In the majority of cases, δ is then much less than the diameter of the conductors, so that the assumption of planar geometry made above is justified. However, another assumption has also been made, namely, that the magnetic field intensity is uniform around the conductor periphery. This is obviously true when the conductor system has circular symmetry, for example, in the case of a coaxial line, but requires investigation in other cases. We will return to this later, but for the moment we will go on to consider the effects of the conductor impedance on the overall line characteristics.

6.2.1 Frequency-dependent line parameters

Dielectric losses are usually small, and can often be safely ignored, but in the majority of cases the effects of conductor losses will be significant. Provided the skin depth δ is small compared with the conductor diameter d, the high-frequency resistance will be identical to the d.c. resistance of a cylindrical shell of thickness δ, so that

$$Z_{C1} = \frac{Z_s}{\pi d} \tag{6.19}$$

The true d.c. resistance of the conductor is

$$R_{dc} = \frac{4}{\pi d^2 \sigma} \tag{6.20}$$

and it will be convenient to denote the ratio of the a.c. impedance to the d.c. resistance by a factor S. S may be regarded as the correction factor by which the d.c. resistance has to be multiplied to obtain the high-frequency impedance. In this present case

$$Z_{C_1} = R_{dc}S_1$$

where

$$S_1 = \frac{d}{4\delta}(1+j) \tag{6.21}$$

Table 6.3   Characteristics of conductors

Material	$\mu \times 10^{-6}$	$\sigma \times 10^{8}$	Temperature coefficient	Skin depth δ			Surface resistivity		
				1 kHz	1 MHz	1 GHz	1 kHz	1 MHz	1 GHz
Aluminium	1·256	0·382	0·0039	$2\cdot57 \times 10^{-3}$	$8\cdot12 \times 10^{-5}$	$2\cdot57 \times 10^{-6}$	$1\cdot02 \times 10^{-5}$	$3\cdot22 \times 10^{-4}$	0·0102
Brass*	1·256	0·256	0·002	$3\cdot14 \times 10^{-3}$	$9\cdot92 \times 10^{-5}$	$3\cdot14 \times 10^{-6}$	$1\cdot25 \times 10^{-5}$	$3\cdot95 \times 10^{-4}$	0·0125
Copper	1·256	0·580	0·0039	$2\cdot09 \times 10^{-3}$	$6\cdot6 \times 10^{-5}$	$2\cdot09 \times 10^{-6}$	$8\cdot25 \times 10^{-6}$	$2\cdot61 \times 10^{-4}$	0·0825
Gold	1·256	0·409	0·0034	$2\cdot49 \times 10^{-3}$	$7\cdot87 \times 10^{-5}$	$2\cdot49 \times 10^{-6}$	$9\cdot80 \times 10^{-6}$	$3\cdot1 \times 10^{-4}$	0·098
Iron+	2500	0·103	0·0052	$1\cdot11 \times 10^{-4}$	$3\cdot52 \times 10^{-6}$	$1\cdot11 \times 10^{-7}$	$8\cdot75 \times 10^{-4}$	$2\cdot76 \times 10^{-2}$	0·875
Lead	1·256	0·046	0·004	$7\cdot4 \times 10^{-3}$	$2\cdot34 \times 10^{-4}$	$7\cdot4 \times 10^{-6}$	$2\cdot95 \times 10^{-5}$	$9\cdot3 \times 10^{-4}$	0·0295
Tin	1·256	0·088	0·0042	$5\cdot34 \times 10^{-3}$	$1\cdot69 \times 10^{-4}$	$5\cdot34 \times 10^{-6}$	$2\cdot13 \times 10^{-5}$	$6\cdot72 \times 10^{-4}$	0·0213

Typical values in MKS units

*Characteristics of alloys are dependent on their composition

+ μ varies widely with impurity content and processing

The total conductor impedance of a line formed by the two conductors is then

$$Z_C = \frac{Z_{s1}}{\pi d_1} + \frac{Z_{s2}}{\pi d_2} \tag{6.22}$$

Inserting this expression into eqns. 6.5 and 6.6, and making the further assumptions that G is negligible and L is large, we find that

$$Z_0 = Z_\infty + \frac{A}{\sqrt{\omega}}(1 - j) \tag{6.23}$$

$$p = K\sqrt{\omega} + j\,(K\sqrt{2\omega} + N\omega) \tag{6.24}$$

where

$$Z_\infty = (L/C)^{\frac{1}{2}} \qquad A = \frac{B}{(2LC)^{\frac{1}{2}}}$$

$$N = (LC)^{\frac{1}{2}} \qquad K = \frac{B}{2}(C/L)^{\frac{1}{2}}$$

$$B = \frac{1}{\pi d_1}\left(\frac{\mu_1}{2\sigma_1}\right)^{\frac{1}{2}} + \frac{1}{\pi d_2}\left(\frac{\mu_2}{2\sigma_2}\right)^{\frac{1}{2}} \tag{6.25}$$

These expressions are found to be good approximations to the measured characteristics of real lines. Examination of them reveals a number of important features.

(i) The attenuation of a line is proportional to the square root of the frequency, and is inversely proportional to the conductor dimensions. Since other characteristics are dependent on the *ratios* of dimensions this means that doubling the size of a line will halve its attenuation while leaving other characteristics unaltered.

(ii) The normal assumption that the characteristic impedance of a line is real and independent of frequency is not strictly correct. At low frequencies it increases and includes a capacitive reactance.

(iii) The propagation velocity is also slightly frequency dependent, thus introducing a further contribution to the distortion of transmitted signal waveforms. However, it may be shown that in conjunction with the attenuation term the $K\sqrt{\omega}$ component in the phase characteristic forms a minimum-phase function (see Bode[25] for discussion of minimum-phase characteristics) so that the transmission line may be considered to be equivalent to a lossless linear-phase (constant-delay) structure in series with a lossy minimum-phase network. Since, for commonly used circuit configurations, the characteristics of amplifiers, filters and equaliser networks also conform to the minimum-phase constraint, any attempt to compensate for the line-attenuation variations by passive networks or by frequency-dependent amplifier characteristics will automatically introduce a corresponding correction for the phase characteristics, and the resulting overall response will also be a minimum-phase function.

6.3 Conductor impedance in more complex situations

Conductors having a surface coating of some other conducting material, for example, silver-plated steel or tin-plated copper, are quite common. Expressions for the impedances of such composite conductors are given by Ramo, Whinnery, and Van Duzer,[7] and a derivation of the resulting expressions for a coaxial line are given by Nahman.[8] The exact solution for homogeneous circular conductors is given by Schelkunoff.[9] The resulting expressions involve complex Bessel functions, and as a result become very complicated. Expansions and tabulations of these functions will be found in Dwight,[10] and Schelkunoff gives some simplified approximate expressions. (The reader wishing to pursue this should note that there is an error in eqn. 82 in Schelkunoff's paper.) The case of a hollow cylindrical conductor is of particular importance, since in this case there are two conductor impedances, corresponding to the inner and outer surfaces, and a transfer impedance expressing the voltage appearing on one surface due to a current flowing on the other. The situation is shown in Fig. 6.3 and the resulting expressions are

$$E_i = Z_{ii}I_i + Z_{io}I_0$$
$$E_0 = Z_{io}I_i + Z_{00}I_0 \tag{6.26}$$

Details of the evaluation of the impedance functions are given by Schelkunoff. The transfer impedance Z_{io} is of particular importance, since this is the major factor in determining crosstalk coupling in coaxial cables.

Proximity effects. In the case of balanced pair lines using solid dielectrics the conductor spacing is small, and the assumption that the magnetic field is uniform around the surface of the conductors is no longer valid. An analysis of this effect

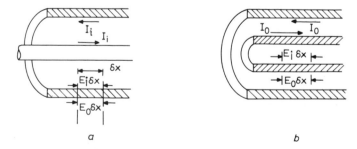

a   *b*

Fig. 6.3 Cylindrical conductor
    a With inner return path
    b With outer return path

has been made by Arnold,[11] who shows that for solid homogeneous conductors, the resistance derived from eqn. 6.21 must be multiplied by another factor T_1 where

$$T_1 = \frac{1}{\left\{1 - \left(\dfrac{d}{s}\right)^2 a(x)\right\}^{\frac{1}{2}}} \tag{6.27}$$

and

d = wire diameter

s = centre to centre spacing between the wires

$$x = \frac{d}{\sqrt{2}\,\delta}$$

A useful approximation is

$$a(x) = 0.526\left\{1 + \tanh\left(1.14 - 3.078/x\right)\right\} \tag{6.28}$$

A somewhat similar approach may be used to allow for the effects of surrounding pairs in the same cable. As suggested by Eager,[12] these may be simulated by a surrounding conducting pipe, whose radius r is equal to the average distance to the surrounding wires.

6.4 Dielectric losses

Up to the present, the conductance term G has been assumed to be negligible. While the dielectric losses in modern plastic insulation materials are extremely low, it is found that these losses do become significant at very high frequencies, and in any case, some of the older insulating materials, notably the paper insulation used in audio pair cables, have appreciable dielectric losses at frequencies above about 100 kHz. Examination of dielectric data (see Table 6.2) shows that, in many cases, the dissipation factor or loss angle remains substantially constant over a wide range of frequencies. One is therefore tempted to assume that the conductance G is proportional to frequency, and on inserting this in eqn. 6.5 it is found that, to a first approximation, an attenuation component proportional to frequency has been introduced. Allowing for the skin effect, the total attenuation is then given by an expression of the form

$$\alpha(f) = \alpha_0 + \alpha_1 \sqrt{f} + \alpha_2 f \tag{6.29}$$

This is found to be quite a good working approximation, and is often employed in published work. It does, however, lead to difficulties in formal analysis. For example, it violates the minimum phase constraints discussed above, and difficulties are encountered when one attempts to obtain the time-domain characteristics by applying the Laplace transform.

A better approach is to model the dielectric by an equivalent RC network of the form shown in Fig. 6.4. This is found to be mathematically more tractable, and it can be shown (Von Hippel[13]) that this sort of model is valid for a wide range of dielectric materials. The admittance of an n section model of this type is

$$G + j\omega C = \sum_{i=1}^{n} \left(\frac{1}{R_i} \frac{X_i}{1 + X_i} + j \frac{\omega C_i}{1 + X_i} \right) + j\omega C_0 \qquad (6.30)$$

where

$$X_i = (\omega R_i C_i)^2$$

A three- or four-section model will usually be found to be adequate. An example of such a dielectric model for polyethylene is given in Fig. 6.4.

6.5 Multiconductor lines and crosstalk

It is quite straightforward to generalise the basic transmission-line eqns. 6.1 and 6.2 to cover the case of a system of n conductors, as shown in Fig. 6.5. For each conductor we have a pair of equations of the form

$$\frac{\partial v_j}{\partial x} = -R_{jj} i_j - R_{jk} i_k \dots -L_{jj} \frac{\partial i_j}{\partial t} - L_{jk} \frac{\partial i_k}{\partial t} \dots \qquad (6.31)$$

$$\frac{\partial i_j}{\partial x} = -G_{jj} v_j - G_{jk} v_k \dots -C_{jj} \frac{\partial v_j}{\partial t} - C_{jk} \frac{\partial v_k}{\partial t} \dots \qquad (6.32)$$

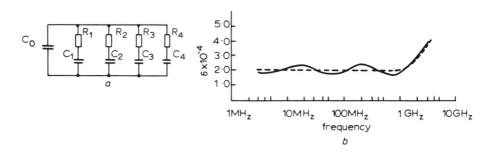

Fig. 6.4 Modelling of dielectric losses of Polythene
    a   Model
    C_0 = 1000 pF R_1 = 60·9 kΩ
    C_1 = 0·383 pF R_2 = 3·22 kΩ
    C_2 = 0·367 pF R_3 = 595 Ω
    C_3 = 0·122 pF R_4 = 6·64 Ω
    C_4 = 1·615 pF
    b Dissipation factor
    ——— model
    --------- target characteristic

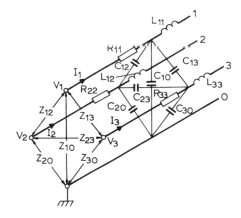

Fig. 6.5 Multiconductor transmission line

The entire system may be conveniently expressed in matrix notation by

$$[v] = -[R][i] - [L]\frac{\partial}{\partial t}[i] \tag{6.33}$$

$$[i] = -[G][v] - [C]\frac{\partial}{\partial t}[v] \tag{6.34}$$

Symbolically, the solutions proceed as before yielding

$$[v] = [A]e^{[p]t} + [B]e^{-[p]t} \tag{6.35}$$

$$[i] = [A][Z_0]^{-1}e^{[p]t} - [B][Z_0]^{-1}e^{-[p]t} \tag{6.36}$$

where

$$[Z_0] = (\{[R] + j\omega[L]\}\{[G] + j\omega[C]\}^{-1})^{\frac{1}{2}}$$

$$[p] = (\{[R] + j\omega[L]\}\{[G] + j\omega[C]\})^{\frac{1}{2}}$$

The main problem is to interpret expressions such as $e^{[p]t}$. This can be done by first expanding $e^{[p]t}$ as a power series in $[p]t$ and then using Sylvester's theorem. (See Pipes[14] for details of this.) A detailed discussion of multiconductor systems from this point of view is given by Hayashi,[15] and, in principle, this offers a solution to all multiconductor problems. However, the resulting expressions are formidable, and, in many cases of great practical interest, their evaluation is still quite impossible. Alternative approaches must then be used.

Crosstalk. We are particularly interested in the coupling, or crosstalk, between different transmission circuits, and there are two situations of special interest in practice. These are

A pair of lines coupled only through a common conductor impedance. This covers the case of crosstalk between coaxial cables, and the simplest conditions arise when the outer conductors of the two coaxial pairs are in metallic contact, as shown in Fig. 6.6. In this case, the voltages and currents existing in the contact between the two conductors must be equal. These were defined by eqn. 6.26 for the case of a single conductor, in terms of an impedance matrix. Extending this, the coupling between the two lines is then given by

$$\begin{bmatrix} E_a \\ I_a \end{bmatrix} = [A_a] [B_b] \begin{bmatrix} E_b \\ -I_b \end{bmatrix}$$ (6.37)

where $[A_a]$ is the transmission matrix for the outer conductor of cable A and $[B_b]$ is the inverse transmission matrix for the outer conductor of cable B (see Shea[5] for the required matrix conversions). With some rearrangement the transfer impedance Z_t between the two transmission circuits can then be obtained. For an incremental length Δx we then have

$$\frac{E_b}{I_a} = Z_t \Delta x$$

Now in terms of the sending-end voltage

$$I_a - \frac{E_s}{Z_0} e^{-px}$$

and the signals produced in the near end and far end terminations of the second pair, due to the induced voltage in element Δx are

$$\Delta E_n = \frac{E_b}{2} e^{-px}$$

$$\Delta E_f = \frac{E_b}{2} e^{-p(1-x)}$$

Integrating to sum the effects of all elementary segments of the cables, the final values for the near end and far end crosstalk ratios are found to be

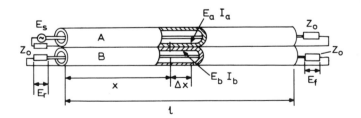

Fig. 6.6 Coupling between coaxial pairs

$$N = \frac{E_s}{E_n} = \frac{2Z_0}{Z_t} \left[\frac{2p}{1 - e^{-2\,pl}} \right] \tag{6.38}$$

$$F = \frac{E_s}{E_f} = \frac{2Z_0}{Z_t} \frac{e^{pl}}{l} \tag{6.39}$$

If, as is frequently the case, the outer conductors of the coaxial pairs are insulated from each other, these outers form an auxilliary transmission line, whose characteristics must be taken into account in evaluating the coupling. The resulting solutions, which are very complex, are discussed by Croze[16] and Schelkunoff.[17]

A pair of balanced lines coupled only through their electromagnetic fields, with no common conductor impedance. This covers the case of crosstalk between pairs in multipair cables. Such cables are made up using different stranding lays (rate of twist) for the different pairs, so that ideally the coupling between pairs will balance out over any appreciable length of line. In practice, this balancing is imperfect, and so, between two pairs within any elementary section of cable, we can define a coupling capacitance C_μ, representing the unbalance in the electrostatic coupling, and a mutual inductance L_m representing the residual unbalance in the magnetic coupling. As shown in Fig. 6.7a, a current I_1 in pair 1 will then cause a crosstalk current I_2 to appear at each end of the elementary section of pair 2. The far-end crosstalk F will then be given by the ratio of I_1 to I_{2f} at the far end of the cable, and the near-end crosstalk by the ratio of I_1 to I_{2n}, when both are measured at the sending end, as shown in Fig. 6.7b. In both cases, the crosstalk for an elementary length is given by

$$X = \frac{I_2}{I_1} = \left(\frac{j\omega C_\mu Z_0}{8} \pm j\, \frac{\omega L_m}{Z_0} \right) \tag{6.40}$$

The total crosstalk in a long cable will then be the sum of a large number of such elementary contributions, modified by the effects of cable attenuation. In the case of far-end crosstalk, it can be seen from Fig. 6.7b that the contributions from all elementary sections traverse the same total path length and so undergo the same attenuation. The individual couplings may be considered to be random, and to add on an r.m.s. basis, so that if the attenuation of the elementary length l is α, the total far-end crosstalk for a cable of length nl is

$$F_l = \overline{X}\, \sqrt{n}\, e^{-n\alpha} \tag{6.41}$$

It follows that if a far end crosstalk ratio F_l is measured on a cable of length l_1 the far-end crosstalk for a length l_2 will be

$$F_2 = F_1 \left(\frac{l_2}{l_1} \right)^{\frac{1}{2}} e^{-\alpha(l_2 - l_1)} \tag{6.42}$$

In the case of near-end crosstalk each contribution is transmitted over a path of different length with a different attenuation. The near-end-crosstalk ratio for a cable of length nl is then

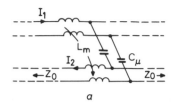

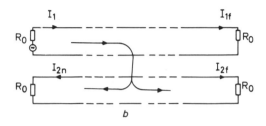

Fig. 6.7 Crosstalk coupling in balanced pair cables
   a Crosstalk in an elementary length
   b Crosstalk in a long cable

Near-end crosstalk ratio $N = \dfrac{I_{2n}}{I_1}$

Far-end crosstalk $\qquad F = \dfrac{I_{2f}}{I_1}$

Far-end crosstalk ratio $F' = \dfrac{I_{2f}}{I_{1f}}$

$$N = \overline{X}\left(\frac{1 - e^{-4n\alpha}}{4\alpha}\right)^{\frac{1}{2}} \tag{6.43}$$

and the relationship between the near-end crosstalk for a cable of length l_1 and a cable of length l_2 is

$$N_2 = N_1 \left(\frac{1 - e^{-4\alpha l_2}}{1 - e^{-4\alpha l_1}}\right)^{\frac{1}{2}} \tag{6.44}$$

It can be seen from eqn. 6.40 that the elementary crosstalk coupling increases in direct proportion to the frequency. In the case of far-end crosstalk we are usually interested in the ratio of the *received* signal to the *received* crosstalk, and since both of these components suffer the same line loss, this loss term cancels out, and the crosstalk/signal ratio then increases in direct proportion to the frequency, namely, at 6 dB per octave. In the case of near-end crosstalk, the effects of line loss are not eliminated. As we have seen in Section 6.2, the attenuation is proportional to the square root of the frequency, and so near-end crosstalk/signal ratio varies in proportion to $f^{\frac{3}{4}}$, that is, at a rate of 4·5 dB per octave. It should be emphasised that both of these figures represent smooth mean trends, and that there are in

practice considerable random deviations from these mean characteristics, particularly in the case of near-end crosstalk.

6.6 Time-domain characteristics

Since, for all practical purposes, transmission lines are linear elements, their time-
domain responses may be derived from their frequency-domain characteristics
through the use of the Laplace transform. Formally, the transform of the input
signal is combined with the transform of the transmission-line function, and inversion
of the result leads to the required solution (see Carslaw and Jaeger[18] for details). In
practice, it is frequently found that one or more of these steps involve integrals that
cannot be evaluated, and it is then necessary either to introduce some simplifying
approximations, or to resort to numerical solution. We will first consider a number
of examples in which analytic solutions are possible.

The RC line. This case is of great historical interest because of its association with
the early work on long submarine telegraph cables, and is still important when
dealing with audio pair cables (see Chapter 7). The submarine cable was briefly
considered in Chapter 1, where the solution originally derived by Lord Kelvin was
given. Kelvin obtained his solution by first defining the steady-state voltage and
current distribution along the line, and then representing this by a sum of sinusoidal
components. This is, in effect, a process of Fourier analysis of the spatial distribution of the signal. The individual sinusoidal components can be considered to
represent the possible resonant states of the line (analogous to the possible modes
of vibration in a stretched string) and are therefore referred to as the natural modes
of the system. The solution obtained through this approach (eqn. 1.3) has the
disadvantage that it is expressed in terms of a specified length of line l, and does not
readily show how the signal varies along the line. By using Laplace transform
methods we can obtain solutions in a more convenient form. For example, for a
semi-infinite line with an input voltage step e_s applied at time $t = 0$, the transforms
derived from eqns. 6.3 and 6.4 are

$$V(x,s) = \frac{e_s}{s} e^{-x\sqrt{RCs}}$$

$$I(x,s) = \frac{e_s}{s} \left(\frac{C}{R}\right)^{\frac{1}{2}} s\, e^{-x\sqrt{RCs}} \tag{6.45}$$

Which, on inversion, give for the current and voltage at a point x

$$v(x,t) = e_s \operatorname{erfc}\left(x\sqrt{\frac{RC}{4t}}\right)$$

$$i(x,t) = e_s \left(\frac{C}{\pi R t}\right)^{\frac{1}{2}} e^{-\frac{x^2 RC}{4t}} \tag{6.46}$$

where

$$\text{erfc}(x) = 1 - \text{erf}(x) = 1 - \frac{2}{\sqrt{\pi}} \int_0^x e^{-y^2} \, dy \tag{6.47}$$

Since, at the point x, the semi-infinite line is indistinguishable from a finite line of length x terminated in its characteristic impedance Z_0, this latter case is also covered by these expressions.

The RC line with arbitrary resistive termination. This can be approached either by an extension of Kelvin's methods, or through Laplace-transform techniques. The latter case is similar to examples given by Carslaw and Jaeger,[18] and the final solution is found to come out as an infinite series, the terms of which are related to the natural modes. For a line of length l, terminated in a resistance R_l, with an applied input voltage step e_s, the solution is

$$e_l = e_s \left[\frac{R_l}{Rl + R_l} + \sum_x \frac{2R_l e^{-\frac{x^2 t}{CRl^2}}}{Rl \left(\cos x - \frac{\sin x}{x}\right)} \right] \tag{6.48}$$

$$i_l = \frac{e_l}{R_l}$$

where the values of x are given by

$$\tan x = \frac{R_l}{Rl} x$$

Note that Kelvin's original solution is the limiting case of this when R_l becomes zero.

Lines with no dielectric losses. Lines of this type were discussed at length in Section 6.2, and it was shown that at high frequencies their behaviour was dominated by the variation in conductor impedance caused by skin effect. For most cases, eqns. 6.23–6.25 give an adequate account of the line characteristics. Using these, the transforms of the step response of a semi-infinite line are then

$$V(x,s) = \frac{1}{s} e^{-(Nxs + Kx\sqrt{2s})}$$

$$I(x,s) = \frac{1}{s(Z_\infty + \sqrt{2}A/\sqrt{s})} e^{-(Nxs + Kx\sqrt{2s})} \tag{6.49}$$

which lead to

$$v(x,t) = \text{erfc}\left[\frac{Kx}{\{2(t-Nx)\}^{\frac{1}{2}}}\right]$$

$$i(x,t) = \frac{1}{Z_\infty} e^{2\left(\frac{K}{N}\right)^2 (t-Nx)} \text{erfc}\left[\frac{K}{N}\{2(t-Nx)\}^{\frac{1}{2}} + \frac{Kx}{\{2(1-Nx)\}^{\frac{1}{2}}}\right] \quad (6.50)$$

(See Bohn[19] for a derivation but note error in his final expression.) Referring back to eqn. 6.33, we see that in practical cases $A \ll Z_\infty$ and hence only a small error results from assuming

$$i(x,t) \approx v(x,t)/Z_\infty \qquad (6.51)$$

It is rather surprising to find that this solution is very similar to that obtained for the semi-infinite RC line. As a result of this, the development in the next two Sections may be applied to both of these cases.

Response to rectangular pulses. Since a rectangular pulse of duration t_0 can be considered as the superposition of two step functions separated by the same time interval, we are now in a position to investigate the propagation of rectangular pulses over the types of line considered in the previous Sections. As an example, consider a semi-infinite line with no dielectric losses. From eqn. 6.48 it can be seen that the voltage waveform at a point x resulting from the input of a rectangular pulse of duration t_0 is

$$v(x,t) = \text{erfc}\left[\frac{Kx}{\{2(t-Nx)\}^{\frac{1}{2}}}\right] \qquad t \leqslant t_0$$

$$= \text{erfc}\left[\frac{Kx}{\{2(t-Nx)\}^{\frac{1}{2}}}\right] - \text{erfc}\left[\frac{Kx}{\sqrt{\{2(t-t_0-Nx)\}^{\frac{1}{2}}}}\right] \qquad t > t_0 \quad (6.52)$$

An evaluation of this expression for the case of a typical coaxial cable is given in Fig. 6.8. This illustrates the way in which distortion builds up during transmission, and important distinctions between the behaviour of short and long lines can be deduced. In the case of short lines the main effect of cable loss is to increase the rise and fall times of the pulse, and under these conditions the width at half height remains substantially constant, even though the shape of the pulse has undergone appreciable distortion. This is illustrated in Fig. 6.9. For these conditions, inter-symbol interference between successive pulses is not serious, and it is possible to regenerate rectangular pulses by a simple slicing circuit set to operate at the 50% amplitude point. On long lines the situation is totally changed. The received waveform is now elongated to have significant amplitude over a very large number of successive time slots, and it is then essential to carry out equalisation before any digital processing can be attempted. Referring back to eqn. 6.24, it is seen that the

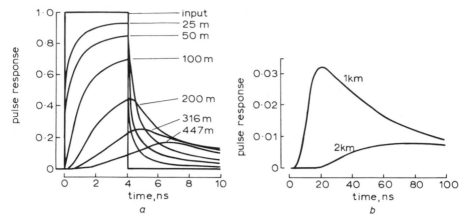

Fig. 6.8 Attenuation of a rectangular pulse in a coaxial line
Curves refer to 4 ns input pulse and 4·4 mm telephony type coaxial cable

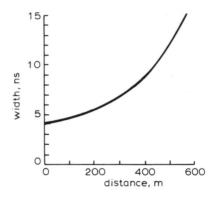

Fig. 6.9 Variation in pulse width during transmission along a coaxial cable
Width measured at 50% amplitude for a 4 ns pulse on 4·4 mm telephony coaxial cable

distortion is primarily caused by the loss of the high-frequency components in the signal, and one might therefore expect that if the transmitted pulse could be predistorted in some manner, so as to emphasise its high-frequency components, the shape of the received signal would be improved. A simple method of achieving this is by the transmission of a dipulse, that is, a positive pulse followed immediately by a negative pulse. Signals of this type may be analysed by a simple extension of the technique used in eqn. 6.50. An example is shown in Fig. 6.10 and it is seen that this is effective in improving the received waveform. It was, in fact, used in some of the early telegraph systems, but has tended to fall out of favour since electronic amplifiers have become available. This is because it involves processing the signal at a high-level point, which is usually more difficult, more expensive, and

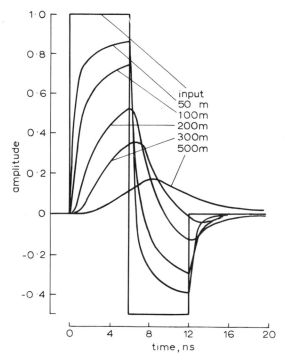

Fig. 6.10 Transmission of a dipulse through 4·4 mm coaxial cable

consumes more power than the comparable equalisation and amplification at the receiving end of the line.

Response to an arbitrary input waveform. If the transient response for a step or impulse input can be obtained this may then be used to derive the response to an arbitrary input waveform by convolution. Physically, this process may be pictured as replacing the arbitrary input by an equivalent sequence of step waveforms or impulses, and then summing the responses to these signals. Mathematically, convolution is described by Duhamel's theorem (see Carslaw and Jaeger,[18] p. 252 for details). This theorem states; if $x_1(s)$ and $x_2(s)$ are the transforms of $x_1(t)$ and $x_2(t)$, then $x_1(s)x_2(s)$ is the transform of $\int x_1(\tau)x_2(t-\tau)d\tau$. As an example, consider the transmission of a cosine squared pulse, duration t_0 at half height, over a line whose losses are due to skin effect. The Laplace transform of this pulse is

$$X_p(s) = \frac{\pi^2(1 - e^{-2t_0 s})}{2s(s^2 t_0{}^2 + \pi^2)} \tag{6.53}$$

and so the transform of the required response is

$$V(x, s) = X_p(s)\, e^{-(Ns + Kx\sqrt{2s})} \tag{6.54}$$

Putting

$$x_1(s) = \frac{1}{s} e^{-(Ns + Kx\sqrt{2s})}$$

$$x_2(s) = \frac{\pi^2(1 - e^{-2t_0 s})}{2(s^2 t_0^2 + \pi^2)}$$

The required solution becomes

$$v(x,t) = \tfrac{1}{2} \int_0^t \text{erfc}\left(\frac{Kx}{\sqrt{2\tau}}\right) \sin\left\{\frac{\pi}{t_0}(1 - \tau)\right\} H(t - 2t_0 - \tau)\, d\tau \qquad (6.55)$$

Reflections in the time domain. While we are primarily concerned with the transmission of digital signals over lines, there are still cases in which the reflected signals
are of interest. One such case arises when the impedance matching at the input and
output of digital regenerators is considered. An understanding of the nature of the
reflections is also of great assistance in interpreting the displays obtained when time-
domain reflectometers are used to test cable systems. In this Section, attention will
be restricted to the case of lines with frequency-dependent conductor losses and
negligible dielectric losses, as defined by eqns. 6.21 and 6.25. More complex situations must normally be handled by numerical methods.

Response to an input current step. Referring back to eqn. 6.23, the Laplace transform of the voltage appearing at the input to a line when a current step is applied is
seen to be

$$H(s) = \frac{1}{s} Z_\infty + \frac{A}{\sqrt{s}} \qquad (6.56)$$

and inversion then yields the corresponding time response as

$$H(t) = Z_\infty + \frac{2A}{\sqrt{\pi}} \sqrt{t} \qquad (6.57)$$

This waveform, shown in Fig. 6.11, is the basic response seen on time domain-
reflectometer displays, and the second term in eqn. 6.57 explains why a slow rise is
often seen after the initial step on such displays. The response to a rectangular pulse
of width t_0 can be obtained as the difference of two such step responses. This is
also shown in Fig. 6.11.

Cable with a resistive termination. The voltage reflection coefficient between a cable
and a terminating resistance R can be obtained from eqn. 6.7. The transform of the
corresponding step response is then given by

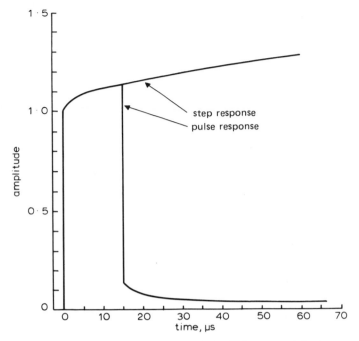

Fig. 6.11 Time-domain reflections from the input of a 2·6/9·5 mm coaxial pair

$$R(s) = \frac{R - \left(Z_\infty + \dfrac{A}{\sqrt{s}} \right)}{s \left(R + Z_\infty + \dfrac{A}{\sqrt{s}} \right)}$$

$$= \frac{R - Z_\infty}{\sqrt{s}\{\sqrt{s}(R + Z_\infty) + A\}} - \frac{A}{s\{\sqrt{s}(R + Z_\infty) + A\}} \tag{6.58}$$

The time response is then found to be

$$R(t) = 1 - \frac{2 Z_\infty}{R + Z_\infty} e^{\left(\frac{A}{R + Z_\infty} \right)^2 t} \operatorname{erfc}\left(\frac{A\sqrt{t}}{R + Z_\infty} \right) \tag{6.59}$$

For $R = Z_\infty$ this reduces to

$$R_0(t) = 1 - e^{\left(\frac{A}{2 Z_\infty} \right)^2 t} \operatorname{erfc}\left(\frac{A\sqrt{t}}{2 Z_\infty} \right) \tag{6.60}$$

The form of this response is shown in Fig. 6.12, and it can be seen that it resembles the step response discussed in the previous Section. It demonstrates that it is impossible to correctly terminate a real cable by a resistor. It is also of interest to note that if

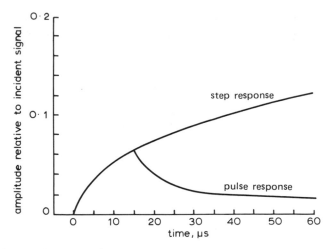

Fig. 6.12 Reflection from a 2·6/9·5 mm coaxial pair terminated in a resistance equal to its asymptotic characteristic impedance

$$A\sqrt{t} \ll (R + Z_\infty)$$

$\text{erfc}\left(\dfrac{A\sqrt{t}}{R + Z_\infty}\right)$ may then be replaced by $1 - \dfrac{2A\sqrt{t}}{\sqrt{\pi}(R + Z_\infty)}$ and $e^{\left(\frac{A}{R+Z_\infty}\right)^2 t}$ by 1.

After some manipulation it is then found that

$$R(t) \approx \frac{R - Z_\infty}{R + Z_\infty} - \frac{4A Z_\infty}{(R + Z_\infty)^2} \left(\frac{t}{\pi}\right)^{\frac{1}{2}} \tag{6.61}$$

Since the reflection coefficient cannot exceed unity, it is clear that this approximation cannot be valid for large values of t. However, by using the asymptotic expansion

$$\text{erf}(x) = 1 - \frac{e^{-x^2}}{x\sqrt{\pi}}\left(1 - \frac{1}{2x^2} + \frac{1 \cdot 3}{(2x^2)^2} \cdots\right) \tag{6.62}$$

a second expression that is valid for large t can be obtained. This is

$$R(t) = 1 - \frac{2Z_\infty}{A(\pi t)^{\frac{1}{2}}} \tag{6.63}$$

As in previous cases, the response to a rectangular pulse can be expressed as the difference of two expressions of the above type, and the response for other pulse waveforms may be obtained by convolution.

Lumped capacitive loads. In this case, we will confine our attention to lossless lines having real characteristic impedances. If a capacitor is connected across such a line at some point remote from its end, the conditions will be equivalent to a line terminated at the tapping point in a resistance R_0 shunted by a capacitor C. The transform of the voltage-reflection coefficient is then

$$R(s) = \frac{\dfrac{R_0}{1 + sCR_0} - R_0}{\dfrac{R_0}{1 + sCR_0} + R_0}$$

$$= \frac{-s}{s + \dfrac{2}{CR_0}} \tag{6.64}$$

and the resulting reflection due to a unit step input is

$$V_r(s) = -\frac{1}{s + \dfrac{2}{CR_0}} \tag{6.65}$$

yielding

$$V_r(t) = -e^{-2t/CR_0} \tag{6.66}$$

This is a case in which the theoretical result must be treated with some caution, since it predicts that the peak amplitude of the reflected signal will be equal to the amplitude of the incident step, and it is known from experience that this does not occur in practical cases. The reason for this discrepancy is that the analysis has been performed for an ideal step with an infinitely fast rise time, and a corresponding infinite frequency spectrum, and any practical case is concerned with signals having finite rise times and bandwidths. Hence, as a more practical example, let us consider the reflection from an incident raised cosine pulse of duration t_0 at half height. The transform of the pulse is

$$C(s) = \tfrac{1}{2}\left(\frac{1}{s} - \frac{s}{s^2 + \left(\dfrac{\pi}{t_0}\right)^2}\right)(1 - e^{-2t_0 s}) \tag{6.67}$$

and the corresponding reflection is

$$V_c(s) = \tfrac{1}{2}\left(\frac{\left(\dfrac{\pi}{t_0}\right)^2}{s^2 + \left(\dfrac{\pi}{t_0}\right)^2}\right)\left(\frac{1}{s + \dfrac{2}{CR_0}}\right)(1 - e^{-2t_0 s}) \tag{6.68}$$

which after some algebraic manipulation leads to

$$V_c(s) = \frac{1}{2\left(1 + \left(\dfrac{2t_0}{\pi CR_0}\right)^2\right)}\left(\frac{-s + \dfrac{2}{CR_0}}{s^2 + \left(\dfrac{\pi}{t_0}\right)^2} + \frac{1}{s + \dfrac{2}{CR_0}}\right)(1 - e^{-2t_0 s})$$

$$V_c(t) = \frac{1}{2\left\{1 + \left(\frac{2t_0}{\pi CR_0}\right)^2\right\}} \left(\frac{2t_0}{\pi CR_0} \sin \pi \frac{t}{t_0} - \cos \pi \frac{t}{t_0} + e^{-\frac{2t}{CR_0}}\right)$$

$$\times \; \{1 - H(t - 2t_0)\} \tag{6.69}$$

Some typical examples of this response are plotted in Fig. 6.13.

Approximate solutions; Extension to the case of lines with composite conductors, and with dielectric losses, becomes very complex, and only a few solutions are known. Nahman[8,20] gives a result for the case of plated conductors, and he also suggests that dielectric losses may be allowed for by altering the exponent in the attenuation term, leading to propagation constants of the form

$$p = Nxs + Kxs^m \tag{6.70}$$

Ignoring the delay component Nxs, the step response is then given by

$$v(x,t) = 1 - \frac{1}{\pi} \int_0^\infty \frac{1}{r} e^{-(rt + ar^m)} \sin(br^m) \, dr \tag{6.71}$$

where

$$a = Kx \cos m\pi$$

$$b = Kx \sin m\pi$$

An alternative method of including dielectric losses can be derived from Section 6.4. Using the network model of the dielectric derived in that Section we may put

$$p = \frac{Z_c}{2Z_\infty} + \frac{Z_\infty}{2} \sum_{i=0}^n \frac{j\omega C_i}{1 + j\omega R_i C_i} \tag{6.72}$$

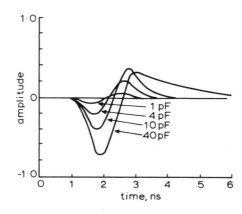

Fig. 6.13 Reflection of a raised cosine pulse from a shunt capacitance in a 75 Ω transmission line
Pulse duration = 1 s at half height

for small dielectric losses. The transform of the step response is then

$$V(\lambda,s) = \frac{1}{s} e^{-K r_0 \sqrt{7 r}} e^{-\left(\frac{Z_\infty}{2} \sum_{i=1}^{n} \frac{sxC_i}{1+sxR_iC_i}\right)} \tag{6.73}$$

which may be evaluated by convolution to give

$$V(x,t) = \operatorname{erfc}\left(\frac{Kx}{\sqrt{2t}}\right) \left(1 - \frac{xZ_\infty}{2} \sum_{i=1}^{n} \frac{1}{R_i}\right)$$

$$+ \frac{xZ_\infty}{2} \int_0^t \operatorname{erfc}\left(\frac{Kx}{\sqrt{2t}}\right) \left\{\sum_{i=1}^{n} \frac{1}{R_iC_i} e^{-\frac{(t-\tau)}{R_iC_i}}\right\} d\tau \tag{6.74}$$

6.7 Irregular lines

Up to the present, we have assumed that the primary line parameters $R, L, G,$ and C remain constant along the length of the line. There are however two other possible situations (i) the parameters vary with position in some regular systematic manner, and (ii) the parameters vary in an irregular manner. The former case arises, for example, in microwave impedance matching networks, and while this is of considerable interest, and has received extensive attention in the published literature, we will not consider it in the present work. The second situation arises in all practical transmission lines, for a variety of reasons. These include

(a) discontinuities in the line geometry, caused by sharp bends, dimensions steps at joints and connectors etc. These produce disturbances in the electromagnetic field, that appear as equivalent lumped impedance discontinuities (see Somlo[21]).

(b) irregularities caused by imperfections in fabrication. Variations in the geometry of lines, such as small changes in conductor spacings, and small changes in stranding pitch can be caused by imperfections and irregularities in the operation of the cable-making machines. There will also be slight variations between different manufactured lengths of cable, which will give rise to further discontinuities when these lengths are joined to make a long line.

(c) mechanical constraints on the line structure. For example, in semi-air-spaced coaxial lines insulating supports must be introduced periodically to support the inner conductor.

The major effect of irregularities is to introduce reflected signals, as determined by eqns. 6.7 and 6.8. These reflected signals will in turn suffer reflection from other irregularities, and so produce secondary reflections propagating in the same direction as the main signal. These secondary reflections, while small, are not always entirely insignificant. Viewed in the time domain they appear as a distortion following any step waveform, or as a 'tail' following a pulse, as shown in Fig. 6.14. In the

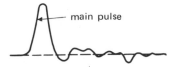

Fig. 6.14 Tail caused by secondary reflections following a transmitted pulse
Effect has been exaggerated for clarity

frequency domain they produce small deviations from the smooth transmission characteristics predicted by theory. The primary reflections are, of course, much larger than the secondary components, and it is usual to employ a measure of these as a check on the uniformity of the line, and then to use this to estimate the likely transmission disturbances. Reflections may be measured in the time domain by standard pulse techniques, or by special-purpose time-domain reflectometers. In the frequency domain, the reflections are most conveniently measured in terms of their effects on the input impedance of the cable. This fluctuates about the smooth mean characteristic, as shown in Fig. 6.15, and the amplitude of these fluctuations is a measure of the irregularities. Recently, a hybrid technique using a pulse-modulated carrier as a test signal has become popular (see Rosman[22]).

The reflected signals represent a loss of energy from the transmission, and so, by analogy with the situation which arises in any network when a mismatch occurs between a source and load, these effects can be described as a return loss. The term 'structural return loss' is now commonly employed in this particular case.

Because the frequency and time-domain responses form a Fourier transform pair, we may equate the pulse response and impedance irregularities by

$$\{\overline{s(t)^2}\}^{\frac{1}{2}} = \{\overline{\delta z(\omega)^2}\}^{\frac{1}{2}} = K \tag{6.75}$$

where
$s(t)$ is the response to a unit pulse
and $\delta z(\omega)$ is the impedance deviation.

The prediction of the resulting transmission degradations has been investigated by a number of workers, but unfortunately much of the literature has appeared in

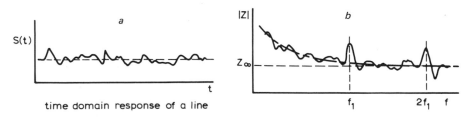

Fig. 6.15 Responses of an irregular line
a Time domain response
b Frequency domain impedance characteristics
Note that fluctuations are superimposed on the mean characteristics and that periodic irregularities show up as large peaks in the response

French and German sources and does not appear to be well known in this country. The best papers in English are by Mertz and Phleger[23] and by Fuchs.[24] It can be shown that if K is the r.m.s. impedance deviation measured on a short length of cable of attenuation α dB per unit length, then for a length l.

rise in attenuation $= K^2 \alpha L$ dB

standard deviation of attenuation $= K^2 (4 \cdot 343 \, \alpha L)^{\frac{1}{2}}$

delay distortion $= \pm 4 \cdot 42 \dfrac{K^2}{Z_0} \tau \left(\dfrac{L}{\alpha}\right)^{\frac{1}{2}}$

attenuation sinuosity $= 0 \cdot 0256 \dfrac{K^2}{Z_0} \pi \tau \left(\dfrac{L}{\alpha}\right)^{\frac{1}{2}}$ (6.76)

where τ = delay per unit length.

6.8 References

1 WILLIS JACKSON: 'High frequency transmission lines' (Methuen, 1947)
2 GRIVET, P.: 'The physics of transmission lines at high and very high frequencies' (Academic Press, 1970) vol. 1
3 GOLDMAN, S.: 'Transformation calculus and electrical transients' (Prentice Hall, 1949)
4 ITT: 'Reference data for radio engineers' (Howard W. Sams, 1968) 5th edn.
5 SHEA, R.F.: 'Principles of transistor circuits' (Wiley, 1953)
6 VON HIPPEL, A.R.: 'Dielectric materials and applications' (MIT, 1954)
7 RAMO, S., WHINNERY, J.R., and VAN DUZER, T.: 'Fields and waves in communication electronics' (Wiley, 1965)
8 NAHMAN, N.S.: 'A discussion on the transient analysis of coaxial cables considering high frequency losses' *IEEE Trans.*, 1962, CT-9, pp. 144−152
9 SCHELKUNOFF, S.A.: 'The electromagnetic theory of coaxial transmission lines and cylindrical shields', *Bell Syst. Tech J.*, 1934, **13**, pp. 532−579
10 DWIGHT, H.B.: 'Tables of integrals and other mathematical data' (McMillan, 1947)
11 ARNOLD, A.H.M.: 'Proximity effects in solid and hollow round conductors', *Proc. IEE*, 1941, **88**, Pt. II, pp. 349−359
12 EAGER, E.S., JACHIMOVICZ, L., KOLODNY, I., and ROBINSON, D.G.: 'Transmission properties of polythene insulated telephone cables at voice and carrier frequencies' *Trans. AIEE*, Nov. 1959, p. 618
13 VON HIPPEL, A.R.: 'Dielectrics and waves' (Wiley, 1954)
14 PIPES, L.A., and HARVIL, L.R.: 'Applied mathematics for engineers and physicists' (McGraw Hill, 1970) 3rd edn.
15 HAYASHI, S.: 'Analysis of the special problems of travelling wave phenomena on multi-conductor transmission systems by Hayashi's new analytic method', Technical Reports of the Engineering Research Institute, Kyoto University, 1952, **2**, p. 138
16 CROZE, R., and SIMON, L.: 'Transmission telephonique; theorie des lignes', Editions Eyrolles, Paris, 1952
17 SCHELKUNOFF, S.A., and ODURENKO, T.M.: 'Crosstalk between coaxial transmission lines', *Bell Syst. Tech. J.*, 1936, **16** pp. 144−164
18 CARSLAW, H.S., and JAEGER, J.C.: 'Operational methods in applied mathematics' (Oxford University Press, 1948) 2nd edn.
19 BOHN, E.V.: 'The transform analysis of linear systems' (Addison Wesley, 1963)

20   NAHMAN, N.S.: 'A note on the transition, (rise time) versus line length in coaxial cables', *IEEE Trans.*, 1973, **CT-20**, pp. 165–167
21   SOMLO, P.I.: 'The computation of coaxial line step discontinuities', *ibid.*, 1967, **MTT-15**, pp. 48–53
22   ROSMAN, G.: 'Assessment of coaxial cable for frequency division multiplex transmission by means of a carrier wave burst test signal', *Proc. IEE*, 1970, **117**, (■), pp. 45–50
23   MERTZ, P., and PFLEGER, K.W.: 'Irregularities in broadband wire transmission circuits', *Bell Syst. Tech. J.*, 1937, **16**, pp. 541–559
24   FUCHS, G.: 'Reflections in a coaxial cable due to impedance irregularities', *Proc. IEE*, 1952, **99**, (4), pp. 121–136
25   BODE, H.W.: 'Network analysis and feedback amplifier design' (Van Nostrand, 1945)

Chapter 7

Practical transmission lines

7.1 Introduction

After the rather intensive theoretical survey of the last Chapter, we must now return to the real world, and consider what happens on practical transmission lines. There are two main areas of interest, the long lines used as the transmission paths in communication systems, and the short lines used for interconnections within and between pieces of equipment. In the case of long lines, as well as studying the electrical characteristics of the transmission line itself, we must take account of geographical features, such as the spacing of joints and the total length of each cable section. As we have seen in Chapter 3, a complete trunk transmission link may be several thousand kilometres in length, and even the individual sections of this link between switching centres may be over a hundred kilometres long. However, the repeater amplifiers or regenerators effectively isolate the different sections of line, and we can normally restrict our attention to the characteristics of single repeater sections. These sections are usually between 1 and 25 km in length. In the local network we are concerned with the link between the subscriber and the local exchange. This is normally a single cable section without repeaters and will be between a few hundred metres and a few kilometres long.

The cable sections in both the trunk and local network are not homogeneous, because it is not practical to make and lay cable lengths of several kilometres. In the United Kingdom the standard practice is to install cables in ducts, and the primary cable lengths are then determined by the practical limits on the lengths that may be drawn into the ducts, usually between 100 and 400 metres. To permit drawing in and jointing, manholes and jointing chambers must be provided, and the exact positioning of these is restricted by the local geography of the route, so that the primary cable lengths are not uniform. Similarly, repeater section lengths vary. Climatic factors are also important, because open-wire lines and pole-mounted cables are affected by changes in temperature, by rain, snow, and such like, and even underground cables exhibit slow seasonal changes as the ambient temperature varies between winter and summer.[1-3]

Interconnection cables can vary in length from a few centimetres up to several hundred metres, in the case of interconnections in large exchange buildings. Temperature changes are not so extreme as for outside plant, but the environment is often electrically noisy and so induced interference has to be taken into account.

7.2 Audio pair cables

As discussed in Chapter 1, cables of this type were first introduced about 85 years ago. The fundamental problems at that time were to select a suitable combination of a good conductor and a satisfactory insulating material, and the combination of copper conductors and dry paper insulation became the accepted standard. The wires were arranged as twisted pairs, or as groups of four conductors (star quads). Crosstalk was controlled by varying the stranding lays between the different pairs or quads.[4] The complete cable was assembled by combining the pairs or quads in concentric layers, or by first forming units of a number of pairs and then assembling the units to make up a complete cable, as in Fig. 7.1. An outer lead sheath was used as protection. This basic design of dry core cable has remained almost unchanged up to the present day. In about 1950, plastic insulation, principally polyethylene (polythene), began to be introduced, and the increasing cost of copper and lead encouraged attempts to economise in the use of these metals. As a result, cables employing aluminium conductors are now common, and the lead sheath is often replaced by a plastic one. Unlike lead, however, plastic sheathing is

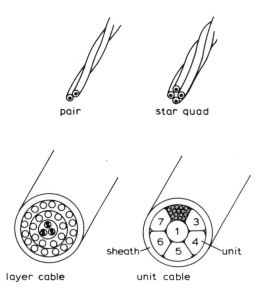

Fig. 7.1 Configurations for audio pair cables

not completely impervious to the ingress of water, and so cables are provided with either an aluminium-foil moisture barrier or with their core interstices completely filled with petroleum jelly. This filling has the added advantage of preventing the longitudinal passage of water in the event of a sheath puncture. Advantage has been taken of the increased sensitivity of modern terminal equipment (i.e. telephone hand sets) to allow higher line losses, and so the conductor sizes employed in audio cables have tended to be reduced in recent years. Star quad cables provide compact assemblies (up to 40% more conductors in a given cross-section than for pair cables) but pair cables offer advantages in flexibility of interconnection. Both types are in common use, but there is a tendency to favour unit pair cables at the present time.

Electrical characteristics. As we shall see later, digital signals at rates up to 2 Mbit/s may be transmitted over audio pair cables, which implies that the characteristics of these cables up to frequencies of about 10 MHz will be significant in digital applications. An account of the characteristics of dry-core cables of the types in common use before 1950 is given by Hebbett.[5] Detailed information on more recent plastic-insulated cables as used in America is given in a series of papers by Jachimowicz and his colleagues[6,7] and information for Japanese cables is given by Matuda.[8] No comparable information has been published for British cables, but it is known from unpublished work that the characteristics of these cables are very similar to those of comparable American and Japanese products. As an example, measured parameters of a British aluminium/polyethylene unit pair cable are given in Figs. 7.2 and 7.3. These results are reproduced by permission of the British Post Office, and illustrate a number of features which are found to be typical of all pair cables. First, the characteristic impedance (Fig. 7.2a) is seen to be split into two regions. At low frequencies it has equal real and imaginary components, which are both inversely proportional to $\sqrt{f}$ and at high frequencies the impedance is real and constant at about 100 Ω. The transition occurs in the region of 100 kHz. This behaviour can be explained by remembering that the leakance G is always small, and at low frequencies the conductor resistance R is much larger than the series reactance ωL so that

$$Z_0 \approx \left(\frac{R}{j\omega C} \right)^{\frac{1}{2}} \qquad (7.1)$$

while, at high frequencies where $\omega L \gg R$

$$Z_0 \approx \left(\frac{L}{C} \right)^{\frac{1}{2}} \qquad (7.2)$$

Obviously, the transition region is defined by

$$f = R/2\pi L \qquad (7.3)$$

Note that these expressions are only first approximations, since other factors, such

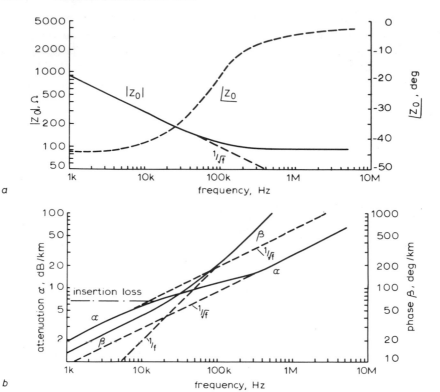

a

b

Fig. 7.2 Parameters of 0·5 mm aluminium-polyethylene pair cable
a Characteristic impedance ⎞
b Propagation characteristics ⎠
[Reproduced by permission of the Post Office]

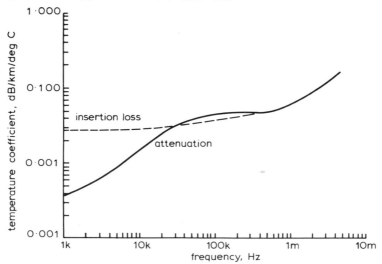

Fig. 7.3 Temperature coefficient of attenuation of a pair cable

as proximity effects (Chapter 6, Section 3) and variations in dielectric characteristics (Chapter 6, Section 4) will have some effect.

Corresponding to these impedance characteristics we would expect the low frequency propagation constant to be given by

$$p = (j\omega RC)^{\frac{1}{2}} = \left(\frac{\omega RC}{2}\right)^{\frac{1}{2}} (1 + j) \qquad (7.4)$$

This exhibits phase and attenuation variations proportional to $\sqrt{f}$ translating to a constant attenuation

$$\alpha = \frac{R}{2}\left(\frac{C}{L}\right)^{\frac{1}{2}} \qquad (7.5)$$

and a linear phase frequency characteristics when ωL becomes comparable with R. It can be seen from Fig. 7.2*b* that this transition does tend to occur, but that at a slightly higher frequency the attenuation returns to a $\sqrt{f}$ dependance, due to the intrusion of skin effect (Chapter 6, Section 2). While this has a major effect on the attenuation it only slightly modifies the phase and impedance characteristics. Dielectric losses also have their greatest effect on the attenuation and as a result it is commonly found that in paper-insulated cables the attenuation is beginning to increase at something above a $\sqrt{f}$ rate, for frequencies in excess of 100 kHz. Dielectric losses are negligible in plastic-insulated cables.

It should be noted that, under practical operating conditions, the attenuation constant α does not give a true indication of the low frequency loss of the line. α is a measure of the signal loss between the source and the load impedances equal to Z_0, the characteristic impedance of the line which, as we have seen, increases at low frequencies. On the other hand, the input and output impedances of the equipments connected to the line are real and constant, and are typically chosen to be equal to the high-frequency value of Z_0. Thus, there are significant losses due to mismatch at the lower frequencies. The low-frequency insertion loss of the line is then very much higher than its attenuation under matched conditions. This is shown in Fig. 7.2*b*. For the terminations employed in this case the insertion loss is constant below the frequency defined by $R = \omega L$.

The importance of temperature variations has already been mentioned. In the case of pair cables, these manifest themselves mainly through variations in the conductor resistance. Thus, in the low and high-frequency regions, where $\sqrt{f}$ frequency dependence occurs, an associated term in $\sqrt{R}$ or $\sqrt{\sigma}$ appears in the expression for attenuation. The temperature coefficient of attenuation will then be half the temperature coefficient of resistance of the conductor material, while in the intermediate region, where attenuation is dependent on R it will be equal to the t.c.r. This is shown in Fig. 7.3. It should be noted that if the frequency spectra of the line signals are confined to either the high-frequency or the low-frequency regions, the effects of temperature variations are identical to the effects of changes in line lengths, and so the same equalisation scheme can be used to cover both types of variation. If operation covers the transition region this introduces some small errors. In the case of paper insulated cables there is also some contribution from temperature variations in dielectric loss, and these can become irregular at high frequencies.[5] This has been attributed to the effects of residual moisture trapped in the insulation.

Typical crosstalk-attenuation characteristics are given in Fig. 7.4. The actual examples shown are for pairs in fairly close proximity within the cable, and so fall somewhere near the lower limit for the range of crosstalk loss experienced in this particular type of cable. In detail the characteristics are highly irregular, but the general trends of a 4·5 dB per octave slope for near-end crosstalk, and a 6 dB slope for far-end crosstalk ratio (see Chapter 6) can be detected. The irregularity of the crosstalk characteristics is associated with irregularities in the cable construction, and, for audio pair cables, this is the major manifestation of production variations.

7.3 High-frequency pair cables

In the period 1930–55, before the widespread introduction of coaxial cable systems, medium-capacity f.d.m. systems employing pair cables were installed in

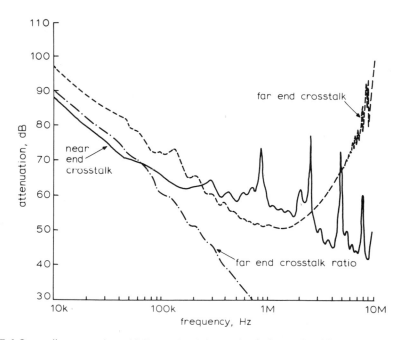

Fig. 7.4 Crosstalk attenuation of 0·5 mm aluminium-polyethylene pair cable
[Reproduced by permission of the Post Office]

many countries. These offered capacities of up to 60 telephony channels and employed frequencies up to about 250 kHz. (Some 120-channel systems extending to 500 kHz were introduced at the end of this period.) Special high-frequency pair cables were developed for these applications, and a number of countries have extensive networks of this type of line. While in the majority of cases these lines are still in use with the original f.d.m. equipment, the possibility of re-equipping them with digital equipment has been considered, although, at the present time (1975) no such conversions have actually been carried out. More recently, new designs of pair cable have been produced, intended explicitly for digital operation.[9-11]

Mechanical design. The main problem in using pair cables at high frequencies is centred on the control of crosstalk. As discussed in Chapter 6, this coupling is largely due to mechanical irregularity of the pairs which manifests itself as unbalanced coupling, and so the major effort in designing h.f. pair cables has been concentrated on this aspect. The basic capacitance has been reduced in the early designs by using semi-air-spaced insulation formed by paper 'string', and, in more recent designs, by use of foamed-plastic insulation. Careful attention to regularity of manufacture and to the choice of the stranding lays reduces the residual unbalance. In some recent designs, thin metal-foil screens have been introduced to separate different sections of the cable.[9,10]

Electrical characteristics. The reduced line capacitance leads to a higher characteristic impedance and reduced attenuation. Rather heavier gauge conductors are used than for audio cables, resulting in a further reduction in attenuation. Thus, while the general form of the line characteristics are similar to those described in the previous Section, we find that the high-frequency asymptotic impedance of carrier cables is rather greater typically in the range 140–170 Ω, and the attenuation is significantly lower, typically 2·2 dB per km at 100 kHz. In unscreened cables crosstalk attenuations are of the order of 10–20 dB better than the figures quoted in Section 7.2. This is sufficient to meet the far-end-crosstalk requirements for carrier transmission, but to meet near-end-crosstalk requirements it has normally been necessary to employ separate cables for the two directions of transmission. Data on established cable designs is given by the CCITT (recommendation G 321) and by Cohen.[12] Information on the new designs being developed will be found in Okamoto,[10] Murata,[9] and Setzer.[11]

Geographical features. As mentioned above, crosstalk requirements dictated the installation of cables in pairs, one for each direction of transmission. Systems were originally designed for repeater spacings of the order of 20 km, and as each repeater location required a number of repeaters, which, in the earliest designs were rather bulky, it is usual to find substantial repeater buildings along pair cable routes.

7.4. Open-wire lines

To the best of our knowledge no use has so far been made of open-wire lines for direct digital transmission, but since some such system will almost certainly appear somewhere at some time, a few words on the characteristics of this transmission medium seem called for. Open-wire lines are most commonly encountered in undeveloped countries or in sparsely populated rural areas, where comparatively long low-capacity transmission links are required. In these situations, the cost of buried cables is usually unacceptable, and the compromise of a more economical, albeit less reliable, connection is then chosen. Data on open-wire lines is given by Boyce,[13] Cohen,[12] and ITT,[14] and the reader is referred to these sources for numerical information. Typical lines use heavy conductors (diameters of 2·5 mm or more) spaced 8–12 inches apart. Thus, skin effects appear at quite low frequencies (10–20 kHz), and the combination of a low conductor impedance and a high characteristic impedance (typically 600–700 Ω) leads, under dry conditions, to very low attenuations. However, wet weather produces leakage losses at the insulators which can increase the line losses by up to 50%, and such changes can occur very rapidly. Even more spectacular increases in loss can arise when lines become coated with wet snow, and, in passing, it is interesting to note that dry snow can induce static charges by frictional effects. Crosstalk between lines on the same pole route is high. Continuous balancing by stranding techniques is not practical, and crosstalk is suppressed by periodic line transpositions.

7.5 Trunk coaxial cables

The first carrier systems utilising coaxial cables were introduced in the late 1930s, and by 1948 a standard design of coaxial cable had emerged; which has since been used extensively in both Europe and America. It is still being manufactured, and the current product differs only slightly from the original version; a striking testimony to the quality of the design. The structure is shown in Fig. 7.5. In the current design, the inner conductor is solid copper, 2·64 mm (0·104 in) diameter, and is supported by polyethylene discs, 1·8 mm thick, at intervals of 33 mm. Two techniques are in common use for providing these supporting discs.[15] In one, they are moulded on the conductor, in the other, they are prepunched from polyethylene tape, and are slotted so that they can be pushed on to the conductor. (See Fig. 7.5a.) The outer conductor consists of a copper tape, 0·254 mm (0·01 in) thick, which is shaped over the supporting discs to form an outer conductor of 9·5 mm (0·375 in) internal diameter with a longitudinal joint. In the first designs, the edges of the tape were serrated as shown in Fig. 7.5a, to form a series of interlocking tabs, but the punching operations needed to form these tabs were found to slow up the manufacturing process. This was partially overcome by introducing a rotary punch, although in more modern versions a butt joint is used, with

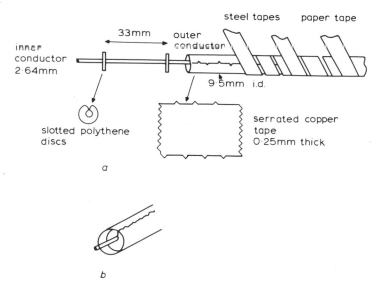

Fig. 7.5 Construction of 9·5 mm trunk coaxial cables
    a Serrated seam 9·5 mm cable
    b Indented seam cable

the edges of the tape indented to prevent them sliding over one another (Fig. 7.5*b*). The longitudinal jointing scheme was adopted to inhibit any tendency for the current in the outer conductor to flow in a spiral path (which could occur with wrapped tapes) since this would increase the low-frequency magnetic coupling between pairs. To provide mechanical support, and to increase low-frequency cross-talk attenuation, the outer copper conductor is covered by two mild steel tapes, each 0·13 mm (0·005 in) thick. These tapes are wound in the same direction with a gap of about 2·5 mm between turns, and with the outer tape covering the gap in the lower tape. The coaxial pair is finally covered by paper tape to give an overall outside diameter of about 12 mm. The complete cable is made up by stranding a number of pairs together. Eight and twelve pair assemblies are common, and in the UK up to 18 pairs are used.

The 9·5 mm cable has been the only type used in America until very recently, but in Europe a smaller size, the 4·4 mm cable, was introduced in about 1960.[16] This employs an almost identical construction to the 9·5 mm cable (Fig. 7.6). The inner conductor is 1·18 mm diameter (0·0465 in) and in most forms of this cable disc supports are used (4·1 mm diameter, 1·0 mm thick, spaced 11·0 mm) as in the 9·5 mm cable, but, as an addition to reduce the risk of breakdown under high power feeding voltage, a 0·1 mm thick polyethylene tape is wound over the discs with a 30% overlap, to form a complete covering under the outer conductor. As an alternative, in some designs a rather thicker polyethylene tube is used, which is pinched at intervals to support the inner conductor (see Fig. 7.6*b*). A third variant,

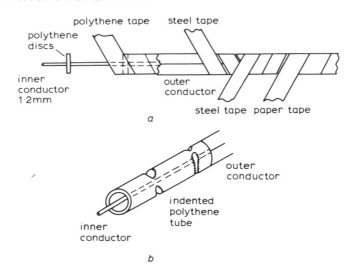

Fig. 7.6 Construction of 4·4 mm trunk coaxial cable
    a Cable with disc supports
    b Cable with indented tube support

not shown, uses an integrated polyethylene disc and tube assembly. The outer conductor is a copper tape, 4·42 mm (0·174 in) internal diameter and 0·18 mm (0·007 in) thick. This is covered by two steel tapes, 0·1 mm thick. In this case, the tapes have opposing lays, the lower tapes being wound with a small gap, and the upper tape being wound with a small overlap, because it is now believed that this configuration offers improved crosstalk attenuation. As in the case of the larger design, multiple assemblies of up to 12 tubes are common.

In the very recent past, new designs have appeared, but these have not yet become established. In America, a modified 9·5 mm cable has been developed. This uses a laminated corrugated outer conductor consisting of a thin copper foil attached to a steel supporting tape by a copolymer adhesive. In Europe, proposals for an even smaller coaxial cable having an outer diameter of 2·8 mm have been made.[19] Both of these designs have been prepared with digital applications in mind.

Electrical characteristics. The characteristics of these cables may be derived to a high degree of precision from the relations given in Section 6.42. However, it is convenient to slightly modify these to express the results in units of decibels and megahertz giving.

$$Z_0 = Z_\infty + \frac{A}{\sqrt{F}}(1-j) \text{ ohms}$$

$$\alpha = a + b\sqrt{F} + cF \text{ dB/km}$$

$$\beta = NF \text{ deg/km}$$

(7.6)

The values of these parameters for the 9·5 mm, 4·4 mm and 2·8 mm* cables are given in Table 7.1. The expressions given above are valid for frequencies down to about 200 kHz, but below this the skin depth is becoming significant compared with the thickness of the outer conductor, so that low-frequency variations of a similar form to those described for pair cables begin to appear. Because coaxial cables are not normally used for low-speed or low-frequency applications these effects can be ignored. As mentioned in Chapter 6, slight deviations in the attenuation characteristics occur when the thickness of the outer conductor is about 1·8 times the skin depth, this being in the region of 0·5 MHz for 9·5 and 4·4 mm cables.

Due to its composite outer conductor the American LC cable exhibits a more complex behaviour, and its attenuation deviates significantly from the $\sqrt{F}$ law below 3 MHz. Apart from this, it is very similar to the standard 9·5 mm cable. In all of these cables the attenuation is dominated by conductor resistance contributions, and so the temperature coefficient of attenuation is derived directly from the

Table 7.1   Parameters of trunk coaxial cables

Cable	9·5 mm	4·4 mm	2·8 mm*	Units
D.C. resistance of inner conductor	3·1	15·2	49	Ω/km
Capacitance	46·5	49	56·7	pF/m
Dielectric constant	1·074	1·17	1·5	
Velocity ratio	0·966	0·926	0·817	
Impedance Z_∞	74·4	73·1	71·8	Ω
A	0·915	1·92	3·06	Ω at 1 MHz
Attenuation a	0·013	0·066	–	dB at 1 MHz
b	2·305	5·15	9·55	per km
c	0·003	0·0047	–	
Phase shift N	1243	1296	1469	deg at 1 MHz per km
Delay	3·45	3·59	4·08	ns per m (at high frequencies)

These are typical figures which do not relate to any specific country or manufacturer.

* Figures for the 2·8 mm cable are derived from current work and may be subject to some amendment before the design is finalised.

temperature coefficient of resistance of the conductors, and leads to a $\sqrt{F}$ variation. (Data on the temperature fluctuations in buried cables will be found in Prache.)[20]

Crosstalk attenuation between coaxial pairs is high. In the absence of the steel binding tapes the coupling would begin to become significant at frequencies below about 1 MHz, but with the tapes present the loss remains high down to much lower frequencies. Calculation of this loss is difficult, since the conditions are complex, and some of the parameters, for example, the permeability of the steel tapes, are not very accurately defined; but following from the analysis in Chapter 6 we can expect that the coupling will vary in inverse proportion to $\sqrt{f}$. Typical limits set in cable specifications are:

(i) For 10 km repeater sections of 9·5 mm cable

near-end crosstalk $> 80 + 73\sqrt{F}$ dB
or 160 dB

far-end crosstalk $> 73 + 73\sqrt{F}$ dB
or 133 dB

(ii) For repeater sections of x km of 4·4 mm cable

near-end crosstalk $> 89 + 24\sqrt{F}$ dB
or 160 dB

far-end crosstalk $> 56 + 5·3 \times \sqrt{F} + 20 \log_{10} \dfrac{282}{x}$ dB.

Measurements made at frequencies below 1 MHz (Fig. 7.7) show that these limits are easily met by existing cables, and it would appear from this that at higher frequencies crosstalk will be completely negligible.

Despite their apparently rather flimsy mechanical construction, air-spaced coaxial cables achieve a surprisingly high degree of uniformity and stability. Because, up to the present time, these cables have been used almost exclusively for f.d.m. transmission, specification and testing of their regularity has been restricted to frequencies below 100 MHz, and, although the primary interest has been in the frequency-domain performance, pulse testing has been commonly employed. To utilise these tests it has been necessary to devise some simple measure of the overall degree of regularity of the cable. Clearly it is not sufficient to consider only the peak amplitude of the pulse echo signal, and in any case allowance must be made for the attenuation of echoes from points remote from the ends of the cable. As an example of the type of assessment in use we cite the CCITT proposals for tests for 9·5 mm cables employed for 60 MHz systems (recommendation G 332).

The CCITT suggest that individual factory lengths should be tested, using a sine squared pulse of 10 ns duration at half height, tests being made from each end of the cable. Then the attenuation of the largest echoes, corrected for cable attenuation, should not be less that 46 dB in all cases, and not less than 50 dB in 95% of the tests. The uncorrected attenuation of the largest echo should be not less than

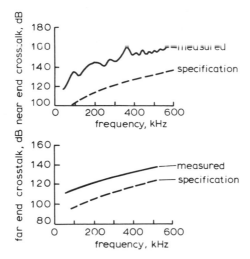

Fig. 7.7 Crosstalk performance of 9·5 mm coaxial cables
Measurements for a 6 mile cable section
[Reproduced from *ATE Journal*, July 1954]

54 dB, and the mean square of the corrected attenuation of the three largest echoes in each test should be greater than 51 dB. Translated into terms of impedance deviations these limits are equivalent to variations in the characteristic impedance of about 0·3 ohms, which, by reference to Table 7.1 is seen to be equivalent to dimensional changes of less than 0·5%. A typical pulse response of a 4·4 mm cable is shown in Fig. 7.8 and it is seen that the specification is easily met.

However, time-domain tests do not show up periodic variations in the cable impedance. These effects may be detected by steady-state measurements of the input impedance, preferably using a burst-carrier technique.[21,22] An example of such a measurement is given in Fig. 7.8. The general level of the frequency-domain reflection is about 20 dB higher than that of the time-domain reflections, which is in agreement with the rough estimate given in Chapter 6, and, in addition, there are large peaks at about 200 MHz and 400 MHz. These reveal periodic variations in the characteristic impedance of the cable, which have been found to be introduced during the stranding operations.[15,23] As discussed in these references, existing stranding machines tend to apply cyclic tensioning and torsion to the tubes, which causes periodic variations in their dimensions.

Geographical features and jointing. Since a high degree of regularity is achieved in the manufacture of coaxial cables, the disturbances introduced during installation assume a much greater importance than in the case of other types of line. These will vary in different countries according to the installation techniques employed. In the UK, the standard practice is to draw the cable into ducts in lengths of up to about 800 m, the exact lengths depending on the type of cable, the route

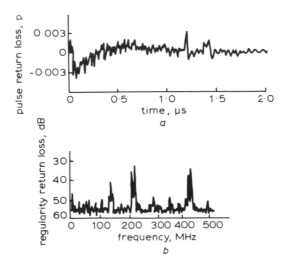

Fig. 7.8 Impedance regularity of 4·4 mm coaxial cable

    a Time domain response: 10 ns pulse/
    Note: 0·003 is equivalent to 50·5 dB return loss/
    b C.W. burst response
    [Reproduced by permission of the Post Office]

geography and the planning rules employed in laying out the duct route. Histograms for the joint distributions on typical cable routes are given in Fig. 7.9. Jointing introduces two types of disturbance; first, there will be a change in cable characteristics across the joint, since the two adjacent lengths will almost certainly have slightly different mean characteristic impedance, and any cyclic variations in impedance will undergo an abrupt phase change at this point. (The significance of this is discussed by Rosman.)[22] Secondly, the joint itself presents a disturbance. The majority of joints are of the end-to-end butt type, and are made after the manner shown in Fig. 7.10. When done properly, this type of joint introduces a negligible additional disturbance.[24]

A rather more complex situation arises at the end of each repeater section. In a typical system, the repeaters are installed in sealed boxes, which for convenience are located at the side of the road, offset from the main cable. A tail cable of some metres in length therefore runs from the main cable joint in the manhole to the repeater box, where it is terminated in a sealed coaxial connector. In present designs, this connector presents a larger discontinuity than a normal cable joint, and at the main cable, while one connection to the tail is made by a normal butt joint, the other must involve a reversal of direction, and so a special U joint is employed. Because of its more complex geometry this exhibits a greater discontinuity than a normal joint. Thus, we typically have two relatively large impedance discontinuities in close proximity, and while these are not sufficiently large to cause difficulties in existing systems, it is possible that they could become significant in any future ex-

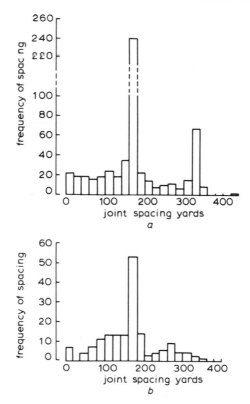

Fig. 7.9 Joint spacing on coaxial cable routes
    a 9·5 mm cable
    b 4·4 mm cable
    [Reproduced by permission of the Post Office]

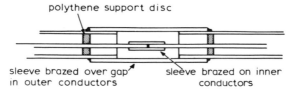

Fig. 7.10 Joint design for 9·5 mm coaxial cable

tensions to higher digit rates. A typical configuration, and its time domain reflecto-meter response, are shown in Fig. 7.11. Note that owing to stranding the distance indicated by electrical measurements exceeds the physical distance by about 7%.

As we shall see in a later Chapter, the distribution of the lengths of the repeater sections is of importance in designing digital systems. A histogram of the repeater section lengths found on a typical British cable route is given in Fig. 7.12.

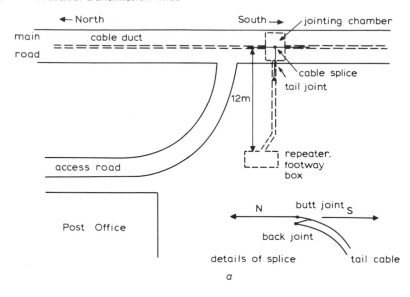

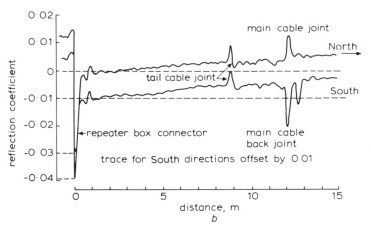

Fig. 7.11 Repeater arrangements on a 4·4 mm coaxial cable
    a Layout of cable end
    b Response to 1 ns risetime step
    True distance = Quoted distance/1·07
    [Reproduced by permission of the Post Office]

7.6 Interconnection cables

It may not be immediately appreciated that fairly long interconnections can occur between the various pieces of equipment within terminal buildings. One example of this is the case of a microwave link, where the radio equipment may be located at the

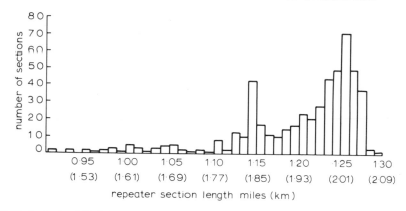

Fig. 7.12 Analysis of 32 4·4 mm cable routes
[Reproduced by permission of the Post Office]

top of a tower, perhaps some 300 m above its associated multiplexing equipment. Thus, in terms of length there is no clear-cut distinction between the trunk transmission lines and internal interconnecting cables. At low speeds, balanced pair cables are used, which are similar to the trunk cables described in Section 7.2. At higher speeds, coaxial cables are employed, which differ from the semi-air-spaced trunk designs. Coaxial cables used for interconnections are generally solid polyethylene dielectric types, with braided outer conductors. The solid dielectric leads to a lower propagation velocity and some decrease in impedance regularity. The use of braided outer conductors introduces a number of 2nd-order effects. The losses are slightly higher than for a solid copper conductor of similar dimensions (see Benson[25]), and the inhomogeneous nature of the outer conductor leads to an increase in the crosstalk coupling with adjacent lines. It has also been reported that if braided conductors are left undisturbed for lengthy periods there is some tendancy for attenuation and crosstalk to increase, due to oxidation of the wires in the braid producing insulating layers between wires which cause the current to flow in helical paths rather than longitudinally. Mechanical disturbance will break down these oxide layers and lead to a fall in crosstalk and attenuation.

7.7 References

1 TATELMAN, P.I.: 'World frequency of high temperatures', US Government report AD 696094, Dec. 1969
2 ITT: 'Reference data for radio engineers', 5th edn, chap. 41
3 SHARMA, K.L.S., and MAHALANABIS, A.K.: 'Modelling and prediction of the daily maximum temperature', *IEEE Trans.*, 1974, **SMC-4**, pp. 219–221
4 DUNSHEATH, P.: 'A history of electrical engineering' (Faber & Faber, 1962) chap. 14
5 HEBBETT, C.M.: 'Transmission characteristics of toll cables', *Bell Syst. Tech. J.*, 1941, **20**, pp. 293–330

6   JACHIMOWICZ, L., OLZEWSKI, J.A., and KOLODNY, I.: 'Transmission properties of filled thermoplastic insulated and jacketed telephone cables at voice and carrier frequencies', *IEEE Trans.*, 1973, **COM-21**, p. 203

7   JACHIMOWICZ, L., EAGER, E.S., KOLODNY, I., and ROBINSON, D.E.: 'Transmission properties of polythene insulated telephone cables at voice and carrier frequencies', *AIEE Trans.*, 1959, **COM-E**, p. 618

8   MATUDA, S.: 'High frequency transmission properties and crosstalk characteristics of intercity and toll cables applicable to p.c.m. transmission design', *Rev. Electr. Commun. Lab.*, 1965, **13**, pp. 906–930

9   MURATA, H., ISHIKAWA, H., INAO, S., and ITO, M.: 'Balanced pair cable for 100 MBit/s system'. Proceedings of the IEEE International Conference on 'Communications' Philadelphia, June 1972, pp. 19.1–19.6

10   OKAMOTO, K., NODA, H., and ONISHI, M.: 'High frequency performance of Z type shielded cable.' Proceedings of the IEEE International Conference on 'Communications', Philadelphia, June 1972, pp. 19.8–19.18

11   SETZER, D.E.: 'Low capacitance cable for 6·3 MBit/s transmission systems'. Proceedings of the IEEE International Conference on 'Communications', Philadelphia, June 1972, p. 19.19

12   COHEN, B.S.: 'A handbook of telecommunication' (Pitman, 1947)

13   BOYCE, C.F.: 'Open-wire carrier telephone transmission' (MacDonald & Evans, London, 1962)

14   ITT: 'Reference data for radio engineers', 5th edn., chap. 30

15   PRITCHETT, J., and BARTLETT, G.A.: 'The 60 MHz transmission system cable design and manufacture', *Post Off. Electr. Eng. J.*, 1973, **66**, pp. 158–166

16   ALLAN, A.F.G.: 'Small diameter coaxial cable developments', *ibid.*, 1964, 57, pp. 1–6

17   NUTT, W.E., WARGOTZ, B., and BISKEBORN, M.C.: 'New long distance broadband coaxial with corrugated outer conductor'. Proceedings of the 17th International Wire and Cable Symposium, Atlantic City, N. J., Dec. 1968

18   IYENGAR, R., McCLEAN, R., and McMANUS, T.: 'Sophisticated coaxial cable for digital duties', *Telesis (Canada)* 2, pp. 9–14

19   PALADIN, G.: 'Transmissione digitale a 8·448 Mbit/s su cavi microcoassiali', *Telecomunicazioni.*, Guigno, 1971, 39, pp. 67–76

20   PRACHE, P.M.: 'The influence of temperature variations on the attenuation of communication cables.' *Cables & Transm.*, 1947, **1**, p. 185

21   ROSMAN, G.: 'Assessment of coaxial cable for frequency-division-multiplex transmission by means of a c.w.-burst test signal', *Proc. IEE*, 1970, **117**, (1), pp. 45–50

22   ROSMAN, G.: 'Sweep frequency measurement of structured return loss by a carrier burst method', *Electron. Lett.*, 1971, 7, pp. 216–217

23   MASTERS, C.H., GARRETT, W.H., and ATKINS, D.H.: 'Manufacturing cloax cable', *Western Electric Eng.*, 1971, **15**, pp. 65–73

24   STENSON, D.W., and SLAUGHTER, W.G.F.: 'The 60 Mhz. f.d.m. system-cable jointing', *Post Off. Electr. Eng. J.*, 1973, **66**, pp. 170–173

25   BENSON, F.A., and TEPEREK, R.J.: 'Losses in braided coaxial cables in the frequency range 100 kHz–100 MHz', *Proc. IEE*, 1973, **120**, (12), pp. 1465–1468

Chapter 8

Other transmission media

8.1 Introduction

In contrast to the very extensive treatment of transmission lines given in the last two Chapters, we propose to present only a brief account of other transmission media. Our justification for this is twofold; first, that as yet no significant use of other transmission media has been made, and as a result there is not much information available on any peculiarities relevant to digital transmission that may occur. Secondly, other transmission media are theoretically much less complex than lines, and their characteristics may usually be considered as simplified variants of the line-transmission expressions. It will also be convenient to append a discussion of the problems encountered in digital transmission over the switched analogue telephone network, since this presents problems rather similar to those encountered in these other transmission media.

At the present time, three transmission media are receiving attention. These are

(a) microwave radio
(b) waveguides
(c) optical fibres.

Of these, only microwave radio has reached the stage of actual use in the communication network. All three media share the common feature of employing very high-frequency carrier signals, and as a result the bandwidths of the modulated signals, expressed in octaves, are very small. Most of the frequency dependent effects encountered in line transmission then vanish.

8.2 Microwave radio

Over the past 20 years extensive use has been made of the frequency bands between 2 GHz and 10 GHz for analogue transmission systems, and as a result this part of the radio spectrum is now almost fully occupied in some countries. The development of digital microwave systems has therefore been concentrated on the use of higher frequencies. At present, systems operating in the 11 GHz band and the 19 GHz band are being considered. Now, while at the lower frequencies losses in the atmosphere can be neglected, this is no longer true above 10 GHz. The situation is shown in Fig. 8.1. There is an absorption peak due to water vapour just above 20 GHz and a second due to oxygen at about 60 GHz. The basic path losses are therefore significantly higher. At lower frequencies, the main concern has been with variations in the path loss caused mainly by atmospheric disturbances, leading to multipath transmission and consequent interference between the signal contributions from the different paths. These produce signal 'fades' which for a small but significant part of the time, can become severe. The radio engineer has therefore tended to design his systems on the basis of the expected depths of fade and the time that the transmission performance, in analogue terms, may be allowed to fall

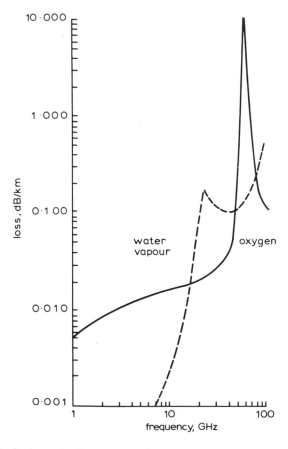

Fig. 8.1 Atmospheric absorption by oxygen and water vapour

below a minimum acceptable level. We do not propose to go into details of the techniques used to plan a radio system, and we refer the reader to the excellent accounts given by Bullington[1] and in the ITT Radio Engineers Handbook, for information on this topic.

In extending the use of microwave radio to digital systems two factors must then be taken into account. First, the multipath fading effects encountered at lower frequencies will still occur, and be somewhat more pronounced, but in addition, further fading caused by heavy rainfall will now also be a significant contributor. Secondly, owing to the very rapid variation in error rate with changes in signal/noise ratio, the effects of fades will become very much more abrupt. Digital systems will not deteriorate gradually, but will in effect switch from nearly perfect operation to an unusable state almost instantly. The type of effects that we may expect to encounter are shown in Fig. 8.2, which is derived from tests made by the British Post Office Research Department.

A radio path may therefore be treated as a frequency-independent loss, that is slightly unstable, and for short periods may increase to very high values.

8.3 Waveguide transmission

In one sense, waveguide transmission can be regarded as radio transmission confined within a controlled environment. It then becomes possible to eliminate the atmospheric absorption and instabilities that are evident in free-space transmission, but this is in part offset by the introduction of losses due to the resistance of the walls of the confining guide. For most waveguide configurations the losses are

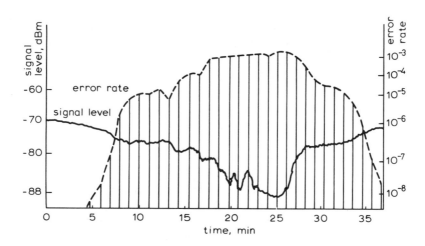

Fig. 8.2 Example of a fade due to a thunderstorm
[Reproduced by permission of the Post Office]

appreciable, the one exception being the H_{01} mode in circular guides. Transmission losses for this mode can be very low, the main difficulty being that to achieve this, the guide dimensions must be large compared with the wavelength employed, so that it becomes easy for the transmitted signal to be converted to other modes having higher losses. Over the past decade considerable attention has been given to waveguide transmission, and it appears that the technical problems have now been largely solved. The introduction of waveguide systems is now restricted by the combination of economic factors, and the traffic requirements, since these systems are applicable only to very high capacity routes. Detailed information will be found in Karbowiak[2] and IEE Conference Publication 71. It appears likely that any practical system will employ waveguides of about 5 cm diameter and carrier frequencies in the range 40–100 GHz. Under these conditions the losses will be about 2–3 dB per km so that repeater spacings of 15–30 km will be used. Each carrier will be phase modulated by digital signals at digit rates between 240 and 500 Mbit/s. This implies that the required bandwidths for each signal will be under 1 GHz and it can be seen that over this range the frequency dependence of the waveguide losses will be insignificant. Hence, the transmission medium itself does not present any special problems.

8.4 Optical transmission

Light being an electromagnetic wave, it is in theory possible to transmit it along a suitable guide in much the same way as the radio signals described in the previous section. In recent years a number of developments have brought this to the verge of technical feasibility, and, although no working system using this transmission medium has yet appeared, it seems highly probable that such a system will be introduced within the next few years. Two main developments have been responsible for this situation. First, the development of semiconductor light sources that can be modulated at high digit rates. These appear in two categories, the first being light-emitting diodes (LEDs) which produce noncoherent radiation, and the second being semiconductor lasers. The former have reached a satisfactory stage of development, but the latter have so far tended to have rather short working lives, and the reasons for this are only just becoming understood. The second main development has been the production of low-loss glass fibres that can be used as the optical waveguides. It is now possible to produce these with losses of under 10 dB/km, which means that systems employing repeater spacings of the order of 2–4 km should be practicable. The exact limits for transmission distances remain to be established, and it may turn out that the loss in the fibres will not be the limiting factor. In addition to being absorbed, the optical radiation tends to be spread out or dispersed during transmission. This sets a limit to the maximum digit rate that can be employed. It appears that it will be of the order of 50 Mbit/s for noncoherent transmission, and above 100 Mbit/s for the coherent case.

8.5 Transmission over switched analogue systems

In recent years, a demand for low-speed data transmission capability over the switched telephone network has arisen. This has been associated with the growing use of remote terminals connected to central computing installations. Typically, data rates in the range 1·2 Kbit/s to 4·8 Kbit/s have been employed. Although, strictly speaking, the switched telephone network cannot be considered as a simple transmission medium, since it involves a variety of transmission systems and intermediate equipment, a single audio channel can be so treated, and in many respects its characteristics parallel those of the radio paths previously discussed. Because of band limitations it is necessary to employ carrier techniques similar to those used in the microwave systems. However, in this case, the bandwidths are significant in relation to the carrier frequency, and amplitude and delay distortions over the transmitted band are quite large. Further, while these distortions will be relatively constant for any one connection, they will differ widely between different switched circuits, and as a result it is necessary, for the higher-speed systems, to carry out equalisation of the transmission path every time a new connection is established. This has led to intensive development of automatic adaptive equalisers for these systems. We will not pursue this topic here, we refer the reader to the article by Westcott[3] for an introductory survey, and to the texts by Bennett[4] and Lucky[5] for more extensive information. A useful account of the characteristics of audio-frequency channels is given by Broad.[6]

8.6 References

1 BULLINGTON, K.: 'Radio propagation fundamentals', *Bell Syst. Tech. J.*, 1957, **36**, 3 pp. 593–626
2 KARBOWIAK, K.E.: 'Trunk waveguide communication' (Chapman & Hall, 1965)
3 WESTCOTT, R.J.: 'An experimental adaptively-equalized modem for data transmission over the switched telephone network', *Radio & Electron. Eng.*, 1972, **42**, pp. 499–507
4 BENNETT, R.W., and DAVEY, J.R.: 'Data transmission' (McGraw-Hill, 1965)
5 LUCKY, R.W., SALZ, J., and WELDON, E.J.: 'Principles of data communication' (McGraw-Hill, 1968)
6 BROAD, E.R., *et al.*: 'Networks to simulate attenuation frequency and group delay frequency characteristics of carrier telephone channels', Post Office Engineering Department Research Report 20962

Chapter 9

Regeneration and waveform transmission

9.1 Basic function of regeneration

Digital transmission systems exist in a bewildering variety of forms, principally determined by the nature of the channel medium over which the system is intended to work. There are, however, certain generic features which may be, to a large extent, dissociated from the channel properties. It is our intention in the present Chapter to select these common features as a basic introduction to the specific applications covered in Chapters 12 and 13. The process which unifies all types of digital transmission is regeneration. To clarify what is meant by regeneration we refer to a functional model of a transmission link, Fig. 9.1, consisting of a chain of digital repeaters, inserted at intervals along a channel of unspecified characteristics. For the sake of simplicity, it is convenient to consider the case of an attenuating baseband channel transmitting binary symbols. At any given repeater input port the incoming signal arrives attenuated and dispersed by the channel and perturbed by noise. The first operation at the receiving repeater must therefore be preamplification and shaping of the weakened signal to a level and form from which a reliable threshold detection may be performed. Conceptually, the threshold detector is a decision circuit, the output of which may be assumed to be in a 'hard' digital logic form. This decision output is then used to control an output stage which produces a sufficient signal level to drive the following channel section. The output signal is normally a pulse as well defined in shape as the relevant circuit technology allows.

The above description could, as far as it goes, be said to form a rudimentary regenerative digital transmission system. However, it may be argued that a truly digital system is quantised both in amplitude and time, and, inasmuch as the regeneration process described above does not exploit the time parameter, it cannot be truly efficient. This supposition may be justified when it is noted that, given that the repeater input signal has finite transition times, the effect of noise is to modulate the time interval between transitions of the threshold detector. These arguments lead to the provision of a timing facility as part of the regenerative process, see Fig. 9.1. In principle, timing may be supplied by means of a separate

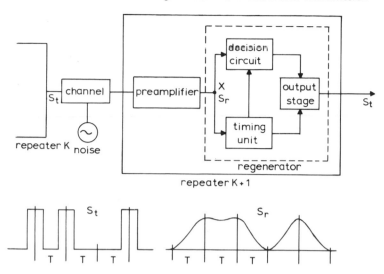

Fig. 9.1 Functional model of a repeatered link

transmission channel conveying a timing wave, which would normally be a sinusoid at a frequency equal to, or a submultiple of, the symbol rate. In practice, however, such externally timed systems cannot be realised without some waste of information capacity and, for this reason, self-timed systems, as shown in Fig. 9.1, are normally preferred. The wide topic of timing extraction will be considered in some detail in later Chapters.

In isolating the regenerator as a self-sufficient functional block we must examine the signal conditions at its input and output ports. Fig. 9.1 indicates that the input to the regenerator $s_r(t)$ differs significantly in shape from the pulse transmitted by the previous repeater. At first sight this may seem surprising, since it would seem reasonable to provide sufficient gain and frequency shaping in the preamplifier to equalise exactly for the loss characteristic of the channel. This, however, would be an unnecessarily generous procedure, because the decision circuit can perform 'hard' decisions on a nearly sinusoidal pulse shape. Furthermore, such a pulse shape implies less preamplifier bandwidth than would be required for complete equalisation; this, in turn, implies a lower noise level at the preamplifier output-decision circuit input port. Clearly, the equalisation design involves a compromise; on the one hand, the noise level may be reduced by broadening $s_r(t)$ until a limiting condition of pulse overlap is reached, on the other hand, a better formed $s_r(t)$ eases the decision function but increases the noise level. Judicious choice of the equalisation characteristic is a central feature of all digital transmission systems, and assumes crucial significance in the design of digital line transmission systems. We defer full discussion of this topic to Chapter 12. It would also be premature at this stage to specify any particular form for the output pulse shape $s_t(t)$ as this is clearly dependent on the form of the channel. The shape indicated in Fig. 9.1 should therefore be taken as arbitrary.

9.2 Signal/noise ratio and error rate

The presence of random noise at the decision circuit input contributes the fundamental mechanism for the generation of random errors in digital transmission. We now examine this mechanism to derive a relation between the signal/noise ratio and error rate.

We consider the most basic form of digital signal, a stream of random binary symbols with permissible states A and B with a separation of V, see Fig. 9.2. At the receiver, in the absence of any disturbing signal, certain detection may be effected by setting a threshold level somewhere between A and B and deciding whether A or B was sent on the basis of the input either exceeding or falling below the threshold. The above ideal situation changes in the presence of any form of disturbance added to the signal. The most basic disturbance is random thermal noise which is closely modelled by a process having Gaussian statistics. Assuming the noise affects both states equally, the threshold must now be set halfway between A and B. This assumes that the cost of an error in either state is equal, and that no specific signal correlations exist to warrant a more refined detection rule. Because of the unbounded nature of the Gaussian distribution there will always be a finite probability that a noise value will exceed $V/2$, thus causing an error. The probability of error P_e is therefore defined by

$$P_e = P(A \mid B)\,P(B) + P(B \mid A)\,P(A) \tag{9.1}$$

By symmetry $P(A \mid B) = P(B \mid A)$, and for equiprobable levels $P(A) = P(B) = 1/2$

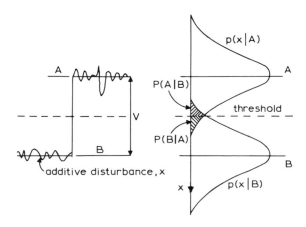

Fig. 9.2 Error mechanism in binary transmission

Hence

$$P_e = P(\Delta \mid R) = \int_{V/2}^{\infty} p(x)\, dx \tag{9.2}$$

If $p(x)$ is a Gaussian distribution, then from eqns. 2.9 and 2.10,

$$P_e = \frac{1}{\sigma\sqrt{2\pi}} \int_{V/2\sigma}^{\infty} e^{-x^2/2\sigma^2}\, dx$$

$$= \frac{1}{\sqrt{2\pi}} \int_{V/2}^{\infty} e^{-x^2/2}\, dx = 1/2 \; \mathrm{erfc}\left(\frac{V}{2\sigma}\right)$$

$$= \overline{\Phi}\left(\frac{V}{2\sigma}\right), \; \text{say} \tag{9.3}$$

This expression relates P_e to the peak signal/r.m.s. noise ratio for a binary baseband signal subject to additive Gaussian noise. The relation is plotted over a range of $P_e = 10^{-5}\text{--}10^{-12}$ in Fig. 9.3. This involves extremal values of the Gaussian distribution obtainable, for instance, from Reference 1. The scales are adjusted to linearise the relation artificially by the well known technique of taking multiple logarithms. In this case, as we have a decibel measure for s.n.r. and, conventionally, an exponential expression for P_e ($P_e = 10^{-x}$), we can take logarithms while still preserving a form meaningful to the reader. In particular, consider a plot of s.n.r. in decibels against $\log_{10} x$. In effect, $\log_{10} x$ is a double logarithm $-\log \log P_e = \log x$, and it is seen from the graph that the relation is nearly linear in the range presented. The equation of the straight line fitted to the graph of Fig. 9.3 is

$$\text{s.n.r., dB} = 20 \log_{10} \frac{V}{\sigma} = 10\cdot65 + 11\cdot42 \log_{10} x \tag{9.4}$$

where

$$P_e = 10^{-x}.$$

Estimated accuracy over the range $4 < x < 15$ is better than $0\cdot05$ dB. The above is a useful slide-rule approximation where comprehensive plots of the function are not readily to hand.

We note a remarkable feature of the relation between P_e and s.n.r., namely the rapid variation in P_e for relatively small variations in s.n.r. This emerges clearly by differentiating eqn. 9.4

$$\frac{d}{dx}(\text{s.n.r.}_{\text{dB}}) = 4\cdot96/x \simeq 5/x \tag{9.5}$$

The approximation is a mnemonic sufficiently accurate for most purposes. We see, therefore, that at $P_e = 10^{-5}$, ($x = 5$), a change of 1 dB causes an order of magnitude

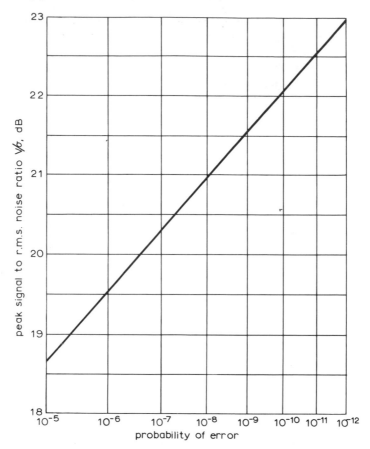

Fig. 9.3 Relation between probability of error and signal/noise ratio for binary baseband transmission

change in error rate, while at $P_e = 10^{-10}$, a one order of magnitude change results from a change of s.n.r. of only 0·5 dB. The quoted values are somewhat inaccurate since they exceed the small increments for which eqn. 9.5 is valid, and are used solely for illustration; Fig. 9.3 should be consulted if better accuracy is desired. Values of P_e for s.n.r. in the range 0 to 25 dB are given in Table 9.1.

9.3 Accumulation of errors in a repeatered link

Because a transmission link may comprise a large number of repeatered sections the dependence of the overall probability of error on the error behaviour of individual sections is of interest. The normal assumption made in assessing the overall error probability of the link is that errors in individual sections are uncorrelated and

Table 9.1   Relation between s.n.r. = 20 log$_{10}$ V/σ and error rate = $A \times 10^{-x}$ for binary baseband transmission with additive Gaussian noise

s.n.r., dB	Error rate $A \times 10^{-x}$		s.n.r., dB	Error rate $A \times 10^{-x}$	
	A	x		A	x
0	3·09	1	13·0	1·28	2
1·0	2·87	1	14·0	6·11	3
2·0	2·65	1	15·0	2·46	3
3·0	2·40	1	16·0	8·03	4
4·0	2·14	1	17·0	2·00	4
5·0	1·87	1	18·0	3·57	5
6·0	1·59	1	19·0	4·17	6
7·0	1·32	1	20·0	2·87	7
8·0	1·05	1	21·0	1·01	8
9·0	7·94	2	22·0	1·55	10
10·0	5·69	2	23·0	8·22	13
11·0	3·80	2	24·0	1·16	15
12·0	2·33	2	25·0	3·06	19

hence that their error probabilities may be summed. Analogue f.d.m. systems offer an interesting comparison here; in these, noise power is the relevant performance criterion and is again summed over the number of repeater sections involved. In a digital link, however, the form of the relation between s.n.r. and P_e causes a surprising feature which is most easily demonstrated by means of a simple example. Consider a digital link consisting of 100 sections each operating with an s.n.r. of 22 dB. From Table 9.1 the error rate for the link is then $1·55 \times 10^{-8}$. If now the s.n.r. *of a single section* deteriorates by 2 dB the overall error rate increases to $99 \times 1·55 \times 10^{-10} + 2·87 \times 10^{-7} = 3·02 \times 10^{-7}$, being now primarily determined by the contribution of the single offending section. In comparison an f.d.m. link containing 100 sections will only suffer a 1% increase in overall noise power if the noise contribution of one section increases by 3 dB!

The above illustrates a fundamental feature of digital transmission systems, namely, that their performance is *critically dependent on worst-case conditions*, in direct contrast to the averaging tendency of analogue transmission.

9.4 Intersymbol interference and eye patterns

From the above we see that the performance of a digital transmission system is dependent on the quality of the regeneration process in each and every repeater in the system. This in turn depends on the quality of the decision circuit and the form of the signal at its input, node X of Fig. 9.1. In assessing the performance of a digital system either at the design stage or in service, it is the waveform at the decision node X which requires critical examination. In practice, the waveform may be viewed on an oscilloscope in one of two modes, either displaying the whole pattern over some basic pattern period or frame, or by synchronising the instrument time base to the signalling rate. The latter has the effect of superposing all the waveform combinations over adjacent signalling intervals, and thus provides a condensed representation of the signal states and transitions between states. The superposition of waveforms so obtained often assumes the lozenge shape of an eye, and is for this reason called an *eye pattern*. Fig. 9.4 shows eye patterns generated from the basic symmetrical waveform $s_r(t)$, for binary and ternary alphabets.

The eye opening associated with the eye pattern defines a boundary within which no waveform trajectories can exist under any condition of code pattern, and its complete shape is important in the assessment of various system imperfections. The eye opening is clearly dependent on the number of code levels, and its size E can be deduced by superposing the interfering tails of $s_r(t)$. For a symmetrical waveform whose value one signalling interval from the waveform centre is a, $\{s_r(\pm T) = a\}$, and zero at subsequent signalling instants ($t = \pm kT, k = 1, 2, ..$)

$$E = 1 - 2a(L - 1) \qquad (9.6)$$

where L is the number of code levels. The second term defines the so-called *intersymbol interference* which may be considered as the effect of telescoping together consecutive symbols. The eye opening is clearly the guaranteed minimum difference between code levels from which a decision has to be made, and it is seen from eqn. 9.6 that intersymbol interference reduces this by a factor $1/E$ of the value obtained in its absence.

The concept of eye patterns leads naturally to the identification of regeneration as essentially a 2-dimensional process involving amplitude setting and timing, analogous to setting a target, here the centre of the eye opening, against cross-hairs. An interesting geometric feature, which may be easily proved analytically, is that, in a multilevel system containing more than one eye, the extremities of different eyes occur at the same points in the signalling interval.

One measure of system performance, described by the relation between s.n.r. and error probability, may be deduced directly from the eye opening. All that is required is the modification of the separation between adjacent levels V in Section 9.2, by the impairment factor $1/E$ due to intersymbol interference. In other words, to guarantee a given error probability the s.n.r. must be increased by $1/E$ of its value in the absence of intersymbol interference. This reasoning, in fact, leads to a somewhat conservative design, since the eye opening defines the worst possible combination of code pattern, and hence the design then caters for continuous transmission of such a pattern. This is clearly a pessimistic approach on the basis that all

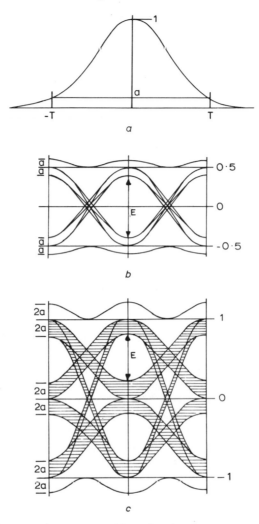

Fig. 9.4 Unit waveform *a*, and eye patterns; binary *b* and ternary *c*

other code-pattern combinations will be subject to a better s.n.r. and hence a lower
error probability, with the result that the average error probability will be lower
than that anticipated by the worst-case design. This, however, may also be used as a
powerful argument in favour of the worst-case approach, in that actual performance
is likely to be better than predicted, which represents a design error in the right
direction. Whether or not the worst-case design is acceptable depends largely on the
'tightness' of various system parameters, in particular power limitation. We consider
alternative approaches in Section 9.7.

9.5 Received waveform criteria

Earlier, in Section 9.1, we touched on the choice of the received waveform $s_r(t)$ on which the regeneration process is based. We remarked that normally the choice is influenced towards bandwidth constraint to minimise noise. We now propose to examine the situation as closely as is possible within the context of general applicability, while deferring the noise minimisation problem to Chapter 12. We may formulate the problem by considering a sequence of discrete values $a_1 a_2 \ldots a_k, \ldots$, comprising the message. These are transmitted as some specific signal at intervals of T seconds, to be converted by the channel and preamplifier response shaping (see Fig. 9.1) into a sequence of waveforms $s_r(t)$ whose coefficients, apart from a constant loss, are the original values a_k. Thus,

$$s_r(t) = \sum_{-\infty}^{\infty} a_k \, r(t - kT) \tag{9.7}$$

where $r(t)$ is a shaping or weighting function and is the impulse response of the filter network formed by the overall response $R(f)$, the two being Fourier transform related. Since the signalling interval T implies the generation of independent message values spaced T seconds apart, a first approach might reasonably be to restrict $r(t)$ wholly within an interval $2T$ so that adjacent symbols of $s_r(t)$ cannot mutually interfere. However, such a restriction in the time domain leads to oscillatory lobes in the frequency domain capable of passing noise.

Instead of attempting to restrict the interval of $r(t)$ as above, we can specify that $r(t)$ be zero at all sampling instants kT except for $k = 0$. With this restriction in time, is it possible to impose additional restrictions in spectrum? The sampling theorem (see Chapter 2) furnishes one set of functions which clearly satisfy these conditions. In the transmission context the theorem may be reformulated as follows: independent values may be transmitted at a signalling rate $1/T$ per second using a response whose bandwidth is restricted to W herz, provided that $W \geqslant 1/2T$. The shaping function which satisfies the limiting condition $W = 1/2T$, is the function given by

$$r(t) = (\sin 2\pi \, Wt)/(2\pi \, Wt) \tag{9.8}$$

This is the familiar $\sin x/x$ form shown in Fig. 9.5a. Its use in practice is, however, to be avoided for several reasons. First, the rectangular spectrum is physically unrealisable, even more so with the linear-phase response required by the sampling theorem. Secondly, the waveform converges towards zero very slowly implying that a given symbol value is dependent on the remote tails of many other symbols. From this it follows that timing and amplitude imperfections affect large groups of symbols and are therefore critical.

This is forcefully illustrated in Fig. 9.5b and c showing worst-case eye boundaries for the limiting case of $W = 1/2T$ and for the case of $W = 0.95/2T$. It is

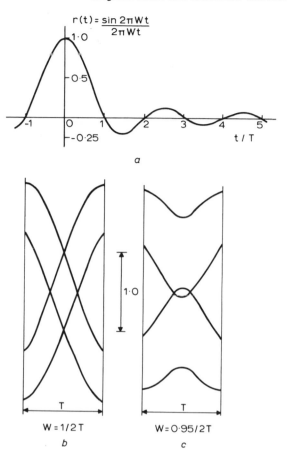

Fig. 9.5 sin *x/x* impulse response *a*, and corresponding binary eye patterns *b* and *c*

seen that for $W = 1/2T$ the eye is fully open, while a pass band which falls slightly short of the limiting case results in nearly complete eye closure.

The sampling theorem, as we recall from Chapter 2, is also satisfied given linear phase and a skew amplitude characteristic in the cutoff region of the frequency response. This clearly offers further scope for the choice of suitable waveforms and spectra. One such set involves the so-called raised-cosine amplitude spectrum and is described by

$$R(f) = 1, \quad B - W < f < W - B$$

and

$$R(f) = 1 + \cos \pi \left(\frac{|f| - W + B}{2B} \right), \quad W - B < |f| < W + B$$

$$R(f) = 0 \text{ elsewhere} \tag{9.9}$$

The corresponding impulse response may be shown to be (Haber)[11]

$$r(t) = \frac{\sin 2\pi (B + W)t \, \cos 2\pi Bt}{\pi t\{1 - (4Bt)^2\}} \qquad (9.10)$$

The skirt region of width $2B$, having sinusoidal roll-off, is symmetric and hence the limiting signalling rate $1/T = 2W$ is obtainable with zero intersymbol interference. The raised cosine spectrum and impulse response are shown in Fig. 9.6 for two values of the skirt width. The corresponding worst-case binary eye openings are also illustrated.

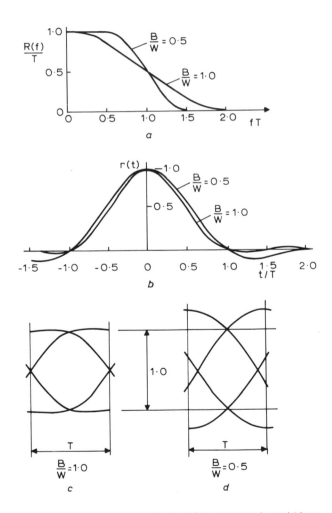

Fig. 9.6 Raised cosine spectra a, corresponding impulse responses b, and binary eye patterns c and d

Nyquist's conditions for zero intersymbol interference at signalling instants have been even further extended by Gibby and Smith.[12] Although mathematically simple, the conditions involve the quantities $A(f) \cos \alpha(f)$ and $A(f) \sin \alpha(f)$, where $R(f) = A(f) \exp j \; \alpha(f)$ which entail a certain amount of computation, and hence the intrinsic simplicity of the original linear phase, skew attenuation condition is lost.

The condition of zero intersymbol interference at signalling instants is a good starting point in the choice of received signal waveform; it is, however, by no means a prerequisite. It does, of course, produce a completely open eye $E = 1$ which simplifies the regeneration process, particularly in the presence of unavoidable circuit imperfections. However, in many instances, as was pointed out in Section 9.1, a well defined waveform and eye pattern are accompanied by increased noise due to the more extensive spectral shaping required. It follows that, in theory, there should be an optimum design which, for any given channel characteristic minimises the error probability. This has been shown to be the case by Tufts[15] who has derived optimum transmit and receive filters which, in general, take the form of a matched filter followed by a tapped delay line or transversal filter. A useful survey of related theoretical work has been prepared by Lucky.[16] The theoretical approach has hitherto suffered from several major shortcomings. First, there is no guarantee that the optimum filter is realisable. Secondly, no general results exist for the case where jitter is present. Thirdly, practical imperfections add a sizeable set of extra parameters which in some cases are capable of overwhelming the idealised theoretical model.

As no unique solution can hitherto claim to be generally optimum in practice, it is pertinent to consider some waveforms which may be considered suitable candidates for the received signal, despite finite intersymbol interference.

One possible waveform is the Gaussian function with its inherent mathematical simplicity. Specification of a standard deviation σ_a to give a value of a at a time T from the origin leads to the following expression for the Gaussian waveform:

$$r(t) = \exp \left(\frac{-t^2}{2\sigma_a^2} \right)$$

or

$$r(t) = \exp \left(-kt^2/T^2 \right) \tag{9.11}$$

For values of a of 2% and 5%, the equations become

$$r(t) = \exp \left(-3{\cdot}912t^2/T^2 \right), \quad a = 2\%$$

$$r(t) = \exp \left(-2{\cdot}996t^2/T^2 \right), \quad a = 5\%$$

The normalised spectrum is given by

$$\frac{R(f)}{R(0)} = \exp \left(-\pi^2 f^2 T^2/k \right) \tag{9.12}$$

If the value of interference a is considered excessive, it is possible to reduce it by synthesising a new waveform $r_c(t)$, by the subtraction of identical Gaussian waveforms $r(t)$ of amplitude a at adjacent timing instants

$$r_c(t) = \frac{1}{1 - 2a^2}\, 2\{r(t) - a\, r(t - T) - a\, r(t + T)\} \tag{9.13}$$

This compensated Gaussian waveform is, for small values of a, approximately zero at $t = \pm T$ and unity at the origin, and is shown together with the basic Gaussian waveform in Fig. 9.7 for $a = 5\%$. The corresponding spectrum is given by

$$R_c(f) = R(f)\,(1 - 2a \cos 2\pi f T) \tag{9.14}$$

Because of their Gaussian composition, this set of waveforms retains analytical simplicity.

Recent studies of optimum response band-limited filters may also lead to a new appreciation of suitable received waveforms. Temes[13] has shown that a filter exists which gives the fastest monotonic step response for a given bandwidth. This filter is based on a complex mathematical function and has been named the prolate filter, after it. Finally, for a fine tutorial treatment of the general problem and criteria underlying the choice of signalling waveforms, the reader is referred to Lerner.[14]

The criteria discussed above are based on the objective of minimising intersymbol interference, and we should point out at this stage that alternative approaches exist. One such approach which will be considered later is known as partial response signalling, where intersymbol interference is deliberately allowed, so generating intermediate received code levels which can be uniquely decoded by virtue of specific encoding rules at the transmitter.

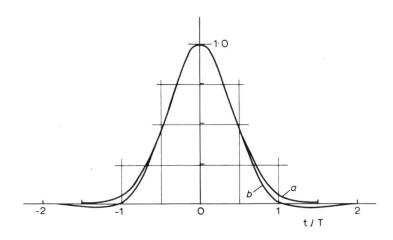

Fig. 9.7 Gaussian waveform *a*, and compensated Gaussian waveform *b*

9.6 Basic imperfections

Intersymbol interference is not the only degradation in a practical system. There are many factors which contribute to the deterioration of the eye pattern, many of them peculiar to specific transmission systems. We can, however, distinguish certain general classes of imperfection to which all other types belong. The most easily identified source is that associated with imperfect decision making at the regenerator. Recalling the concept of regeneration as the setting of the eye pattern against cross-hairs, it is clear that misadjustment of the cross-hairs will degrade the operation. In terms of regenerator quantities, this misadjustment will be caused by decision-threshold inaccuracy in the vertical plane of the eye pattern, and by timing inaccuracy in the horizontal plane. A less obvious imperfection arises from tolerances involved in the realisation of the equalisation response $R(f)$, and hence the received waveform at the decision node. Unfortunate combinations of circuit tolerances can obviously cause greater pulse spill-over, and consequently, intersymbol interference in excess of the nominal design value, causing a reduction of the eye opening. These effects are illustrated in Fig. 9.8, which attempts to convey that the design cannot expect the decision threshold or the timing instant to be defined any more precisely than lying somewhere within the regions of width d and h, respectively. Imperfect equalisation further reduces the eye opening to the limits shown. The implications are, therefore, that a realistic design must not base the

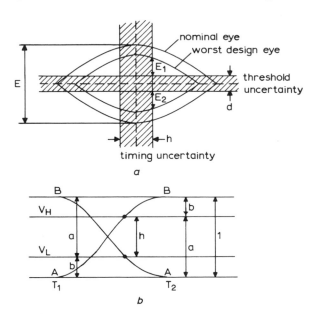

Fig. 9.8 Eye-pattern imperfections
  a General tolerances
  b Hysteresis

prediction of error probability on the nominal peak eye opening E, but must rely on what remains of the eye after allowing for imperfections, leaving the two portions E_1 and E_2. Eqn. 9.3 is thus slightly modified to give

$$P_e = \tfrac{1}{2} \, \overline{\Phi} \left(\frac{E_1}{\sigma} \right) + \tfrac{1}{2} \, \overline{\Phi} \left(\frac{E_2}{\sigma} \right) \tag{9.15}$$

If $E_1 = E_2$ eqn. 9.15 reduces to eqn. 9.3; however, in most practical cases, especially in multilevel transmission, the assumption of complete eye symmetry does not apply.

The above simple description of the practical situation is open to attack from a more probing formulation. Thus, the causes of the regions of uncertainty in time and amplitude require examination, and again, we find that a more complete examination must entail statistical descriptions of these regions. Furthermore, the statistics involve awkward correlations because, for instance, the timing uncertainty or so-called jitter is correlated closely with the signal. It is clear that the problem can be taken to any desired degree of complication; this, in the main, leads to results which do not match the required effort, and we feel that any sophistication is only warranted in specific examples.

Regenerator hysteresis. The effect of hysteresis on regenerator performance is by no means obvious, and we have found sufficient disagreement on this topic to warrant a brief diversion. Basically, we wish to show whether hysteresis has a beneficial or adverse effect on error performance. Consider, for simplicity, a random binary signal with states A and B; Fig. 9.8b illustrates the two thresholds V_H and V_L applicable to rising and falling waveforms.

Consider also signal conditions for two adjacent signalling instants T_1 and T_2. Then the probability of an error due to thermal noise may be written as

$P(e) =$ Prob [given signal AA, noise $\geqslant |a|$, or, given BB, noise $\geqslant |a|$ or, given AB, noise $\geqslant |b|$; or, given BA, noise $\geqslant |b|$]

$$= P(AA) \, P \left(|N| \geqslant \frac{1+h}{2} \right) + P(BB) \, P \left(|N| \geqslant \frac{1+h}{2} \right)$$

$$+ P(AB) \, P \left(|N| \geqslant \frac{1-h}{2} \right) + P(BA) \, P \left(|N| \geqslant \frac{1-h}{2} \right).$$

If $P(AA) = P(BB) = P(AB) = P(BA) = 1/4$, then

$$P(e) = \tfrac{1}{2} P \left(|N| \geqslant \frac{1-h}{2} \right) + \tfrac{1}{2} \, P \left(|N| \geqslant \frac{1+h}{2} \right)$$

and this, for Gaussian noise, will always exceed the zero-hysteresis probability $P(|N| \geqslant 1/2)$. We must conclude, therefore, that for a random signal hysteresis tends to degrade performance.

In summary, all imperfections in regeneration may be described by uncertainties in time, amplitude and pulse shape. We shall examine the causes of these uncertainties in later Chapters

9.7 Statistical assessment of error performance

As indicated in Section 9.4, estimates of error probability based on worst-case eye-opening considerations offer the easiest guide to digital system performance, but they are overpessimistic. Simple reasoning can show the approximate extent of the inaccuracy involved. For any eye pattern, the eye opening is defined by the two enclosing waveform trajectories each corresponding to a specific code sequence. Both these sequences, and only these, are responsible for the worst case error condition. Clearly, if the probability of occurence of these sequences were known then an average error probability could be computed by making a 1st-order assumption, based on the relation between s.n.r. and error probability, that all other error probabilities associated with other sequences will be very significantly lower, and can be ignored.

Proceeding on these terms, for any eye pattern the number of sequences which define eye openings equals twice the number of eyes, or $2(L-1)$ where L is the number of code levels. If, for simplicity, we assume that intersymbol interference is caused solely by adjacent impulses, then the eye pattern is largely determined by sequences of three symbols having a total number of L^3 combinations. With equiprobable independent symbols, the probability of worst-case trajectories is thus $2(L-1)/L^3$, and when multiplied by the worst-case error probability gives an estimate of the *average* error probability. The above factor is $1/4$, $4/27$ and $3/32$ for systems using binary, ternary and 4-level transmission, respectively.

If pulse tails spill over more than the adjacent signalling intervals, so that N intervals are involved, then in general the number of sequences which must be taken into account increases from L^3 to L^N, and the computation of intersymbol interference components becomes embarrassingly laborious for even moderate N. However, in principle it is possible to arrive at the average error probability by computing the intersymbol interference and eye opening for each code sequence, and hence the associated error probability, finally averaging over all combinations. A great deal of recent work has been devoted to deriving estimates for the average probability of error without the amount of computational effort required in the exhaustive method.

Saltzberg[2] and Lugannani[3] by assuming that the intersymbol interference term is the sum of independent random variables were able to apply an inequality relation to the average error probability, referred to as the Chernoff bound.[4] This upper bound was typically a decimal order lower than the simple worst-case error probability. Further work by Ho and Yeh[5,6] and Shimbo and Celebiler[7] revealed that the intersymbol interference term could be expressed as a convergent series resulting in very significant savings in computation. Ho and Yeh indicate that, for a particular 4-level system, the computation was 10^4 times faster, while the error probability was two decimal orders lower than that derived from the Chernoff bound. The same authors[8] report a further improvement in computational technique leading to rapidly approached tight upper and lower bounds. A typical example shows the average error probability lying between $1 \cdot 1 \times 10^{-8}$ and $1 \cdot 2 \times 10^{-8}$, which represents an accuracy far in excess of practical system tolerances.

The results described hitherto all assume that the symbols comprising the digital signal are generated independently. In general, however, it is likely that real signals exhibit a certain degree of correlation which clearly alters the statistics of the intersymbol interference. More significantly, very distinct correlative properties are often imposed on the digital sequence by particular types of codes (see Chapter 11). In some cases, such as in the bipolar code where successive marks are prohibited, the permitted eye pattern is obvious, allowing a simple adjustment of the error probability estimate. The issue becomes complicated with block codes involving coding translation of complete blocks of symbols, exemplified by the 4B-3T code. The problem is beginning to receive attention (Glave,[9] Prabhu,[10]) but it is difficult to derive any definite conclusions of general use.

Summarising, methods have been developed for the efficient computation of average error probability; we must leave the reader himself to decide whether his system is adequately specified by average or worst-case error criteria.

9.8 Optimal regeneration in theory and practice

In the above we have introduced basic concepts of regeneration, and proposed rather obvious and reasonable means of realising it. Noticeably, however, we have not considered alternatives, nor have we indicated the existence of theoretical optimal solutions; to lend perspective to our treatment we must briefly do so.

For our starting point we have to ask the reader to study the first part of Section 13.2 dealing with signal representation and optimum detection. Following the general approach used in Section 13.2 we first consider individual messages $s_i(t)$ of binary symbols, say, each of duration T

$$s_0(t) = a_0\, r(t)$$

$$s_1(t) = a_1\, r(t - T)$$

$$s_2(t) = a_2\, r(t - 2T)$$

$$- \; - \; - \; - \; - \; - \; -$$

$$s_i(t) = a_i\, r(t - iT)$$

This is no more than an expansion of eqn. 9.7 where the a_i are binary coefficients. Assume that the unit waveforms $r(t-iT)$ do not mutually overlap or, in mathematical terms, are orthonormal. Following the discussion of Section 13.2 we could directly implement an optimal detection strategy as shown in Fig. 13.1 by setting $\phi_i(t) = r_i(t)$. Since, however, the $r_i(t)$ exist strictly in mutually exclusive time intervals then it is intuitively clear that the set of local correlator waveforms can be pooled together to produce the optimal detector of Fig. 9.9.

In general, this detector is awkward to realise in practice. However, in exceptional circumstances it may, in fact, be closely approached. In particular consider the case where $r(t)$ = rect T in which case the product modulator is superfluous and the optimal detector reduces to the form of Fig. 9.9b. For obvious reasons this is known as an 'integrate-and-dump' detector, and is optimal in the absence of band-limiting, this condition being a consequence of specifying the received signal as a summation of rectangular functions.

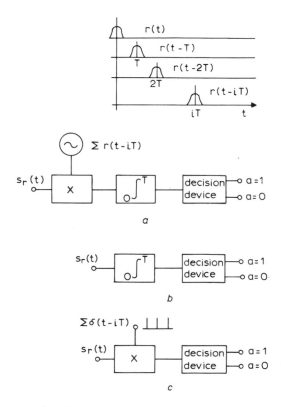

Fig. 9.9 Binary baseband detectors
   a Optimal detector
   b Integrate and dump detector
   c Sampling detector

As our earlier discussion focused on the practical case of regeneration with band-limiting, the question now arises of the degradation relative to optimal detection, first, of a band-limited integrate and dump system, and, secondly, of our previously described 'sample-only' system, Fig. 9.9c. This is analysed by Park[17] assuming rectangular band-limiting; several conclusions may be deduced from his results. First, integrate-and-dump is always better than 'sample-only'. Secondly, with a bandwidth of $0.5/T$ or less, integrate-and-dump is only about 0.2 dB better than 'sample-only'. Thirdly, for a bandwidth of $0.5/T$ the 'sample-only' detector is about 1.5 dB worse than optimal. Finally, the degradation of the 'sample-only' system increases steeply beyond a bandwidth of $1/T$ (1 dB for $B = 1/T$, 4 dB for $B = 1.5/T$, relative to integrate-and-dump). These results should be treated with some caution since they apply to rectangular band-limiting which we rejected in our earlier treatment; however, we feel that the general trend should apply to other band-limiting functions.

In conclusion, we may state that (i) integrate-and-dump regeneration is theoretically optimal in the absence of band-limiting; (ii) in the presence of band-limiting of the type described in Section 9.5, the difference between integrate-and-dump and sample-only detection is small and neither is likely to be more than 1.5 dB worse than optimal.

9.9 References

1   US National Bureau of Standards: 'Tables of the error function', 1954
2   SALTZBERG, B.R.: 'Intersymbol interference error bounds with applications to ideal band-limited signalling', *IEEE Trans.*, 1968, **IT-14**, pp. 563–568
3   LUGANNANI, R.: 'Intersymbol interference and probability of error in digital systems', *ibid.*, 1969, **IT-15**, pp. 682–688
4   CHERNOFF, H.: 'A measure of asymptotic efficiency for tests of a hypothesis based on the sum of observations', *Ann. Math. Stat.*, 1952, **23**, pp. 493–507
5   HO, E.Y., and YEH, Y.S.: 'A new approach for evaluating the error probability in the presence of intersymbol interference and additive Gaussian noise', *Bell Syst. Tech. J.*, 1970, **49**, pp. 2249–2265
6   – 'Error probability of a multilevel digital system with intersymbol interference and Gaussian noise', *Bell Syst. Tech. J.*, 1971, **50**, pp. 1017–1025
7   SHIMBO, O., and CELEBILER, M.I.: 'The probability of error due to intersymbol interference and Gaussian noise in digital communication systems', *IEEE Trans.* 1971, **COM-19**, pp. 113–119
8   YEH, Y.S., and HO, E.Y.: 'Improved intersymbol interference error bounds in digital systems', *Bell Syst. Tech. J.*, 1971, **50**, pp. 2585–2599
9   GLAVE, F.E.: 'An upper bound on the probability of error due to intersymbol interference for correlated digital signals', *IEEE Trans.*, 1972, **IT-18**, pp. 356–363
10 PRABHU, V.K.: 'Intersymbol interference performance of systems with correlated digital signals', *ibid.*, 1973, **COM-21**, pp. 1147–1152
11 HABER, F.: 'Rapidly converging sample weighting functions', *Trans.Comm. Syst.*, 1964, **CS-12**, pp. 116–117
12 GIBBY, R.A., and SMITH, J.W.: 'Some extensions of Nyquist's telegraph transmission theory', *Bell Syst. Tech. J.*, 1965, **44**, p. 1487

13 TEMES, G.C.: 'The prolate filter: an ideal lowpass filter with optimum step-response', *J. Franklin Inst.*, 1972, **293**, pp. 77–103

14 LERNER, R.M.: 'Representation of signals and design of signals', chaps. 10 and 11 of 'Lectures on communication system theory', BAGHDADY, E. (Ed.) (McGraw Hill, 1961)

15 TUFTS, D.W.: 'Nyquist's problem – the joint optimisation of transmitter and receiver in pulse amplitude modulation', *Proc. IEEE*, 1965, **53**, pp. 248–260

16 LUCKY, R.W.: 'A survey of the communication theory literature – 1968–1973', *IEEE Trans.*, 1973, **IT-19**, pp. 725–739

17 PARK, J.H.: 'Effects of band limiting on the detection of binary signals', *ibid.*, 1969, **AES**, pp. 867–870

Chapter 10
Timing extraction and jitter

10.1 Timing irregularity and its effects

While the sources of digital messages almost always emit the signal elements at a uniform rate, subsequent operations during transmission tend to destroy this uniformity. Thus, if we examine the message at some point along the transmission path we find that the signal elements are arriving in a slightly irregular manner. This raises the following questions:

(i) Is this timing irregularity likely to be significant?
(ii) How is it caused?
(iii) How does it accumulate along the transmission path?
(iv) Can anything be done to reduce or eliminate it?

Before attempting to deal with these questions we must first define the timing irregularity, or jitter, as it is usually called, in a quantitative manner. Let us consider a uniform train of impulses, which after transmission arrive in a slightly irregular manner, as shown in Fig. 10.1. The deviations from the regular timing can be depicted by a series of ordinates at the original regular intervals, and these in turn can be taken to be samples of some continuous function $j(t)$. $j(t)$ is taken as our fundamental description of the jitter. We can clearly associate mean and r.m.s. values with this function, and like other functions of time $j(t)$ can be equated to a continuous series of sinusoidal components through the Fourier transform. Hence, we will have an associated function $J(\omega)$ which describes the spectral distribution of the jitter, and multiplying $j(t)$ and $J(\omega)$ by their complex conjugates we can obtain expressions for the 'energy distribution' of the jitter in the time and frequency domains. We must remember, however, that since we are dealing with phase variations this 'energy' and the 'jitter power' derived by integrating it, have no real physical existence. We can also see that the jitter can contain both systematic and random components.

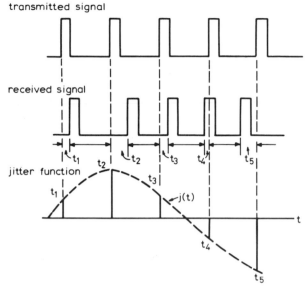

transmitted signal

received signal

jitter function

Fig. 10.1 Definition of timing jitter

10.1.1 Significance of jitter

This has been briefly mentioned at various points in earlier Chapters, and it has been shown that jitter introduces transmission impairment in two ways.

(*a*) During the process of regeneration timing irregularity may displace the effective decision point from the centre of the signal 'eye', and thus lead to a reduction in the signal/noise margin (See Chapter 9). It should be noted, however, that the local timing extraction circuits employed in the regenerators may, at least to some extent, follow the timing variations in the incoming signal, and this must be taken into account when assessing the effects on the error rate.

(*b*) In the case of a digitally encoded analogue signal, jitter in the digit stream arriving at the decoder finally leads to the decoded analogue samples appearing in a slightly irregular manner, as shown in Fig. 10.2. The output may then be considered to consist of samples at the correct instants plus a series of dipulse samples representing the differences between the actual output and the correct output. These samples may be considered to represent a distortion component added to the signal, in an analogous manner to that employed in analysing quantising noise.

Attempts to evaluate the jitter noise described in (*b*) run into severe mathematical difficulties, and the published accounts of work in this area all involve various approximations. One of the most detailed investigations, carried out by Bennett,[1] has already been described in Chapter 2 (Section 2.5). This analysis was aimed at evaluating distortion in digitally encoded f.d.m. channels, and the final result was

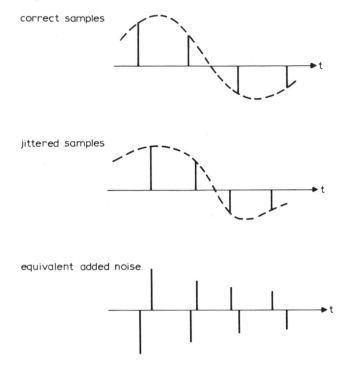

correct samples

jittered samples

equivalent added noise

Fig. 10.2 Introduction of noise by jitter in digital/analogue conversion

$$\frac{\text{jitter distortion power per channel}}{\text{total f.d.m. signal power}} = 4\pi^2 f_c{}^2 \sigma_j^2 B/f_0 \tag{10.1}$$

where, as discussed in Chapter 2

f_c is the channel frequency
σ_j is the r.m.s. jitter
B is the channel bandwidth
f_0 is the f.d.m. signal bandwidth.

Numerical results for f.d.m. systems will be found in Chapter 4.

While, under the assumptions used in its derivation, Bennett's expression can only be legitimately applied to f.d.m. signals, it is of interest to examine the results in other cases. As a first example we take a telephony channel in a standard p.c.m. transmission system, employing an 8 kHz sampling rate. In this case, the total signal power is the power of a single speech signal and is taken as s^2. We apply eqn. 10.1 to evaluate the jitter distortion power $\delta(e^2)$ in a bandwidth δf giving

$$\frac{\delta(e^2)}{s^2} = \frac{4\pi^2 \sigma_j^2 f^2}{f_0} \, \delta f \tag{10.2}$$

Integrating this over the entire channel bandwidth f_0 and rearranging yields

$$u_j - \frac{\sqrt{3}}{2\pi f_0} \frac{e}{s} \qquad (10.3)$$

This is found to be identical to the expression derived by Cattermole[2] using an entirely different approach. In the case of p.c.m. telephony, since jitter distortion can only occur simultaneously with the speech signal, it is largely masked by it, and a comparatively low signal/distortion ratio, say 33 dB, can be tolerated. Inserting this value in eqn. 10.3 leads to a permissible jitter of $1.4 \, \mu s$.

In the case of a 625-line p.a.l. colour television signal tests show that a signal/noise ratio of about 50 dB is tolerable. Using this in Bennett's expression we find that the predicted value for allowable jitter is 0.15 ns, which is in quite good agreement with the value of 0.2 ns derived by direct subjective experiments.

It would therefore appear that Bennett's expression yields a fairly good estimate of the permissible jitter for most cases of practical interest.

10.1.2 Systematic and low-frequency jitter

We shall see later that justification operations can produce systematic jitter contributions, and if the conditions are right these can contain large amplitude low-frequency components. In addition, the devices used to reduce jitter, known as dejitterisers, are most effective in dealing with the higher-frequency components, and tend to leave a residual low-frequency jitter impressed on the signal. The tolerance to such low-frequency jitter can differ significantly from that for the higher-frequency components. Let us first consider one channel in an f.d.m. signal, and represent this by a single tone, amplitude V at an angular frequency ω rad/s. A sinusoidal jitter of r.m.s. value s seconds at an angular frequency m rad/s will phase modulate this f.d.m. signal with a peak deviation of $\sqrt{2} \, \omega s$ radians leading to the signal

$$v = V \sin \{\omega t + \sqrt{2} \, \omega s \sin (mt)\}$$

$$\approx V \sin \omega t + \frac{V \omega s}{\sqrt{2}} \sin (\omega + m) t - \frac{V \omega s}{\sqrt{2}} \sin (\omega - m) t$$

We see that the jitter now appears as two discrete sidebands. If the jitter frequency exceeds about 2 kHz these components will fall into adjacent channels, and analysis shows that the tolerable jitter is then roughly the same as that derived from Bennett's expression. However, if the jitter frequency is less than about 1 kHz the sidebands will fall within the channel, and since the jitter distortion will then only occur in the presence of a signal, a signal/noise ratio of about 33 dB can be allowed. Calculation shows that the permissible r.m.s. jitter levels are then 35 ns, 6·9 ns, and 0·95 ns, for group, supergroup, and hypergroup channels, respectively. Thus, the limits given in Table 4.3 can be relaxed considerably for jitter components below 2 kHz.

In the case of television, phase modulation will produce sidebands of the high-level components which occur at the line frequency and its harmonics. This will be equivalent to single-frequency interference, and the specifications for such interference in television systems suggest that the jitter tolerance will be unaltered at low frequencies. In the case of high-quality programme circuits, the effects of low-frequency modulation can be subjectively objectionable, and one might expect rather more stringent limits for low-frequency jitter than for the higher-frequency components. We do not, however, have any quantitative information on this point.

10.2 Existence of timing signal

For complete regeneration a local source of timing information is required in each regenerator. The preferred means of providing this is by extraction of the timing information from the incoming signal, which presupposes the existence of this information within the signal. As a preliminary to studying the extraction process we must examine this assumption. If, by carrying out a spectral analysis of the signal, we can show that it contains a discrete component at the pulse-repetition frequency f_c we can assume that timing information is available, since this component may be extracted by a suitable filter and used to drive a local clock pulse generator. However, direct spectral analysis of the signal proves to be difficult. We find that, since we are dealing with a nonrepetitive pulse sequence, we cannot employ Fourier series analysis, and equally, since a random pulse train will have finite components over an infinite time interval we cannot establish the convergence of the integrals needed to perform a Fourier transform analysis.

A technique which is used to avoid these difficulties in the case of purely random signals is to work through the intermediary of the autocorrelation function $R(\tau)$. We will defer a detailed discussion of this process to Chapter 11, and for the moment merely note that the autocorrelation function is defined by

$$R(\tau) = \lim_{T \to \infty} \frac{1}{T} \int_{-T/2}^{T/2} f(t) f(t + \tau) \, dt \tag{10.4}$$

In general, the autocorrelation function will be found to consist of a nonperiodic component $R_1(\tau)$ and a periodic component $R_2(\tau)$.

Then, corresponding to the nonperiodic part, the energy spectrum of the signal contains a continuous distribution given by

$$F_1(\omega) = \int_{-\infty}^{\infty} R(\tau) \cos(\omega\tau) \, d\tau \tag{10.5}$$

and corresponding to the periodic part the energy spectrum contains a series of spectral lines given by

$$f(t) = \sum_{n=-\infty}^{\infty} a_n \cos(n\omega t + \theta_n) \tag{10.6}$$

where

$$R(\tau) = \sum_{n=-\infty}^{\infty} a_n^2 \cos(n\omega\tau) \tag{10.7}$$

For the moment we will assume that our pulse trains may be treated as random signals, and apply the above relations to evaluate the energy spectrum of a random sequence of rectangular pulses of amplitude E, duration t_0, and pulse repetition period T, as shown in Fig. 10.3. We assume that the probability of occurrence of a pulse in any time slot is $1/2$ so that

$$R(0) = E^2 t_0/2T$$

For τ lying between t_0 and $T - t_0$ one of the two terms in the product $f(t) \cdot f(t - \tau)$ must be zero, so that

$$R(\tau) = 0$$

When $\tau = nT$ both terms may occur with probability $1/2$ so that

$$R(nT) = E^2 t_0/4T$$

A little thought will show that for intermediate values the autocorrelation function will vary linearly between these limits, so that the final form will be as shown in Fig. 10.3b. Splitting this into its periodic and nonperiodic components, and applying eqns. 10.5–10.7 we find that the continuous energy spectrum is given by

$$F_1(\omega) = E^2 \left(\frac{t_0}{2T}\right)^2 T \frac{\sin^2 x}{x^2} \tag{10.8}$$

where

$$x = \frac{\omega t_0}{2}$$

and the repetitive part yields a line spectrum given by

$$F_2(\omega_1) = E^2 \left(\frac{t_0}{2T}\right)^2 \frac{\sin^2 x}{x^2}$$

where

$$x = \frac{n\omega_1 t_0}{2} = \frac{n\pi t_0}{T} \tag{10.9}$$

These expressions are plotted in Fig. 10.4. We note that due to the $\sin(x)/x$ term the line spectrum must have a null at $\omega = 2\pi/nt_0$ and that in particular, when we have full width pulses with $t_0 = T$ the line spectrum vanishes apart from the d.c. term. Thus, while narrow rectangular pulses fulfil our requirements for the existence of timing information, full width pulses apparently fail to do so.

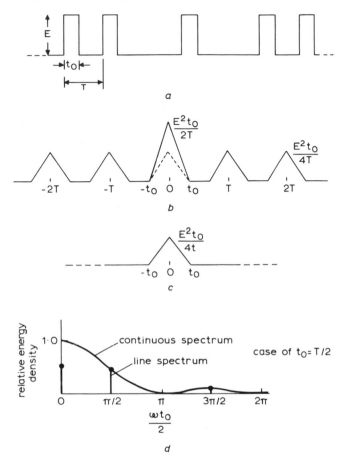

Fig. 10.3 Energy spectrum of a random pulse train
    a Random pulse train
    b Autocorrelation function
    c Nonperiodic part of autocorrelation function
    d Energy spectrum

However, visual inspection of a full-width pulse train suggests that timing information is in fact still present, since we can clearly see the mark to space transitions,
and identify the instants at which these occur. The explanation of this apparent
paradox lies in the fact that the continuous spectral component is not a truly
random signal, and has hidden phase relationships between its different components. This has been discussed in detail by Bennett[1] who introduced the term
'cyclostationary' to describe pseudorandom processes of this type. He also pointed
out that this throws some doubt on the validity of using the autocorrelation relationships in the evaluation of the energy spectrum, and that a rather critical
investigation of the situation is needed to justify their use.

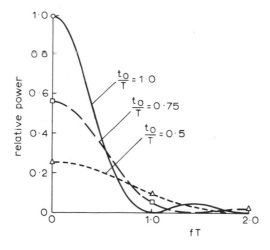

Fig. 10.4 Energy spectrum of a random rectangular pulse train
Pulse duration t_0
Pulse time slot duration T
Curves indicate continuous spectrum and marked points indicate spectral lines
○ for spectral lines with $t_0/T = 1$
□ for spectral lines with $t_0/T = 0.75$
△ for spectral lines with $t_0/T = 0.5$

To extract a timing signal in these circumstances we have to subject the signal to some nonlinear operation. Considering our example of the full-width rectangular pulse train, we can see that if we apply this signal to a threshold detector and pulse generator, arranged so as to produce a short pulse of amplitude E_1 and duration t_1 on every positive-going transition through the threshold level, we can produce a new train of short pulses. In this case, since the probability of a positive transition is only 0·25, the pulse density will be reduced, but the new signal will otherwise be similar to the original pulse train, and, by carrying out a spectral analysis as before, we can show that this new pulse train will possess a line spectrum with a fundamental component at the p.r.f. given by

$$V = \frac{E}{4\pi} \sin\left(\frac{\pi t_1}{T}\right) \tag{10.10}$$

In general, a simple threshold-detection operation will not suffice, and rather more drastic nonlinear processing will be needed. Typically, we may employ a combination of a differentiator, a rectifier, and a threshold detector, as shown in Fig. 10.5. Mathematically, the results of these operations may be considered to be equivalent to generating sum and difference components of the various elements of the signal spectrum. Most of these combine in a random manner, but due to the hidden phase relationships, those falling at multiples of the pulse-repetition frequency tend to add in a systematic way, thus producing the line spectrum.

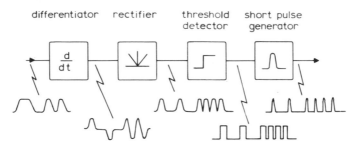

differentiator   rectifier   threshold   short pulse
                               detector   generator

Fig. 10.5 Recovery of timing information by nonlinear operations

The fact that timing information may be derived from the pseudorandom portion of the signal spectrum is very important, since, as we will see in Chapter 12, noise considerations usually lead to a curtailment of the transmitted frequency spectrum which would preclude transmission of a spectral line at the p.r.f., even when this did exist. However, mention of the restricted transmission bandwidth reminds us that in practical systems we do not deal with rectangular pulses, but with signals having finite transition times, and we must next consider the implications of this as far as timing extraction is concerned.

Since the threshold detector is triggered at a fixed level, the exact instant of triggering will be dependent on the signal amplitude, and will be disturbed by any intersymbol interference caused by the finite transition times. This will not eliminate the spectral lines, but will add a further contribution to the continuous spectrum of the derived pulse train. The derived timing pulses will also be non-rectangular. This leads to the question of the relationship between the energy spectrum and the pulse shape. We can depict the pulse train by a derived message sequence that modulates a repetitive pulse generator. The derived message sequence may be represented by the series $\ldots a_{-2}, a_{-1}, a_0, a_1, a_2, \ldots$ where the term a_n takes the value 1 or 0 depending on the presence or absence of a pulse in the nth time slot. If the pulse waveform is $g(t)$ and the time slot duration is T, then the waveform of the derived timing signal is

$$f(t) = \sum_{n=-\infty}^{\infty} a_n g(t - nT) \tag{10.11}$$

Assuming that the probability of a pulse in any given time slot is p, and there is no correlation between terms in the sequence a_n, then it can be shown that the continuous part of the spectrum of $f(t)$ is given by

$$F_1(\omega) = \left\{ \frac{p(1-p)}{T} \right\} |G(\omega)|^2 \tag{10.12}$$

(this being the double-sided version of the spectrum) and the line spectrum is given by

$$F_2(\omega) = \left(\frac{p}{T}\,G(0)\right)^2 \quad \text{for } n = 0$$

$$= \frac{p^2}{T^2}\,|G(n\omega_c)|^2 \quad \text{for } n = \pm 1, \pm 2, \dots. \tag{10.13}$$

where

$$\omega_c = 2\pi/T$$

and $G(\omega)$ is the Fourier transform of $g(t)$ given by

$$G(\omega) = \int_{-\infty}^{\infty} g(t)\,e^{-j\omega t}\,dt \tag{10.14}$$

From this we can see that the ratio of the line-spectrum component at the pulse-repetition frequency f_t to the energy per unit bandwidth of the continuous spectrum in its immediate vicinity is

$$\frac{F_2(\omega_c)}{F_1(\omega_c)} = \frac{p}{T(1-p)} \tag{10.15}$$

Referring back to eqn. 10.11, let us consider the sequence a_n to be a repetitive full-width pulse pattern. The random signal can be considered as the limiting case of this as the repetition period becomes infinite. The energy spectrum may be considered to be the result of modulating a pulse train repeating at $T = 1/f_c$ by this pattern. This spectrum will consist of the base-band spectrum of the pattern plus sidebands about the harmonics of the pulse-repetition frequency f_c, but we can see from eqn. 10.12 that because the sequence employs full-width rectangular pulses, the spectrum of the pattern itself must have nulls at f_c and its harmonics. Thus, in the final energy spectrum of the complete signal, the parts of the continuous spectrum in the vicinity of the spectral lines must originate from the sidebands produced by the modulation process. Since amplitude modulation produces symmetrical sidebands it then follows that the components in the continuous spectrum at $f_c \pm \delta f$ will be highly correlated.

Systematic message sequences. The analysis given above has not emphasised the fact that we have been dealing with *random* processes, and estimating *long-term-average* values of the timing component. Over a short period, the actual value may differ appreciably from the long-term average. In addition, although it is convenient to represent message signals by means of random sequences, the incoming pulse trains are frequently very far from being random. For example, if the system is unloaded, or only carrying a small amount of traffic, the signal may be an almost repetitive pulse pattern, or may contain a very small proportion of marks. This can easily lead to situations where the line component at the p.r.f. falls to a very low level, even when elaborate nonlinear recovery systems are used, and some additional measures must be introduced to cover these situations.

Two techniques are commonly employed.

(i) The message sequence is restricted by using a line code (see Chapter 11) that guarantees a certain minimum content of timing information in the incoming signal.
(ii) At the transmitting end of the system the original message signal is processed by a data-scrambler (Savage[3]) to break up any systematic patterns which may occur. This topic will be discussed later in this Chapter.

10.3 Timing jitter in a single regenerator

Pulse trains derived in the manner discussed in the last Section satisfy our initial criterion for the existence of usable timing information, that is, they contain a strong spectral line at the pulse-repetition frequency. However, these pulse trains cannot be used directly as timing signals, since they still do not contain pulses in every time slot, and further operations are needed in order to generate a regular repetitive train of clock pulses. Most of the timing jitter is introduced during these operations.

A typical extraction scheme is shown in Fig. 10.6. A pulse train derived as in Fig. 10.5 is applied to a high Q resonant circuit tuned to the pulse-repetition frequency f_c. The output of this tank circuit, which approximates to a steady sinusoidal signal at f_c, is applied to a second threshold detector arranged to trigger on positive going transitions through zero level. This, in turn, drives the local clock pulse generator.

We again consider the conditions for a random message input. Because it is not possible to realize a tuned circuit with an infinite Q factor, the signal applied to the threshold detector will consist of the extracted spectral line plus a small segment of the continuous spectrum. At this point the continuous spectrum will consist of the contribution from the spectrum of the transmitted signal, together with a random component due to noise and crosstalk introduced in the preceding transmission section. This latter component will be completely uncorrelated at successive regenerators, and so the resulting jitter contributions will add on a power basis. The message spectrum will tend to introduce similar jitter contributions in every

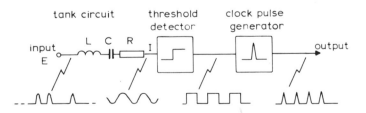

Fig. 10.6 Arrangements for clock pulse generation

regenerator, and there is therefore a high probability that these will add in a systematic manner.

Wo thoroforo rootrict oui attention to this second component. We have already seen that spectral elements equidistant on either side of f_c are almost identical, and thus we may consider the output of the tank circuit as a carrier with associated sidebands, which because of this symmetry, produce amplitude modulation of the carrier. Since amplitude modulation does not disturb the positions of the zero crossings of the waveform, ideally no jitter should be introduced. This happy situation is made unattainable by various practical imperfections that introduce amplitude/phase conversion.

10.3.1 Tank circuit mistuning
The first mechanism of this type that we will consider is mistuning of the resonant circuit. Suppose this is arranged as a series RLC combination, as in Fig. 10.6, and actually resonates at f_0. Then the tank transfer function will be

$$Y_t(\omega) = \frac{I}{E} = \frac{1}{R + j\left(\omega L - \frac{1}{\omega c}\right)} \tag{10.16}$$

and at the p.r.f. the relative transmission of the carrier will be

$$\frac{Y_t(\omega_c)}{Y_t(\omega_0)} = \frac{1}{1 + jQ\left(\lambda_c - \frac{1}{\lambda_c}\right)} \tag{10.17}$$

where

$$Q = \frac{\omega_0 L}{R}$$

$$\lambda_c = \frac{f_c}{f_0} = \frac{\omega_c}{\omega_0}$$

The carrier then experiences a phase shift ϕ defined by

$$\tan \phi = Q\left(\lambda_c - \frac{1}{\lambda_c}\right) \tag{10.18}$$

Let us now consider the modulation of the carrier by the components in the continuous spectrum situated in bands δq at frequencies $f_c \pm f_q$. These components may be considered as discrete terms of amplitude $(2F_1(f_q)\delta q)^{\frac{1}{2}}$ and from eqn. 10.15 we have for $p = 0{\cdot}5$

$$\frac{\text{sideband power}}{\text{carrier power}} = \frac{\delta q}{f_c} \tag{10.19}$$

The relative amplitudes of the two sideband components after transmission through the tank circuit are then

$$\frac{s_1}{s} = \cos(\theta_1) = \cfrac{1}{1 + jQ\left(\lambda_c + \lambda_q - \cfrac{1}{\lambda_c + \lambda_q}\right)}$$

$$\frac{s_2}{s} = \cos(\theta_2) = \cfrac{1}{1 + jQ\left(\lambda_c - \lambda_q - \cfrac{1}{\lambda_c - \lambda_q}\right)} \quad (10.20)$$

where

$$\lambda_q = \frac{f_q}{f_0} = \frac{\omega_q}{\omega_0}$$

It can be seen immediately that the two sidebands are no longer equal, and following standard techniques these can now be split up into a pair of symmetric and antisymmetric components. The latter then introduce phase modulation of the carrier. Routine algebraic operations which we will not reproduce here (the reader will find details in Manley[4]) lead to the final result that for $f_q \ll f_c$ and $|f_c - f_0|$ small, the spectral components at $f_c \pm f_q$ produce timing jitter at frequency f_q given by

$$J = |J_a(f_q)| \sin[2\pi f_q t + \arg\{J_a(f_q)\}] \quad (10.21)$$

where

$$|J_a(f_q)| = \frac{8\sqrt{\dfrac{\delta q}{f_c}}\,(dQ)\,(\lambda_q Q)}{1 + (2Q\lambda_q)^2} \quad (10.22)$$

$$\arg\{J_a(f_q)\} = \tan^{-1}\left\{\frac{\cos\theta_1\sin(\theta_1 - \phi) + \cos\theta_2\sin(\theta_2 - \phi)}{\cos\theta_1\cos(\theta_1 - \phi) - \cos\theta_2\cos(\theta_2 - \phi)}\right\} \quad (10.23)$$

$$d = \lambda_c - 1$$

Eqn. 10.22 then gives an estimate of the magnitude of the jitter caused by tank-circuit misalignment. The jitter is expressed in radians, referred to the timing frequency f_c, and so must be divided by $2\pi f_c$ to obtain a measure in time units.

The spectral distribution of the jitter given by these expressions is plotted in Fig. 10.7. Its most significant feature, which we shall return to later, is that it tends to zero as f_q approaches zero.

10.3.2 Threshold-detector misalignment
If the operating point of the threshold detector is offset from zero, as shown in Fig. 10.8 any amplitude variation in the timing signal will then cause variations in the triggering time. This amplitude modulation can be calculated from the amplitudes of

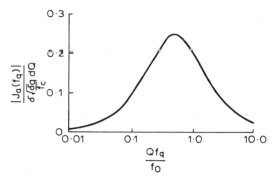

Fig. 10.7 Spectral distribution of jitter caused by tank-circuit misalignment

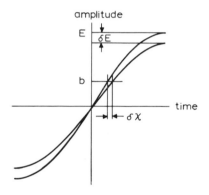

Fig. 10.8 Production of timing jitter by threshold offset
Threshold, normally at zero amplitude, is offset to *b*

the symmetric sideband components appearing at the output of the tank circuit. Since, in practice, the amount of tank mistuning will be small, these will be virtually identical to the amplitudes of the input sidebands, and we may treat threshold misalignment as an independent source of jitter. It should be noted, however, that some form of amplitude limiter is frequently inserted between the tank circuit and the threshold detector, and if this includes any further resonant circuits, it will reduce the effective amplitude modulation by a factor L. As shown in Fig. 10.8, if the amplitude of the timing signal is E, and this undergoes a change δE the resulting timing angle variation $\delta\chi$ for a threshold offset b is given by

$$\delta\chi = \frac{\delta E}{E} \tan\chi \qquad (10.24)$$

where

$$\sin\chi = b/E$$

Then applying an analysis similar to that employed in the previous Section, the final jitter spectrum is given by

$$|J_b(f_q)| = \frac{-2\sqrt{\frac{\delta q}{f_c}}\, L\tan\chi}{\{1 + (2Q\lambda_q)^2\}^{\frac{1}{2}}} \tag{10.25}$$

$$\arg\{J_b(f_q)\} = \tan^{-1}\left(\frac{\cos\theta_1\sin(\theta_1-\phi) - \cos\theta_2\sin(\theta_2-\phi)}{\cos\theta_1\cos(\theta_1-\phi) + \cos\theta_2\cos(\theta_2-\phi)}\right) \tag{10.26}$$

As in the case of eqn. 10.22 $|J_b(f_q)|$ is expressed in radians at the timing frequency, and must be divided by $2\pi f_c$ to convert to time units. This is plotted in Fig. 10.9.

10.3.3 Effects of input pulse shape

We have already indicated that some further timing-distortion contributions can be introduced through the effects of intersymbol interference on the threshold detectors. A more critical examination reveals that the pulse shape introduces further jitter through other mechanisms. Reference back to eqns. 10.12 and 10.13 and to Fig. 10.4 shows that for wide pulses there will be an appreciable slope in the continuous spectral component in the vicinity of the p.r.f. As we have seen, any such slope will lead to phase modulation of the extracted timing signal. Qualitatively, it can be seen that since the differences become small as the modulation frequency is reduced, we will tend to derive a jitter component with a near-zero low-frequency spectrum, similar to that produced by tank-circuit misalignment. As we shall see later, this tends to make the overall effects in long systems unimportant, and we therefore do not propose to carry out a detailed investigation of this effect. We refer the reader to Manley[3] for an account of this mechanism, and for discussion of some further effects introduced by the threshold-detection process.

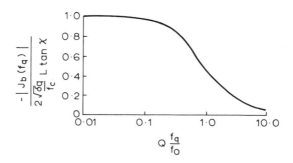

Fig. 10.9 Spectral distribution of jitter caused by threshold offset

10.4 Accumulation of timing jitter

We have now shown that in each regenerator jitter is produced by various amplitude/phase conversion processes. This jitter is then passed on to the next regenerator in the form of time modulation of the transmitted pulse sequence, and ultimately appears at the input to the next tank circuit as a phase modulation of the incoming timing carrier. Since a highly Q-tuned circuit tends to remove the sidebands it will, to some extent, suppress the phase modulation of the incoming signal. Thus each regenerator reduces the incoming jitter and then injects a further component produced by its own amplitude/phase conversion mechanisms.

10.4.1 Response of tuned circuit/phase modulation

Let us suppose that the incoming modulated timing carrier is described by E exp $[\alpha_i(t) + j\{\omega_c t + \beta_i(t)\}]$. Then $\alpha_i(t)$ defines the amplitude modulation and $\beta_i(t)$ the phase modulation. Now if the time domain response of the tank circuit is

$$g_t(t) = \frac{1}{2\pi} \int_{-\infty}^{\infty} Y_t(\omega)\, e^{j\omega t}\, d\omega \tag{10.27}$$

the output of the tank circuit will then be given by the convolution integral as

$$E \exp [\alpha_0(t) + j\{\omega_c t + \beta_0(t)\}] = \int_{-\infty}^{\infty} \exp [\alpha_i(t-x) + j\{\omega_c(t-x)$$

$$+\beta_i(t-x)\}]\, g_t(x)\, dx \tag{10.28}$$

After further manipulation (see Byrne[5] for details) we finally arrive at the result

$$\alpha_0(t) + j\beta_0(t) = \int_0^t \frac{\omega_0}{2Q}\, e^{-\frac{\omega_0 x}{2Q}}\, \{\alpha_i(t-x) + j\beta_i(t-x)\}\, dx \tag{10.29}$$

Because the right-hand side of this equation is still in the form of a convolution integral, it follows that on returning to the frequency domain

$$\beta_0(j\omega) = \frac{\beta_i(j\omega)}{1 + j2Q\dfrac{\omega}{\omega_0}}$$

$$\alpha_0(j\omega) = \frac{\alpha_i(j\omega)}{1 + j2Q\dfrac{\omega}{\omega_0}} \tag{10.30}$$

We can now see that as far as incoming phase modulation is concerned, each tank circuit behaves as a simple RC filter having a transfer function.

$$Y_i(j\omega_q) = \frac{1}{1 + j2Q\dfrac{\omega_q}{\omega_0}} \tag{10.31}$$

10.4.2 Summation of timing jitter in a chain of regenerators

If each regenerator injects a jitter contribution $J(\omega_q)$ we can now see that the resultant jitter at the end of a chain of n identical regenerators is

$$J_n(\omega_q) = J(\omega_q)\,[1 + Y_j(\omega_q) + \{Y_j(\omega_q)\}^2 + \ldots \{Y_j(\omega_q)\}^{n-1}]$$

$$= J(\omega_q)\,Y_j^n\,(\omega_q) \tag{10.32}$$

where

$$Y_j^n(\omega_q) = \frac{1 - \{Y_j(\omega_q)\}^n}{1 - Y_j(\omega_q)}$$

$Y_j^n(\omega_q)$ can be considered as a form of jitter transfer function for the complete regenerator chain. It is plotted for a number of values of n in Fig. 10.10. It can be seen that the low-frequency value of this function is proportional to n, and its 'cut-off frequency' is inversely proportional to n. We find that for jitter of the type discussed in Section 10.3.1 this cut-off frequency is well below the peak frequency of the jitter spectrum, for large n. As a result of this, the jitter contribution from regenerators in the early part of the chain is rapidly attenuated, and it can in fact be shown[4] that the peak jitter in a long chain of regenerators never exceeds four times the jitter contributed by a single regenerator. This type of jitter is thus not a major factor in long systems.

On the other hand, contributions from threshold-detector misalignment (Section 10.3.2) have their peak value at low frequencies and so these are accumulated

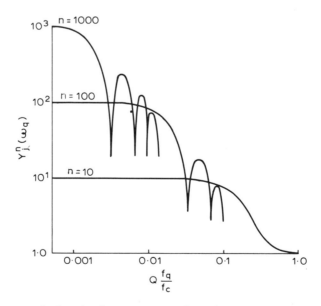

Fig. 10.10 Jitter transfer function for *n* regenerators in tandem

in proportion to the length of the system. Because of the cut-off frequency of $Y_j^n(\omega_q)$ we can ignore the spectral characteristics of the injected jitter and consider this to be a flat spectrum of magnitude

$$J_0 = 2 \sqrt{\frac{\delta q}{f_c}} \, L \tan \chi \quad \text{radians}$$

$$= \frac{2}{\omega_c} \sqrt{\frac{\delta q}{f_c}} \tan \chi \quad \text{seconds} \tag{10.33}$$

The product $J_0 Y_j^n(\omega_q)$ then effectively describes the jitter spectrum at the output of the system, and integrating this we find that the total jitter power is

$$P_j^n = \frac{\omega_0}{Q} J_0^2 \left(n - \frac{(2n-1)!}{2[4^{n-1}\{(n-1)!\}^2]} \right)$$

$$\approx \frac{\omega_0 n}{Q} J_0^2 \tag{10.34}$$

10.4.3 Measurement of jitter contribution of a single regenerator

The theoretical estimates of the jitter contributions from a single regenerator given in the preceding Sections cannot be regarded as entirely satisfactory. They assume that the jitter originates from one or two well defined mechanisms, which is certainly not true in many instances, and in any case it is often difficult to obtain values for the parameters needed to define these mechanisms. It would therefore be much more satisfactory to obtain a direct measure of the jitter contribution of a single regenerator. A technique for carrying out such a measurement has been proposed by Byrne and his colleagues.[5] The regenerator is tested with all possible repetitive patterns of a chosen length N. For each pattern the average digit time delay t_p is measured and the standard deviation σ of this is calculated. (See Byrne for details.) The low-frequency power density of the jitter is then

$$J_0 = 2T\sigma^2 \tag{10.35}$$

where T is the time duration of the N digit block.

10.4.4 Accumulation of jitter for systematic pulse patterns

The estimate of jitter accumulation given in previous Sections has been based on the assumption of random message signals. In practice, we quite frequently encounter situations where the transmitted signal changes from one repetitive pattern to another. As discussed in Section 10.4.3, provided its repetition period is not too long, each such pattern can be associated with an average delay t_p in the timing path through the regenerator. Thus, a change from pattern J to pattern K will cause a relative change in clock timing of $t_k - t_j$, and in a system of N regenerators the total timing change will obviously be $N(t_k - t_j)$. We must next investigate the manner in which this transition takes place. This can most easily be evaluated by interpreting eqn. 10.32 as a Laplace transform. The input jitter at each regenerator is now a unit step whose transform is $(t_k - t_j)/s$, and hence the transform of the resulting jitter after N regenerators is

$$J_n(s) = \left(\frac{t_k - t_j}{s}\right) \left\{ \frac{1 - \left(\dfrac{1}{1 + \dfrac{2Q}{\omega_0} s}\right)^N}{1 - \dfrac{1}{1 + \dfrac{2Q}{\omega_0} s}} \right\}$$

$$= \left(\frac{t_k - t_s}{s^2}\right) \frac{\omega_0}{2Q} \left\{ \left(1 + \frac{2Q}{\omega_0} s\right) - \left(\frac{1}{1 + \dfrac{2Q}{\omega_0} s}\right)^{N-1} \right\} \qquad (10.36)$$

But

$$\left(\frac{1}{1 + \dfrac{2Q}{\omega_0} s}\right)^{N-1} \approx \exp \left\{-(N - 1)\frac{2Q}{\omega_0} s\right\} \qquad (10.37)$$

So on transforming back to the time domain

$$J_n(t) \approx (t_k - t_j) \left\{1 + \frac{\omega_0}{2Q} t - \frac{\omega_0}{2Q} (t - t_n) H(t - t_n)\right\} \qquad (10.38)$$

where

$$t_n = (N - 1)\frac{2Q}{\omega_0}$$

This function is plotted in Fig. 10.11 and we see that, excluding the contribution from the last regenerator, the transition takes place linearly, and rather surprisingly, the rate of transition is independent of the number of regenerators in the chain. Build up of jitter now appears as an increase in the duration of the transition as we progress down the chain.

   In practice, this leads to some secondary effects. The linear transition has the effect of changing the pulse repetition frequency whilst it takes place, and so regenerators towards the far end of the system will operate with an apparent mistuning of their tank circuits for an appreciable period after the pattern change. In addition, the change in pulse pattern and associated change in pulse density will cause an amplitude transient at the output of each tank circuit. The combined effects appear as a long duration transient disturbance in both amplitude and phase.

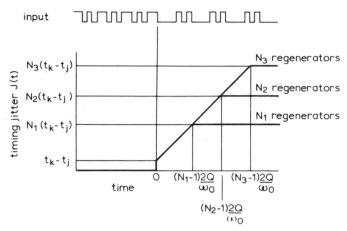

Fig. 10.11 Transient in extracted timing signal for a chain of *n* regenerators subjected to an input-pattern change

The limiter will deal with the amplitude changes, but the phase shift will severely degrade the regenerator operating margins for a period. It has been observed in tests on long regenerator chains that following a pattern change operations continues for a short time, and that there is then an interruption in transmission before stable operation is restored. This has been attributed to the above effect.

10.5 Jitter caused by justification operations

The justification operations which normally form part of the multiplexing process are a second major source of timing jitter. Justification was described in Chapter 5, and we recall that the process is essentially one of inserting additional time slots into the incoming signal, to bring its net digit rate up to the value employed in the synchronous part of the multiplex equipment. At the receiving terminal these added time slots are deleted, with the result that the signal now arrives in bursts, separated by gaps of one time-slot duration. This irregular signal is then retimed by passing it through a buffer store from which it is read out by a local clock. The clock is derived from the average rate of the incoming signal by a phase-locked loop control system. A typical arrangement is shown in Fig. 10.12. This retiming operation is naturally imperfect, and a residual jitter component remains superimposed on the outgoing signal.

10.5.1 Calculation of justification jitter
It is convenient to carry out the analysis of the jitter before the final retiming operation, which can be shown to be equivalent to low pass filtering of the resultant jitter. Now, ignoring any timing jitter that is already present, the incoming

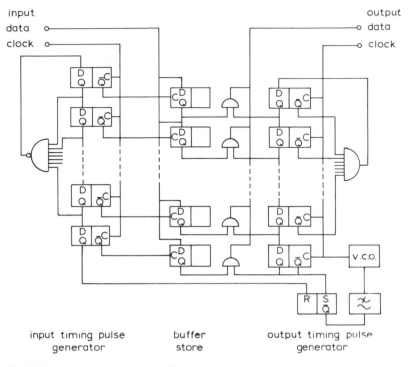

input timing pulse   buffer   output timing pulse
generator   store   generator

Fig. 10.12 Circuit arrangements of a dejitteriser

signal will be arriving at a uniform rate slightly below the required synchronous rate of the multiplex. If we imagine that the incoming and outgoing time slots are initially in alignment, then, as shown in Fig. 10.3, the incoming signal proceeds to fall behind the outgoing signal, until a certain delay is reached at which a decision to justify is made. If this is implemented immediately, the next time slot will be justified, and obviously the justification operations will occur at a uniform rate f_j where

$$f_j = f_m - f_s \qquad (10.39)$$

and

f_j = justification frequency

f_m = multiplex frequency

f_s = incoming signal frequency.

At the receiving terminal the justified time slots are deleted so that the signal arrives in bursts at a digit rate f_m, and every so often there is a pause of one digit period t_m so that the long term average rate remains at f_s. The resulting timing jitter is then as shown in Fig. 10.13. This is the component normally referred to as

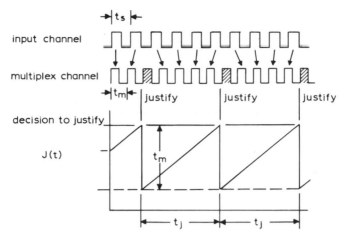

Fig. 10.13 Justification jitter
$$f_s = 1/t_s \qquad f_m = 1/t_m \qquad f_j = 1/t_j$$

justification jitter. It has an r.m.s. value of $t_m/2\sqrt{3}$, which is independent of f_s. By carrying out a Fourier analysis we can show that it is composed of a series of components at harmonics of f_s described by the series

$$j(t) = t_m \left(\frac{1}{\pi} \sin 2\pi f_s t + \frac{1}{2\pi} \sin 4\pi f_s t + \ldots \right) \tag{10.40}$$

Now as far as we are aware, in all systems so far implemented justification does not occur on demand, but is delayed until a specific time slot allocated to justification operations becomes available. The result of this is that there is a variable delay after the decision to justify, before it is implemented, and the resultant timing jitter is then as shown in Fig. 10.14a. The difference can be regarded as being due to an added jitter component shown in Fig. 10.14b, which is defined as the waiting-time jitter.

Qualitatively, it can be seen that this will add further jitter components extending down to direct current, but formal analysis proves to be rather difficult, and as there is no virtue in separating the two components in this manner, other methods of partitioning the jitter are often used. One such approach has been described by Duttweiler.[6] He uses the fact that the justification opportunities occur at a regular rate f_0, so that the jitter can be expressed as the sum of a regular triangular component repeating at this rate f_0 and an irregular component comprising a sequence of rectangular pulses of duration f_0, but of variable amplitude as in Fig. 10.14c. The first component gives a spectrum similar to that of eqn. 10.40, but now at harmonics of f_0. The second can be seen, by reference to our earlier account of pulse spectra, to have an envelope of $\sin (\pi f/f_0)/(\pi f/f_0)$. Since this has nulls at the harmonics of f_0 the two components are independent. The final result (see

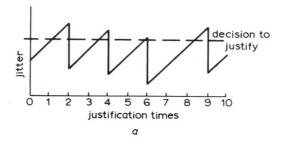

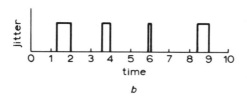

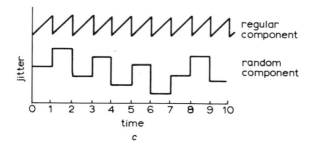

Fig. 10.14 Waiting-time jitter
  a Composite jitter
  b Waiting-time jitter
  c Alternative partitioning of jitter

Duttweiler for derivation) is that the jitter-power spectrum is

$$P(f) = \sum_{n=1}^{\infty} \left(\frac{f_j}{2\pi n f_m}\right)^2 \left\{ \delta\left(\frac{f}{f_0} - n\right) + \delta\left(\frac{f}{f_0} + n\right) \right\}$$

$$+ \left(\frac{\sin(\pi f/f_0)}{\pi f/f_0}\right)^2 \sum_{n=1}^{\infty} \left(\frac{1}{2\pi n}\right)^2 \left\{ \text{rep}_{f_0} \left[\delta\left(f - n\frac{f_s f_0}{f_m}\right)\right] \right.$$

$$+ \text{rep}_{f_0} \left[\delta\left(f + n\frac{f_s f_0}{f_m}\right)\right] \right\} \tag{10.41}$$

where

  $\delta(x)$ is the Dirac impulse function

and

> $\text{rep}_T \, \delta(x)$ then represents a train of impulses at
>
> $x = 0, \pm T, \pm 2T, \pm \ldots$

If the jitter rate is some submultiple of the maximum allowable justification rate so that

$$\frac{f_s}{f_m} = \frac{p}{q}$$

where p and q are relatively prime integers then the second term in eqn. 10.41 may be replaced by

$$\left(\frac{\sin (\pi f/f_0)}{2q \, \pi f/f_0} \right)^2 \sum_{n=1}^{q-1} \text{cosec}^2 \left(\frac{n}{q} \pi \right) \text{rep}_{f_0} \, \delta \left(f - \frac{np}{q} f_0 \right) + \frac{1}{12q^2} \text{rep}_{f_0} \, \delta(f) \quad (10.42)$$

This completes the analysis of the jitter appearing at the output of the multiplexer before the final retiming operations. The phase-lock loop in the output clock system then acts as a low-pass filter that suppresses the high-frequency jitter components. The performance of this will depend on the design of the circuit, and the reader is referred to works on this specialist topic for further information.

The build up of jitter in cascaded justification systems has been considered by Duttweiler, but his results should be treated with some caution, since the input buffer store used in each justification system will serve as a dejitteriser for the incoming signal, and so the transmission of any incoming jitter will be dependent on the detailed performance of this part of the system.

There is also little point in estimating mean square jitter power, since the dominant contributions from justification systems will be the low-frequency systematic components that appear when the input signal rate is some submultiple of the multiplex rate. As discussed in Section 10.1, these must usually be treated as independent sources of impairment.

10.6 Jitter reduction

Two techniques are available for reduction of the magnitude of the jitter appearing at the output of a long system. We may either

(i) take measures to prevent the generation and systematic accumulation of the jitter

or

(ii) reduce the magnitude of the jitter which is already present.

The first approach involves recoding the signal by a device known as a 'data scrambler'. An elementary data scrambler is shown in Fig. 10.15. The incoming binary signal is added to a second sequence in a modulo-2 adder (exclusive-OR gate). This second sequence is in fact derived by further exclusive-OR operations on delayed versions of the gate output. We will not attempt to deal with the theory of this device here, but refer the reader to the paper by Savage[3] on this topic. The effect of this process is to generate a new message sequence that is systematically related to the original input, but in a general sense is 'more random'. In particular, long sequences of marks or spaces, and simple repetitive patterns are broken up. This prevents the appearance of systematic jitter of the type discussed in Section 10.4.4. If further similar scramblers are inserted at points further along the transmission path, the message will be recoded, with the result that the systematic accumulation of jitter by the mechanisms described in Section 4.2 is also curtailed. It remains to show that the original message can be retrieved. Reference to the descrambler shown in Fig. 10.15 illustrates how this is done. More elaborate scramblers are described in Savage[3] (CCITT,[7] and Fracassi[8]).

One disadvantage of data scramblers is that they introduce error extension effects. A transmission error will obviously cause one error when it arrives at the descrambler, but study of Fig. 10.15 reveals that it causes further errors when it reaches the tapping points in the decoder shift register, since it then produces errors in the feedback signal. In the example shown, each transmission error results in a burst of three errors in the decoded signal.

A dejitteriser is essentially identical to the buffer storage arrangements used to smooth the output of a justification system, as described in Fig. 10.12. These circuits are largely ineffective in dealing with very low-frequency jitter.

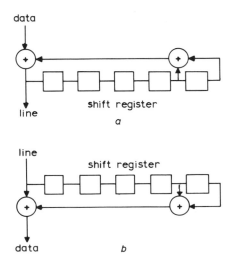

Fig. 10.15 Arrangements for data scrambling
   a Scrambler
   b Descrambler

10.7 References

1 BENNETT, W.R.. 'Statistics of regenerative digital transmission', *Bell Syst. Tech. J.*, 1958, 37, pp. 1501–1542
2 CATTERMOLE, K.W.: 'Pulse code modulation' (Illiffe, 1969) p. 113
3 SAVAGE, J.E.: 'Some simple self synchronising digital data scramblers', *Bell Syst. Tech. J.*, 1967, 46, pp. 449–497
4 MANLEY, J.W.: 'The generation and accumulation of timing noise in p.c.m. systems – an experimental and theoretical study', *ibid.*, 1969, 48, pp. 541–613
5 BYRNE, C.J., KARAFIN, B.J., and ROBINSON, D.B.: 'Systematic jitter in a chain of digital regenerators', *ibid.*, 1963, 43, pp. 2679–2714
6 DUTTWEILER, D.J.: 'The generation and accumulation of timing noise in p.c.m. systems', *ibid.*, 1972, 51, pp. 165–207
7 CCITT Green Book VII, Recommendation V27
8 FRACASSI, R., and FROEHLICH, F.: 'A wideband data station', *IEEE Trans.* 1966, COM-14, p. 648

Chapter 11

Transmission codes

11.1 Function of transmission encoding

In previous Chapters we have examined the sources from which digital signals originate, and the operations involved in combining these signals into a composite digit stream before transmission. We have also seen that the characteristics of the transmission path and the associated transmission equipment impose certain restrictions on the signals that may be transmitted. In general, the source digit stream will not conform to these restrictions, and hence some encoding process has to be introduced between the source and the transmission path. This encoding may be regarded as a signal-matching operation.

As a first step to studying this process we must consider the various features that a transmission code should exhibit.
These are

(a) transparency ⎫
(b) unique decodability ⎬ usually essential

(c) efficiency ⎫
(d) error correction or detection capability ⎪
(e) favourable energy spectrum ⎬ desirable
(f) adequate timing information content ⎪
(g) adequate content of signal level information ⎭

(h) content of framing information ⎫ requirement introduced by
(i) low error extension ⎬ the coding operation

We will consider these in more detail.

11.2 Transparency

A transparent code does not impose any restrictions on the content of the transmitted message. In the case of binary signalling this means that any sequence of marks and spaces may be presented to the coding equipment and will be transmitted faithfully. This requirement must be satisfied in almost all cases since we normally are unable to predict the exact way in which a transmission system will be operated, and we must therefore cater for all possible input messages.

11.3 Unique decodability

It does not automatically follow that a coding scheme yielding a unique output code for each input symbol can be unambiguously decoded to retrieve the original sequence of inputs. A simple example of a possible transmission code will serve to demonstrate this.

Suppose we have

Input symbol	Transmission code
00	1
01	10
10	101
11	110

The decoding of the third and fourth symbols is ambiguous, because the resulting transmission sequences could equally well have been produced by combinations of the first and second symbols. This leads to the requirement that the codes must satisfy the prefix condition, namely, no code word may form the first part, or prefix, of any other code word. However, this still leaves us free to use two or more different transmission codes for the same symbol, or even the same code for two different symbols, providing some information contained in another part of the transmission makes it possible to uniquely decide which choice has been made in any given instance. This leads to the rather stronger property of instant decodability that is exhibited by many codes. This is the property that any incoming code word may be uniquely decoded as soon as its reception is complete. This requirement, while not essential, is useful in many cases.

11.4 Efficiency

The example given above has illustrated the point that the input and output codes need not be of the same length. In fact, we shall find that it is usually essential to make the transmission code 'longer' than the incoming signal, since further code symbols must be included to allow for the addition of the various desirable features we have listed. Any such addition represents, in one sense, a loss of efficiency, since we are now sending symbols that do not contribute to the transmission of the incoming information. We obviously want to minimise this loss of efficiency, and to make a quantitative assessment of the situation and to carry out comparisons between different coding schemes, we require a systematic measure of the efficiency. In cases where both input and output are binary codes, we could do this merely by measuring the relative number of symbols in the input and output messages, but the situation becomes much more complex when the coding operation involves a change in the symbol base, for example, a conversion from a binary input to a quaternary (4-level) output. In this case, the performance is measured by comparing the information capacity (sometimes referred to as the entropy) of the output and input codes. The necessary theoretical basis for this was developed by C.E. Shannon, and we refer the reader to Shannon's original work,[1,2] or to one of the later texts (for example Schwartz[3]) for details. The final results are that if, in each time slot, a code consists of one of m different symbols, and the probability of the ith symbol occurring is p_i, then the rate of transmission of information is

$$H = \sum_{i=1}^{m} p_i \log_2 {}^1/p_i$$

$$= -3 \cdot 332 \sum_{i=1}^{m} p_i \log_{10} p_i \qquad \text{bit/symbol} \qquad (11.1)$$

Now if the input to the transmission encoder is arriving at a rate of s_i symbols per second, the total information within the signal is

$$C_1 = s_1 \sum_{i=1}^{m} p_i \log_2 {}^1/p_i \qquad \text{bit/s} \qquad (11.2)$$

and if the corresponding output code comprises s_2 symbols per second, where each symbol is one of n characters, with probability q_i of the ith character, then the available capacity of the code is

$$C_2 = s_2 \sum_{i=1}^{n} q_i \log_2 {}^1/q_i \text{ bit/s} \qquad (11.3)$$

The efficiency is then

$$\eta = c_1/c_2 \qquad (11.4)$$

For example, if the input is a binary signal at 1 MHz, with a probability of a mark of 0·25, and of a space of 0·75, then

$$C_1 = 0 \cdot 8115 \text{ Mbit/s}$$

If the corresponding output is a sequence of ternary symbols at 0·8 MHz, with all three symbols having the same probability then

$$C_2 = 1 \cdot 268 \text{ Mbit/s}$$

The efficiency is then 64%.

11.4.1 Redundancy

Any coding scheme which is not 100% efficient implies that there is some unused transmission capacity, and so we may associate the efficiency η of a code with its redundancy R, where

$$R = \frac{c_2 - c_1}{c_1} = \frac{1}{\eta} - 1 \tag{11.5}$$

The redundancy is then a measure of the 'unused' capacity. We must, however, recognise that the term 'unused' only applies in a rather narrow sense. In a redundant code we do not make use of the total available capacity to transmit message information, and, as a result, some of the possible sequences of symbols that could be transmitted will not appear. However, as we shall see in later Sections of this Chapter, the absence of these sequences is essential to allow us to introduce the various other desirable features listed in Section 1, and in a sense we are actually employing the redundant part of the code capacity to transmit information relating to timing, error rate, signal levels etc. Clearly, the greater the portion of the available capacity allocated to this purpose, the easier it becomes to include such features. Experience with practical codes shows that to introduce these additions without using complex equipment a high redundancy (e.g. 60%) must be employed, and that even when using fairly complicated implementations, it is difficult to reduce the redundancy below about 20%.

11.5 Error correction, detection and monitoring

This topic has received an immense amount of attention during recent years and would justify a complete volume to itself. We must therefore refer the reader to the specialist sources (*IEEE Transactions* on 'Information theory', *Journal of Information and Control*, text books by Peterson,[4] Berlekamp[5] etc.), and in this Chapter we will merely give a brief introduction to some of the more important aspects of this field.

For a start, we must make it quite clear what is covered by the term 'error'. Since we are only considering digital transmission, we will confine our attention to signals which consist of a series of discrete symbols, each of which may take one of a finite number of distinct values. We therefore have three possible types of error. The first is an error of substitution, in which the true value of a symbol is replaced by some other value during transmission. We next have errors of omission, in which a symbol is completely deleted from the transmitted sequence, and complementing this, errors of commision, in which a spurious symbol, of some arbitrary value, is inserted into the sequence. These latter two classes are associated with synchronisation failures in the transmission path, and tend to be much less common than errors of substitution. We can also introduce the idea of the magnitude of an error. The smallest magnitude of error will be that which leads to the insertion or deletion of a single symbol, or a transition to an adjacent level in a multilevel symbol. Larger errors will insert or delete several symbols, or cause a jump of a number of levels.

The treatment of errors may be carried out at several levels of sophistication. At the highest level, we can modify the transmitted signal so that, within limits, errors may be both detected and corrected by the receiving equipment. Now to accomplish this, it is clear that we must add some symbols to the transmitted information, and, to keep the complexity of the coding and decoding equipment within bounds, the checks must relate to a finite section of the transmission. Thus, practical systems divide the incoming signals into blocks or words, and check each word independently. Two approaches can then be used. First, checks can be appended to each word, such that the receiving equipment can deduce the presence and the location of an error directly from these. This leads to the attribute of 'instantaneous decodability' mentioned earlier. Alternatively, the checks may only reveal that an error has been introduced, and to correct this the receiving terminal then requests a retransmission of the block in question. While this is a simpler process, it introduces two important restrictions. First, there must be a return transmission channel available to carry the requests for retransmission, and secondly, the actual rate of transmission will vary with the number of repeats, and so it will be dependent on the error rate.

If we do not request transmission we are left with an error detecting system. This will be adequate, if the transmission is such that we can accept a certain error content, and only need to know when the error rate is becoming excessive. Alternatively, there are cases where the characteristics of the source signal make it possible to 'conceal' errors without actually correcting them. Examples of this arise in the transmission of digitally encoded analogue signals where the error check words coincide with the word structure employed in the analogue/digital conversion process. In this case, the receiving terminal can associate an error with a particular analogue sample. This sample can then be replaced by an estimate of its probable value, which in most cases will be either the last correctly transmitted value, or an interpolation between the preceding and following error-free samples. This technique has been found to be very effective in the case of transmission of television or high-quality sound channels by digital means.

Correction of this type still requires the detection of every error, but cases also arise where even this is not needed, and it suffices to obtain an indication of the frequency of errors. For example, we may employ an error-monitoring scheme to detect the appearance of an error condition at error rates well below the level which would be significant in terms of transmission impairment. This makes it possible to take remedial action before any service interruption can occur. The advantage of this approach is that error-monitoring arrangements are usually very much simpler than comprehensive error-detection schemes, and can therefore be introduced at a large number of locations in the system, thus making it possible to identify the point at which the error condition has appeared.

11.5.1 Techniques for error detection and correction

The simplest method of error detection is the use a parity check. In this, the incoming message is divided into blocks (these need not correspond to the word structure of the original message), and to each of these an additional time slot is added to contain the parity check. A mark or space is inserted in this time slot so as to make the total number of marks in the word even, if the convention of even-parity checking is employed, or the converse to achieve an odd-parity check condition. This scheme will clearly detect all single errors or any combination of an odd number of errors in a word, but will fail to detect any even number of errors. It will not indicate the number of errors that have occurred in any instance. Thus, it is unsuitable for use in cases where high error rates occur, or where errors may appear in bursts.

In 1950, Hamming showed how this type of system could be extended to form an error-correcting code (Hamming[6]). In his basic scheme m information digits were associated with k parity-check digits to form a word length $n = m + k$. Each parity check was associated with a certain group of digits within the word, such that if any single error occurred the resultant pattern of parity-check violations indicated its position within the word, thereby providing a single-error correcting code. The k check digits then have to indicate $n + 1$ conditions, comprising the n possible error locations and the zero error condition. This requires

$$2^k \geqslant m + k + 1$$

$$2^m \leqslant \frac{2^n}{n + 1} \tag{11.6}$$

The block structure of code words of up to length 16, as defined by these relations is given in Table 11.1.

Hamming allocated the checks so that the parity violations gave the position of the error as a binary number. Thus, check digit k_1 was associated with all digit positions having a mark in the least significant position of the binary number denoting the location, k_2 with locations having a mark in their second location, and so on. This is illustrated in Table 11.2. Finally, Hamming proposed to allocate positions 1,2,4,8,16, ... to check digits, thereby producing a systematic extendable set of code words. The resulting Hamming (7,4) code, that is a code of 7 digit words containing four information digits, is shown in Table 11.3.

Although in this form the process of producing a Hamming code seems relatively straightforward, this simplicity conceals a very sophisticated mathematical background, which must be used to extend the theory and analysis of error correcting codes beyond Hamming's original ideas. In this Chapter, we will only be able to introduce some of the major ideas involved in this theory, and we must leave the reader to pursue these in other works specialising in this topic (Peterson,[4] Berlekamp,[5] Lucky, Salz, and Weldon[7]).

Table 11.1 Hamming single error-correcting codes

Word length, *n*	No. of information digits, *m*	No. of check digits, *k*
1	0	1
2	0	2
3	1	2
4	1	3
5	2	3
6	3	3
7	4	3
8	4	4
9	5	4
10	6	4
11	7	4
12	8	4
13	9	4
14	10	4
15	11	4
16	11	5

11.5.2 Minimum distance

For a start, it is obvious that an error can only be detected if it changes the code word in question into some other word that does not appear at all in the code. Thus, the allowable code words form a restricted set, with the characteristics that all members of the set must differ from one another in at least two digit positions. However, while this will be sufficient to introduce an error-detecting capability, it will not allow error correction, since if two words differ in only two positions, then when the illegal words lying between them arise, it cannot be deduced which of the two possible source words was originally present. If the code words differ in three locations, then any single error will lead to an illegal word that is 'nearer' to one legal word than to any other, and so, on the assumption that only a single error has occurred, it is possible to deduce the word that should have been received. The number of positions in which the code words differ is thus referred to as the 'distance' between them, and in any given code there will then be a minimum distance *m*, which defines the greatest degree of similarity between any two legal words in the code.

In general, when the minimum distance is *m*, up to *m*-1 errors can be detected in any code word, and error correction is possible for up to $(m-1)/2$ errors in a word.

Table 11.2 Allocation of check digits in a Hamming code

Check digit	Positions checked	Binary code
1	1	1
	3	11
	5	101
	7	111
	9	1001
2	2	10
	3	11
	6	110
	7	111
	10	1010
	11	1011
3	4	100
	5	101
	6	110
	7	111
	12	1100
	13	1101
	14	1110
	15	1111

11.5.3 Parity-check matrix

Suppose we consider the 7,4 Hamming code defined in Table 11.4. In this code, the message is allocated to the first four digits, and the error checks to digits 5, 6, and 7. Now, it is clearly very cumbersome to list out the complete code, even in a simple case such as this, and we require some more compact mathematical representations for dealing with this type of problem. We can see that the parity checks can be described by the following three equations,

$$d_1 \oplus d_2 \oplus d_3 \oplus d_5 = 0$$

$$d_1 \oplus d_2 \oplus d_4 \oplus d_6 = 0$$

$$d_1 \oplus d_3 \oplus d_4 \oplus d_7 = 0 \qquad (11.7)$$

where $\oplus$ denotes modulo-2 addition.

These equations can be conveniently expressed in matrix form

$$[H]\,[d]^t = 0 \qquad (11.8)$$

where $[H]$ is the parity check matrix and $[d]^t$ is the transpose of the row vector

Table 11.3 Hamming 7,4 code

Symbol	C_1	C_2	I_1	C_3	I_2	I_3	I_4
0	0	0	0	0	0	0	0
1	1	1	0	1	0	0	1
2	0	1	0	1	0	1	0
3	1	0	0	0	0	1	1
4	1	0	0	1	1	0	0
5	0	1	0	0	1	0	1
6	1	1	0	0	1	1	0
7	0	0	0	1	1	1	1
8	1	1	1	0	0	0	0
9	0	0	1	1	0	0	1
10	1	0	1	1	0	1	0
11	0	1	1	0	0	1	1
12	0	1	1	1	1	0	0
13	1	0	1	0	1	0	1
14	0	0	1	0	1	1	0
15	1	1	1	1	1	1	1

$[d_1, d_2, d_3 \ldots d_n]$ representing the code word. In the present example we have

$$[H] = \begin{bmatrix} 1 & 1 & 1 & 0 & 1 & 0 & 0 \\ 1 & 1 & 0 & 1 & 0 & 1 & 0 \\ 1 & 0 & 0 & 1 & 0 & 0 & 1 \end{bmatrix} \tag{11.9}$$

It can also be seen that if we form another row vector $[d_m]$ consisting of the message digits, then the product $[H].[d_m]^t$ will yield the required check digits. Thus, formally, we can construct an encoder by storing the parity check matrix, making provision to perform the matrix multiplication, and adding the result to the incoming message.

Now suppose that when the word $[d]$ is transmitted we actually receive $[d_r]$ where some of the digits are now in error. This may be considered to be equivalent to the modulo-2 addition of an error vector $[e]$

$$[d_r] = [d] \oplus [e] \tag{11.10}$$

At the receiver we again multiply the transpose of the received vector by $[H]$

$$[H] [d_r]^t = [H] \{[d]^t \oplus [e]^t\}$$
$$= [H] [d]^t \oplus [H] [e]^t$$

Table 11.4 Alternative example of Hamming 7,4 code

Word	Message digits				Check digits		
1	0	0	0	0	0	0	0
2	0	0	0	1	0	1	1
3	0	0	1	0	1	0	1
4	0	0	1	1	1	1	0
5	0	1	0	0	1	1	0
6	0	1	0	1	1	0	1
7	0	1	1	0	0	1	1
8	0	1	1	1	0	0	0
9	1	0	0	0	1	1	1
10	1	0	0	1	1	0	0
11	1	0	1	0	0	1	0
12	1	0	1	1	0	0	1
13	1	1	0	0	0	0	1
14	1	1	0	1	0	1	0
15	1	1	1	0	1	0	0
16	1	1	1	1	1	1	1

It can be seen from eqn. 11.8 that the first term in this expression is zero, and so we are left with

$$[s] = [H]\,[e]^t \tag{11.11}$$

$[s]$ is called the syndrome of the code word. A nonzero value indicates that errors are present, and providing that the number of errors does not exceed the correcting capability of the code, the syndrome will uniquely indicate the location of the errors. Thus, in principle, error correction may be carried out by storing a look-up table of all possible values of the syndrome, together with the corresponding error patterns. In practice, this would prove to be very cumbersome, and much of the work on error correction schemes is aimed at devising codes which permit more elegant and simpler error correction.

Group theory. This branch of mathematical theory has been found to have important applications in the study of coding schemes. From the previous discussion, it follows that if we have two legal code words c_1 and c_2, then

$$[H]\,[c_1 \oplus c_2]^t = [H]\,[c_1]^t \oplus [H]\,[c_2]^t = 0 \tag{11.12}$$

Hence $[c_1] \oplus [c_2]$ must also be a legal code word. In a similar way we find other relationships between the words, and it can be established that all the legal words in a Hamming code satisfy the following conditions:

(i) the sum of any two code words is itself a code word
(ii) addition of words is associative

$$[A] \oplus \{[B] \oplus [C]\} = \{[A] \oplus [B]\} \oplus [C] \tag{11.13}$$

(iii) there is a unique word, consisting entirely of zeros, which when added to any other word leaves it unchanged

$$[A] \oplus [0] = [0] \oplus [A] = [A] \tag{11.14}$$

(iv) every code word has a unique inverse, which when added to it results in the zero word

$$[A] \oplus [A_i] = [0] \tag{11.15}$$

In mathematical terms these four relations form the defining properties of a group under modulo-2 addition, and enable us to apply the theory of groups to coding problems.

Cyclic codes. It is found that codes exist in which each code word is a shifted version of another code word. That is, if $a_1, a_2, a_3, \ldots a_n$ is a member of the code, so is $a_n, a_1, a_2, \ldots a_{n-1}$. This property can lead to a significant simplification in the process of correcting these codes, because it is now only necessary to devise a scheme for dealing with one single-digit position. All other digits can then be dealt with by shifting the word until they occupy the appropriate position. As this suggests, processing of these codes can often be accomplished by shift register systems, and it is found that there are associations between the codes and the repetitive sequences that can be generated by feedback shift registers. These ideas also lead to an algebraic approach to coding theory, that we cannot include in this text. Readers should consult the standard texts on coding for information on this topic. The Bose-Chaudhuri-Hocquenghem (BCH) codes are an important class of cyclic codes that are frequently encountered in works on this topic.

11.6 Spectrum shaping

The need for spectrum shaping arises in two ways. The first of these arises through attempts to match the signal to the characteristics of the transmission path. In general, transmission paths are linear and may be most easily described in terms of their frequency responses. These are almost invariably limited, having both upper and lower limits beyond which no effective transmission takes place. The average energy spectrum of the signal may be estimated by the autocorrelation techniques described in Chapter 10. (Note, however, that some further discussion of this topic will appear later in this Section.) Any failure to transmit the complete spectrum is equivalent to subtracting the missing components from the received signal, and hence appears as a distortion. Clearly, if the total amount of the spectral energy lost is small, the resulting distortion will also be small. The minimum bandwidth required to transmit the signal without degradation is, as is well known, defined by the Nyquist bandwidth criterion (Nyquist[8]). However, it is very important to note what Nyquist did not define, since his work is frequently assumed to imply greater restrictions than actually exist. He did not place any restrictions on the way in which the signal energy is distributed within the band, and so it is quite permissible to employ signals where the energy is zero at the limiting frequencies defining the band edges, is low, or zero outside the band, and may also be low or zero at points within the transmitted band. He did not specify that the signals had to be sampled at a uniform rate, and he did not consider the implications of treating the received waveform as a multilevel signal rather than a binary signal. He only considered the transmission of the signal elements, and did not consider the limits imposed on the transmission of the information carried by these elements. In fact, Shannon's work reveals that the bandwidth required for a given rate of information transmission can be arbitrarily small, provided the noise level is low enough.

The second reason for controlling the signal spectrum arises from the need to minimise interference between different channels. This is almost invariably caused by the high-frequency end of the signal spectrum. One example arises in the case of baseband transmission over balanced pair cables. As discussed in Chapters 6 and 7 the crosstalk between different pairs on such cables increases with increasing frequency. On one given pair, the interference from systems utilising other pairs in the cable will appear as a further noise contribution. The total interference power can be significantly reduced by minimising the high-frequency content of the transmitted signals. In a similar manner, in the case of radio and waveguide systems employing modulated carrier transmission, the high-frequency part of the modulating signal spectrum can contribute to interchannel interference.

11.6.1 Techniques for spectrum shaping

We shall briefly consider three problems:

(i) the reduction of low-frequency spectral energy
(ii) the reduction of high-frequency spectral energy
(iii) the introduction of spectral nulls within the band occupied by the signal.

As a preliminary to this we must first return to the problem of obtaining an estimate of the spectral distribution of a digital signal, which was introduced in Chapter 10. We shall be mainly concerned with the spectral energy density of a signal $w(f)$ which may be defined as the limit of the average power of the signal per hertz within a small band Δf as $\Delta f \to 0$. That is

$$w(f) = \lim_{\Delta f \to 0} \frac{av\, E^2\,(f,\, \Delta f)}{R\, \Delta f} \tag{11.16}$$

where $E(f, \Delta f)$ is the r.m.s. voltage that would appear at the output of a filter of bandwidth Δf centred on frequency f. (For simplicity we will from now on assume that $R = 1\cdot0$.) In many cases $w(f)$ may be derived directly from the Fourier transform of the signal. To avoid difficulties in the formal analysis we first assume that the signal $s(t)$ is zero outside a finite time interval $-T/2$ to $+T/2$. Then the corresponding frequency-domain signal is given by

$$S(f) = \int_{-\infty}^{\infty} s(t)\, e^{-j\omega t}\, dt$$

$$= \int_{-T/2}^{T/2} s(t)\, e^{-j\omega t}\, dt \tag{11.17}$$

By applying Parseval's theorem the average power over the interval $-T/2$ to $+T/2$ is found to be

$$P = \frac{1}{T} \int_{-T/2}^{T/2} s^2(t)\, dt$$

$$= \frac{1}{T} \int_{-\infty}^{\infty} |S(f)|^2\, df \tag{11.18}$$

But from the definition of the spectral density

$$P = \int_{-\infty}^{\infty} w(f)\, df$$

and hence

$$w(f) = \frac{|S(f)|^2}{T} \tag{11.19}$$

As an example, we will consider a random pulse train comprising a sequence of time slots of duration T, in each of which the signal may take up either the value $+V$ or $-V$, with a probability of $0\cdot5$ in each case. We can then use the period T to estimate the spectral density, since in each such period there will either be a pulse of amplitude $+V$ and duration T, or the corresponding pulse of amplitude $-V$. Thus

$$w(f) = \frac{1}{T} \left\{ 0\cdot5 \left(VT\, \frac{\sin \frac{\omega T}{2}}{\frac{\omega T}{2}} \right)^2 + 0\cdot5 \left(-VT\, \frac{\sin \frac{\omega T}{2}}{\frac{\omega T}{2}} \right)^2 \right\}$$

$$= V^2 T \left(\frac{\sin \frac{\omega t}{2}}{\frac{\omega T}{2}} \right)^2 \tag{11.20}$$

As a slightly more complex example let us now suppose that each time slot may contain either a pulse of amplitude V and duration t_1, with probability 0·5, or contain no pulse. To deal with this we first divide the signal into two components, a regular pulse train of amplitude $V/2$ and duration t_1, and a corresponding symmetrical random pulse train with signals of amplitude $+V/2$ or $-V/2$ as shown in Fig. 11.1. The random component may now be handled in a very similar manner to the previous example, yielding the continuous spectral energy density

$$w_1(f) = \frac{1}{T}\left(\frac{Vt_1}{2}\frac{\sin\frac{\omega t_1}{2}}{\frac{\omega t_1}{2}}\right)^2 \tag{11.21}$$

The periodic component may be evaluated by the normal Fourier series analysis as

$$w_2(f) = \sum_{n=-\infty}^{\infty}\left(\frac{Vt_1}{2T}\frac{\sin n\pi\frac{t_1}{T}}{n\pi\frac{t_1}{T}}\right)^2 \tag{11.22}$$

This example has revealed an important feature of the spectra of digital sequences, namely, that they may contain both a continuous energy distribution and a series of discrete frequency components, or spectral lines.

Now, as discussed in Chapter 10, it is often found to be simpler to evaluate the spectral density of a random signal through the intermediary of the autocorrelation function $R(\tau)$.

This is defined as

$$R(\tau) = \lim_{T\to\infty}\frac{1}{T}\int_{-T/2}^{T/2} s(t)\,s(t+\tau)\,dt \tag{11.23}$$

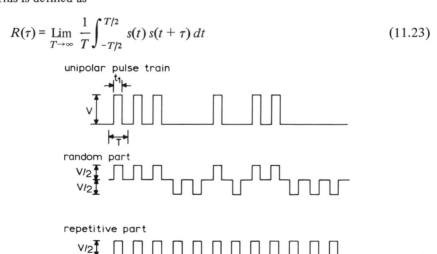

Fig. 11.1 Partitioning of a random unipolar pulse train

and is found to have the following properties:

(i) $R(0)$ = the total signal power P
(ii) $R(0) \geqslant R(\tau)$ for all τ
(iii) $R(\tau) = R(-\tau)$ for a stationary random signal
(iv) If $f(t)$ is periodic $R(\tau)$ is periodic with the same period.
   That is, if $f(t)$ repeats at intervals of nt, then $R(\tau)$ repeats at intervals of $n\tau$
(v) The autocorrelation function and the spectral power-density function form a Fourier transform pair.

This last property is the most significant, and is frequently referred to as the Wiener-Kinchin relations. Remembering, as we have demonstrated above, that the autocorrelation function may be expressed as the sum of a nonperiodic component $R_1(\tau)$ and a periodic component $R_2(\tau)$, repeating at intervals T, we find that the continuous part of the spectral density function is

$$w_1(f) = \int_{-\infty}^{\infty} R_1(\tau)\, e^{-j\omega\tau}\, d\tau \tag{11.24}$$

For stationary signals due to the symmetry of $R(\tau)$

$$w_1(f) = \int_{-\infty}^{\infty} R_1(\tau) \cos \omega\tau\, d\tau \tag{11.25}$$

The periodic part $R_2(\tau)$ then leads to a line spectrum

$$w_2(f) = \sum_{n=-\infty}^{\infty} a_n^2 \cos 2\pi n t/T \tag{11.26}$$

where

$$a_n^2 = \frac{1}{T} \int_{-T/2}^{T/2} R_2(\tau)\, e^{-j2\pi n\tau/T}\, d\tau$$
$$f = n/T$$

In passing, we should note that while so far we have defined the spectral density in terms of f, we could equally express it as a function of the angular frequency ω and eqn. 11.24 indicates that this will in fact be identical to the previous expression. At first sight this is a little unexpected, since we would anticipate that the power density in terms of ω would differ from the density in terms of f. The explanation is that the difference actually arises at the next stage, when the spectral density is integrated over a finite bandwidth to evaluate the actual power. We then have

$$P = \int_{f_1}^{f_2} w(2\pi f)\, df$$
$$= \frac{1}{2\pi} \int_{\omega_1}^{\omega_2} w(\omega)\, d\omega \tag{11.27}$$

As an example of the application of this technique we will evaluate the power density spectrum of sequence of 'top-hat' pulses, as defined in Fig. 11.2, where the probability of occurrence of a pulse in any given time slot of duration T is p.

As in the previous example, we subtract a periodic component of amplitude p, which leaves either a signal of amplitude $(1 - p)$ occurring with probability p, or a signal of amplitude $-p$, occurring with probability $(1 - p)$. This is equivalent to a random signal of amplitude $+1$ scaled by a factor $p(1 - p)$. In this case, the evaluation of the autocorrelation function is simplified by noting that since we are dealing with rectangular pulses, the function $R(\tau)$ will consist of a series of straight-line segments joining values of τ corresponding to coincidence of pulse transitions. Thus, we need only evaluate $R(\tau)$ at $\tau = 0, t_1, 2t_1, 3t_1$.

The corresponding values for $-\tau$ are identical due to the symmetry of the signal. The resulting autocorrelation function is shown in Fig. 11.3. This can be partitioned into three triangular components, as shown, and the spectral-density function is then obtained by summing the Fourier transforms of these components. This yields

$$w_1(f) = V^2 t_1 p(1 - p) \left\{ \frac{5}{4} \left(\frac{\sin \pi t_1 f}{\pi t_1 f} \right)^2 - 2 \left(\frac{\sin 2\pi t_1 f}{2\pi t_1 f} \right)^2 + \frac{3}{4} \left(\frac{\sin 3\pi t_1 f}{3\pi t_1 f} \right)^2 \right\}$$

which simplifies to

$$\frac{V^2 t_1 p(1 - p)}{3} \left(\frac{\sin \pi t_1 f}{\pi t_1 f} \right)^2 (1 - \cos 2\pi t_1 f)^2 \tag{11.28}$$

The repetitive section of the function, also shown in Fig. 11.3, is rather simpler, and transforming this leads to the line spectrum.

$$w_2(f) = \frac{pV^2}{4} \left(\frac{\sin \frac{n\pi}{3}}{\frac{n\pi}{3}} \right)^2 \cos (2n\pi f t) \tag{11.29}$$

The reader may find it interesting to attempt to repeat the above example for the more general case of a top-hat waveform with segments of unequal durations. He will find that the estimation of the autocorrelation function is then far from being a simple matter. It can also be seen that difficulties will arise if we depart from

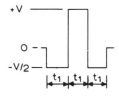

Fig. 11.2 Top-hat pulse

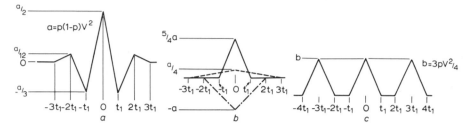

Fig. 11.3 Autocorrelation analysis of a random sequence of 'top-hat' pulses
   a Autocorrelation function of random part of signal
   b Partitioned autocorrelation function of random part
   c Autocorrelation function of repetitive part

rectangular pulse shapes, to more general forms, such as raised cosine pulses. Cases of this type can be handled by treating the random pulse train as the convolution of a random train of impulses with a single pulse of the required waveform $s_0(t)$. That is

$$s(t) = \left\{ \sum_{n=1}^{\infty} a_n \, \delta(t - nT) \right\} * s_0(t) \tag{11.30}$$

where a_n takes the values $0,1$ or $-1, +1$.

Now the convolution of two functions of time is equivalent to forming the product of the corresponding frequency-domain functions, and so we may derive the required spectral density by first calculating the autocorrelation function of the series of impulses $R_d(\tau)$. Secondly, obtaining the corresponding spectral density by the Fourier transform of this and thirdly, multiplying this by the squared modulus of the Fourier transform of the pulse waveform.

This yields,

$$w(f) = \int_{-\infty}^{\infty} R_d(\tau) \, e^{-j\omega\tau} \, d\tau \, |G_0(f)|^2$$

$$= w_d(f) \, |G_0(f)|^2 \tag{11.31}$$

As an example, we take the case of a bipolar train of raised cosine pulses with the probability of a pulse in any one time slot being $0 \cdot 5$. The bipolar signal is formed by taking a random binary pulse train and inverting alternate marks, as shown in Fig. 11.4.

The corresponding impulse train and its autocorrelation function are also shown in Fig. 11.4. It follows immediately that the spectral density corresponding to this autocorrelation function is

$$w_d(f) = \frac{1}{T} \, (1 - \cos 2\pi f T) \tag{11.32}$$

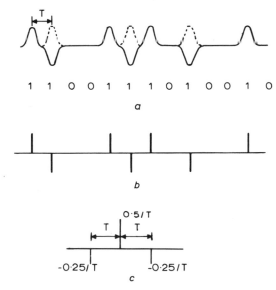

Fig. 11.4 Bipolar coding
 a A bipolar pulse sequence
 b Equivalent impulse train
 c Autocorrelation function of impulse train

and the resulting spectrum for raised cosine pulses is then

$$w(f) = \frac{1}{T}(1 - \cos 2\pi f T)\left\{At_0 \frac{\sin 2\pi f t_0}{2\pi f t_0 (1 - 4f^2 t_0^2)}\right\}^2 \tag{11.33}$$

It should be noted that this simple expression only applies when the probability of a pulse in any time slot is 0·5. In other cases, the situation becomes more complex, and the reader is referred to the detailed analysis given by Bennett and Davey[9] for the solution to this case.

11.6.2 Reduction of low-frequency energy
In the previous Section, it was shown that the spectral-density function of a digital signal could be considered as the product of two terms, one originating from the nature of the pulse sequence as expressed through its autocorrelation function, and the other originating from the waveshapes of the component pulses in the sequence.

It therefore follows that the energy in the low-frequency portion of the spectrum can be reduced either by choice of a suitable pulse shape, or by some modification of the nature of the pulse sequence. Both of these techniques have in fact been introduced in the examples given in the last Section. The 'top-hat' pulse demonstrates one possible pulse shape which would yield zero energy at direct current. The general form of this signal is shown in Fig. 11.5. The energy spectrum of this pulse is

$$w_p(f) = \left\{ \frac{V_1}{\pi f(T - t_1)} \left(T \sin \pi f t_1 - t_1 \sin \pi f T \right) \right\}^2 \tag{11.34}$$

For the case of $V_1 = V_2$, which is particularly easy to implement, this gives

$$w_p(f) = \left\{ \frac{2V_1}{\pi f} \sin \frac{\pi f T}{2} \left(1 - \cos \frac{\pi f T}{2} \right) \right\}^2 \tag{11.35}$$

Another related pulse form is the dipulse shown in Fig. 11.5. The energy spectrum of this for the case of $V_1 = V_2$ is

$$w_p(f) = \left\{ V_1 T \left(\frac{\cos \pi f T - 1}{\pi f T} \right) \right\}^2 \tag{11.36}$$

These spectra are plotted in Fig. 11.6, where they are compared with a random binary pulse train of the same peak-to-peak amplitude. The use of the sequence characteristics to introduce a d.c. null has also been introduced in the example of bipolar (or alternate mark inversion) coding. It is clear that an infinite number of ternary codes of this general type can be devised. For example, we could arrange for n successive marks to have the same polarity before a reversal occurred, or we could devise some more complex reversal convention. Qualitatively, we can see that long runs of marks of the same polarity will lead to an increase in the low-frequency energy, even though the d.c. value remains zero. We can obtain a measure

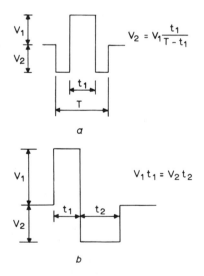

Fig. 11.5 Pulse waveforms giving a d.c. spectral null
    a 'Top-hat' pulse
    b Dipulse

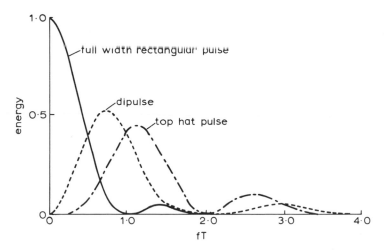

Fig. 11.6 Energy spectra of pulses

of the l.f. performance of a code through its digital sum variation (d.s.v.). This is the peak-to-peak excursion of the algebraic running sum of the code symbol amplitudes.

Partial-response encoding offers another method of introducing a d.c. spectral null. This technique, which has been described in detail by Kretzmer,[10] involves replacing each mark in the incoming binary signal by a sequence of multilevel symbols in a number of time slots. In one sense, this may be regarded as a process of replacing each mark by a new more complex pulse shape, but since the conversion can be carried out by digital operations on the signal sequence, it is also fair to consider it as a recoding operation. The partial response process may be utilised for both high-frequency and low-frequency spectral shaping. The variant of particular interest in the present instance is known as 'class 4 partial response encoding' and involves replacing each mark by the sequence $1, 0, -1$.

11.6.3 Reduction of high-frequency energy

Direct curtailment of the frequency spectrum by modifying the waveforms of the transmitted pulses is not satisfactory, because this simply increases the intersymbol interference. This leaves the possibility of modifying the message-symbol sequence. At first sight, this seems to be an unpromising line of attack, because, superficially, the Nyquist criterion appears to rule out any chance of reducing the required bandwidth. However, as pointed out earlier, the Nyquist limit was devised in terms of binary transmission, and does not fully cover the effects of conversion to multilevel signals. Any code which directly, or indirectly, involves such a conversion may therefore effect some saving in required bandwidth. Direct conversion to multilevel symbols at a reduced rate is obvious and needs no further discussion. Our main interest is in the indirect schemes where the existence of the multilevel conversion may not be readily seen. The best known example of such codes is the Duobinary scheme devised by Lender.[11] In this arrangement, the signal is first recoded, as shown in Fig. 11.7 so that spaces become binary transitions, and marks are represented by no transition. If this signal is now transmitted over a system having a bandwidth approximating to 1/4 of the digit rate (i.e. approaching half the Nyquist bandwidth) it is found that when a space follows a mark a slow signal transition takes place, but if a second space then arises, this is offset by the reverse transition before it can be completed, as shown in Fig. 11.7. The end result is that marks appear at the receiver as positive or negative peak signal excursions, while spaces always appear as intermediate levels. Various other coding schemes of this type have now appeared, notably the partial-response schemes mentioned in the previous Section. Lender has also produced an extended version of his system employing more than three levels, under the name of polybinary coding.[12] An excellent account of the theory of such codes is given in Bennett and Davey,[9] whose treatment lends perspective to the source references and is therefore preferable as an introduction.

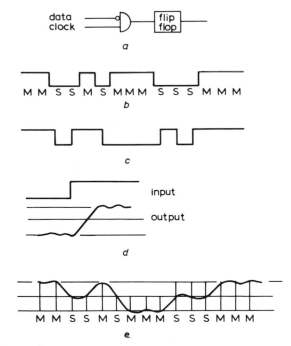

Fig. 11.7 Duobinary coding
    a Coder
    b Input sequence
    c Duobinary sequence
    d Channel step response
    e Output from duobinary sequence

11.6.4 Introduction of spectral nulls

It is also possible to carry out the recoding process in a manner that introduces zeros in the energy spectrum at a finite number of points within the transmitted band. A simple example of this is higher-order bipolar coding.[13] In this scheme, the incoming binary pulse train is treated as N interleaved digit streams, and each of these is independently encoded into the bipolar format. The output is then N interleaved bipolar signals, as shown in Fig. 11.8. From eqn. 11.33 we see that each of these interleaved signals must have the spectral distribution

$$w(f) = f_d(1 - \cos 2\pi N f/f_d)\,\{G(t)\}^2 \qquad (11.37)$$

where $G(t)$ is the spectral distribution of the pulse shape used. Summation cannot introduce any new components, and so the composite signal must have spectral nulls at multiples of f_d/N where f_d is the digit-repetition frequency.

11.7 Timing information

The timing content of digital sequences has already been dealt with in Chapter 10, and requires no further comment. The discussion on spectral shaping in Section 11.6 has, however, revealed one other possibility. Because it is possible to devise coding schemes with zeros in the energy spectrum at points within the transmitted band, it follows that it would be possible to insert sinusoidal tones at these points which could be extracted and used for retiming at the regenerators. As far as we are aware, this arrangement has not yet been employed in any installed system.

11.8 Signal-amplitude information

In the Chapters dealing with transmission media, the variable nature of the transmission loss was discussed, and we will later see, in the discussion of regeneration, that it is highly desirable to present the incoming signal to the regenerator-threshold detector at a constant amplitude. In practice, this requires the use of some variable equaliser in the input preamplifier of the regenerator, this in turn must be controlled by the incoming signal amplitude, and a measure of this amplitude must be available. The code must therefore be designed so that some parameter associated with the signal amplitude remains substantially constant, irrespective of the message sequence being transmitted. One possibility is to design the code so that peak amplitude excursions must occur at relatively short intervals. Another approach would be to arrange that the short-term r.m.s. value remained constant.

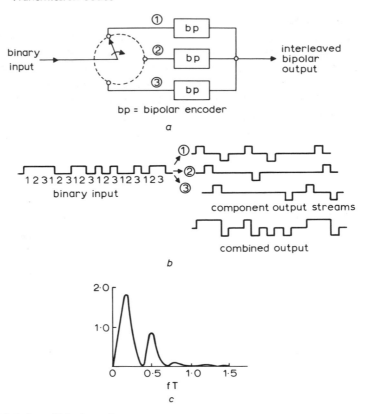

bp = bipolar encoder

a

1 2 3 1 2 3 1 2 3 1 2 3 1 2 3 1 2 3
binary input

component output streams

combined output

b

fT

c

Fig. 11.8 Interleaved bipolar code
    a Implementation
    b Coding process
    c Energy spectrum

11.9 Framing content

It is quite possible that the transmission encoding process will introduce a further frame structure into the system. For example, we shall later encounter transmission codes which convert groups of four binary digits into groups of three ternary digits, and so the beginning of each ternary group must be identified at the decoder. While it would, in principle, be possible to introduce frame-alignment words, as discussed in Chapter 5, when multiplexing was considered, this tends to be an unattractive solution in the case of line coding; and the codes so far used have relied on detecting certain symbol patterns which define the boundaries of the code words. For example, if we arrange the code to avoid using the pattern of three zeros as a ternary word, the occurrence of this pattern must span a word boundary, and so gives an indication of the location of this boundary.

11.10 Error extension

It is possible for a single error arising during transmission to produce a number of errors in the final output, due to interactions in the decoding operations at the receiving terminal. We have already seen one example of this in the case of the data scramblers described in Chapter 10. Similar types of error-extension mechanisms can arise in some types of transmission codes. The block codes mentioned above will also be prone to such effects, because an error in one of the ternary digits is likely to affect more than one of the decoded binary digits. As far as possible, codes should be designed to minimise these effects.

11.11 Practical transmission codes

We conclude this Chapter by briefly describing some of the more important transmission codes that have been developed for digital transmission.

11.11.1 Error-correcting codes

A good example of a general-purpose binary-error detecting or correcting code is given in CCITT recommendation V41 (Green Book vol. VIII). It is intended for use on low-speed data transmission systems, and has various options that cover a wide variety of transmission requirements. It is a block code, the data being split up into blocks of either 220, 460, 940, or 3820 digits, according to the transmission medium and expected transmission path delay. Four service digits are added to the beginning of each block, which is then transmitted and simultaneously fed into a 16-stage shift register, initially set to the all-zero state. The input and the outputs of stages 5, 12 and 16 are combined by modulo-2 addition, as shown in Fig. 11.9. At the end of the block the contents of the shift register are appended as a check word, by closing gates A and B and opening gate C.

At the receiving terminal the signal is fed into a similar shift register. It will be clear that at the end of the data block the contents of this register will be identical to the incoming check digits, provided no transmission errors have occurred. At this point, the contents of the register are added, modulo-2, to the incoming check word. The result will be a sequence of 16 zeros, unless an error has appeared. The first option is to use this as a simple error-detection scheme and demand retransmission of the complete block, if required. As a second option, the error correction may be carried out at the receiving terminal. Full details of this are given by Rowland.[14] The additional circuit arrangements needed are shown in Fig. 11.10. In this case, at the end of the block gate 1 is opened and gate 2 is closed, and the shift register incremented by between 31787 and 32507 shifts, dependent on the data block length being used. The contents of the buffer store B are then cycled from output to input, while the shift register continues to be incremented. Occurrence of the zero state in stages 1–15 of the shift register signifies that the error has been

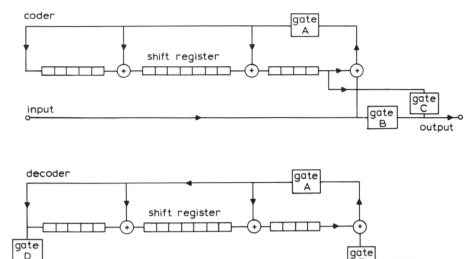

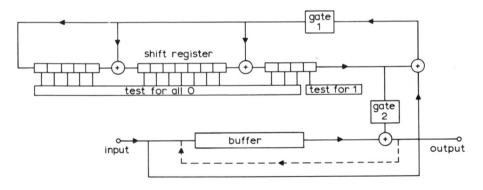

Fig. 11.9 Error detecting code and decoder for V41 system

Fig. 11.10 Single-bit error correcting decoder for V41 system

located. To correct the error gate 1 is closed and gate 2 is opened during the next shift pulse. Any further all-zero state in the shift register before the end of the cycle indicates that more than one error has occurred. Because this scheme is only capable of correcting single errors (although it can detect multiple errors) a retransmission must then be requested.

11.11.2 Restricted binary codes
The error-correcting code described in the previous Section represents one example of the general class of binary codes formed by taking a block of n message digits

and converting this into a new block of m digits for transmission, m being greater than n. In the example described, the message digits were not changed, and the recoding was carried out by systematic logical operations. In the more general case, any m digit pattern may be associated with any of the 2^n possible input patterns, and although some logical conversion process is to be preferred, a look-up table could be used, thus avoiding the need for any systematic choice in the pattern associations. Since 2^{m-n} of the possible 2^m patterns will not be used, it is possible to select the code alphabet to give a balanced mix of desirable properties. One possibility is to choose only code words containing a constant number of marks, thereby producing a code having a constant d.c. component, which for practical purposes means that a low-frequency cut-off may be allowed in the transmission path without introducing significant transmission degradation. The constant mark content also leads to a good timing content, and a fair measure of error-detecting capability. Codes of this type have been described by Cattermole[15] and Neu.[16]

11.11.3 Bipolar codes

The disadvantage of restricted binary codes is that they increase the line-symbol rate, with consequent repercussions on the overall system. During the work on early p.c.m. systems in the period around 1960, workers at the Bell Telephone Laboratories realised that in the case of baseband systems, the use of 3-level (ternary) codes, comprising marks, spaces and negative marks, involved comparatively little added complexity in the equipment, and made it possible to include the required redundancy without increasing the symbol rate. The code they developed, which is still in widespread use, was known as bipolar, or alternate mark inversion (a.m.i.) code. The coding process is simply the choice of alternate positive and negative pulses to represent marks, as shown in Fig. 11.4, and decoding can be performed by rectification. Intermediate regenerators must employ two threshold detectors and two output pulse generators to process both positive and negative marks, but are otherwise identical to simple binary regenerators. The immediate effect of this operation is to eliminate the d.c. component of the signal. The mathematical analysis of the energy spectrum of a.m.i. code has already been given in Section 11.6 for the case of equal probabilities of marks and spaces. In passing we note that in the general case, with a probability p of a mark occurring the energy spectrum is

$$w(f) = \frac{8p(1-p) f_d |G(f)|^2 \sin^2 \pi f/f_d}{1 + (2p-1)^2 + 2(2p-1) \cos 2\pi f/f_d} \tag{11.38}$$

The bipolar code, at least to some extent, satisfies all of the conditions listed at the beginning of this Chapter. It is certainly transparent and uniquely decodable. Its efficiency is perhaps rather low, since each ternary symbol is capable of transmitting $\log_2(3) = 1 \cdot 6$ bits, while the code actually carries one bit per symbol, yielding a redundancy of 60%. There is an inbuilt error-detecting capability, since errors will introduce violations of the sign alternation of the mark. The energy spectrum, as previously discussed, is certainly favourable, and the timing content and peak signal content are no worse than for an unprocessed binary signal.

Compatible high-density bipolar code. Practical experience has shown that the most serious shortcoming of a.m.i. code is the lack of timing information when signal patterns containing a low mark density are transmitted, and, as a result, various attempts to modify a.m.i. to improve this aspect of its performance have been made. The most widely accepted solutions are the compatible high-density bipolar codes (c.h.d.b.) proposed by Croisier.[17] The basic idea is that when a run of more than n zeros occurs, the $n + 1$th zero is replaced by a mark, to increase the timing content. To identify this mark as a substitution it is inserted with the same polarity as the preceding mark, thereby violating the alternate mark-inversion rule. However, this in itself is not sufficient, since it would then be possible for a series of insertions to occur, all having the same polarity, and thus reintroducing a d.c. component into the signal. To deal with this, Croisier proposed a further modification, which forces the a.m.i. violations to alternate in polarity. In this, the coder keeps a running check on the situation, and if two violations of the same polarity are called for, a double substitution is made. The first zero in the run of $n + 1$ zeros is replaced by a mark which obeys the a.m.i. rule, and the $n + 1$th zero is then replaced by a mark of the same polarity as the last *transmitted* mark. The decoder then has to check for two parameters. First, for an a.m.i. violation, and secondly, for the number of zeros preceeding this violation, to determine if the last transmitted mark is also a substitution. We note that as a result of this we have lost the attribute of instantaneous decodability, but the delays involved are small, and in all other respects the characteristics of the code have been improved. An example of c.h.d.b. coding is shown in Fig. 11.11.

It might be thought that the error-detecting capability of the code has been substantially reduced by this modification, because a bipolar violation no longer gives a unique indication of the occurrence of an error. A little thought will show that this is not the case. Any single error will insert a spurious bipolar violation, or will delete one of the deliberate violations. This will become apparent when at the next violation, the alternation of violations does not appear. A check on this feature will therefore provide a single error-detecting capability.

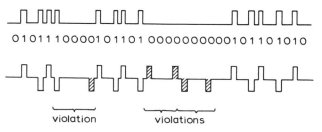

Fig. 11.11 CHDB₃ code

It can be seen that an entire class of these codes exists, the general member being denoted by c.h.d.b.$_n$ code, where n is the maximum allowable run length of spaces. The spectral-energy distributions of these codes will be rather similar to that of simple a.m.i. code, but in detail will be different for each member of the class.

A discussion of this topic is given by Buchner.[18] The energy spectrum for c.h.d.b. with 0·5 probability of a mark is

$$w(f) = \frac{40 - 32 \cos \omega t}{465T(1025 - 64 \cos 5\omega t)}\bigg(7258 \cdot 5 - 1929 \cos \omega t - 1424 \cos 2\omega t$$

$$-160 \cos 3\omega t + 32 \cos 4\omega t$$

$$-\frac{131288 \cdot 5 - 41399 \cos \omega t - 86112 \cos 2\omega t}{85 - 44 \cos \omega t - 24 \cos 2\omega t - 16 \cos 3\omega t}\bigg)|G(f)|^2 \qquad (11.39)$$

11.11.4 4B3T codes

These are codes which attempt to provide similar overall characteristics to the c.h.d.b. codes, while at the same time making more effective use of the available transmission capacity. This is achieved by converting groups of four binary digits into groups of three ternary digits, thereby reducing the redundancy to 20%. The blocks of four binary digits represent 16 possible code words, while the corresponding three ternary symbols can provide 27 combinations. The problem is then to pair the binary and ternary groups so as to provide the desired characteristics, and to do this in a manner that does not present too many difficulties in its practical realisation. As far as we know, all the codes of this type which have received serious attention have employed the important technique of mode alternation. In this approach, a number of alternative ternary alphabets are used to represent the 16 binary signals, the particular alphabet being employed at any time defines the mode of the code, and is determined by some measure of the past history of the signal. Because this past history will be known to the decoder instantaneous decoding is possible. The most common scheme is to organise the mode alphabets to bias the output sequence towards either a positive or negative accumulation of the digital sum variation, and to switch the modes so as to maintain the d.s.v. as close to zero as possible. This has the effect of minimising the low-frequency spectral energy.

As an example, we will consider the MS43 code proposed by Franaszek.[19] This code employs three alphabets, R_1, R_2, and R_3, which are selected according to the state of the digital sum variation of the code. For this code, the running digital sum can vary over 6 values, from 0 up to 5; but the extreme values can only occur during either the first or second elements of a ternary word, so the range of terminal states is restricted to 1–4. If the terminal state is 1 alphabet R_1 is selected for the next code word, if it is 2 or 3 alphabet R_2 is selected, and if it is 4 alphabet R_3 is used.

The composition of these alphabets is given in Table 5. They are arranged so that each group of three ternary symbols is uniquely associated with one group of four

Table 11.5 The MS43 code

Binary words	Ternary alphabets		
	R_1	R_2	R_3
0000	+ + +	− + −	− + −
0001	+ + 0	0 0 −	0 0 −
0010	+ 0 +	0 − 0	0 − 0
0100	0 + +	− 0 0	− 0 0
1000	+ − +	+ − +	− − −
0011	0 − +	0 − +	0 − +
0101	− 0 +	− 0 +	− 0 +
1001	0 0 +	0 0 +	− − 0
1010	0 + 0	0 + 0	− 0 −
1100	+ 0 0	+ 0 0	0 − −
0110	− + 0	− + 0	− + 0
1110	+ − 0	+ − 0	+ − 0
1101	+ 0 −	+ 0 −	+ 0 −
1011	0 + −	0 + −	0 + −
0111	− + +	− + +	− − +
1111	+ + −	+ − −	+ − −

binary digits, and hence it is not necessary for the decoder to identify the mode or alphabet employed by the transmitter. The code allows up to four successive zeros in the ternary sequence, or up to five successive positive or negative marks. The ternary word consisting of three zeros is not used, and so any sequence of three or four zeros must span the boundary between two adjacent words. This feature may be used to achieve frame alignment. Because every word contains one or more marks the timing content is high. A measure of the transmission error rate can be obtained by monitoring the d.s.v. and counting the number of occasions when the theoretical limits of 0 and 5 are exceeded.

A comparison of the spectral-energy distributions for a number of line codes is given in Fig. 11.12. For this comparison the basic signal element is taken as a full-width rectangular pulse. The amplitudes are expressed in terms of the low-frequency spectral density of a random binary signal, as defined by eqn. 11.20.

11.11.5 Other codes
A variety of other binary and ternary codes have been proposed, but as far as we are aware, none of these have been extensively used in practical systems. A list of codes compiled by Mr. R.M. Dorward is given in the Appendix. Extension to codes employing more than three transmitted levels has also received some attention.

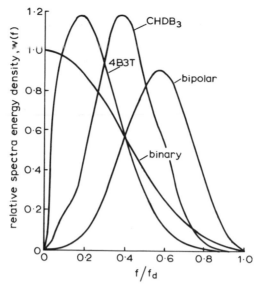

Fig. 11.12 Energy spectra of transmission codes

A scheme of this type has been adopted by the American Bell System for the transmission of high-speed digital signals over analogue transmission plant.[20]

11.12 References

1   SHANNON, C.E.: 'A mathematical theory of communication', *Bell Syst. Tech. J.*, 1948, 27, pp. 379–423 and pp. 623–656
2   SHANNON, C.E.: 'Communication in the presence of noise', *Proc. IRE*, 1949, 37, p. 10
3   SCHWARTZ, M.: 'Information transmission; modulation and noise' (McGraw-Hill, 1970)
4   PETERSON, W.W.: 'Error correcting codes' (MIT and Wiley, 1961)
5   BERLEKAMP, E.R.: 'Algebraic coding theory' (McGraw-Hill, 1968)
6   HAMMING, R.W.: 'Error detecting and correcting codes', *Bell Syst. Tech. J.*, 1950, 29, pp. 147–160
7   LUCKY, R.W., SALZ J., and WHELDON, E.J.: 'Principles of data communication' (McGraw-Hill, 1968)
8   NYQUIST, H.: 'Certain topics in telegraph transmission theory', *Trans. AIEE*, 1928, 47, pp. 617–644
9   BENNETT, W.R., and DAVEY, J.R.: 'Data transmission' (McGraw-Hill, 1965)
10 KRETZMER, E.R.: 'Generalisation of a technique for binary data communication', *IEEE Trans.*, 1966, **COM-14**, pp. 67–68
11 LENDER, A.: 'The duobinary technique for high speed data transmission' *AIEE Commun. & Electron.*, 1963, 82
12 LENDER, A.: 'Correlative digital communication techniques', IEEE International Convention Record, 12, Pt 5 1964
13 AARON, M.R.: 'P.C.M. transmission in the exchange plant', *Bell Syst. Tech. J.*, 1962, 41, pp. 99–141
14 ROWLAND, R., and JONES, A.C.: 'Error detection and correction in commercial systems', IEE Vacation School on 'Signal processing', Hatfield, 1974

15   CATTERMOLE, K.W.: 'Low disparity codes and coding for p.c.m.', IEE Conference on 'Transmission aspects of communication networks', 1964, pp. 179–182

16   NEU, W., and KUNDIG, A.: 'Project for a digital telecommunications network', *IEEE Trans.*, 1968, **COM-16**, p. 633

17   CROISIER, A.: 'Introduction to pseudoternary transmission codes', IBM J. *Res. & Dev.*, 1970, **14**, pp. 354–367

18   BUCHNER, J.B.: 'Ternary line signal codes', International Seminar on 'Digital communications', Zurich, 1974

19   FRANASZEK, P.A.: 'Sequence state coding for digital transmission', *Bell Syst. Tech. J.*, 1968, **47**, pp. 143–157

20   GUNN, J.F.: 'Mastergroup digital transmission on modern coaxial cable systems', *Bell Syst. Tech. J.*, 1971, **50**, pp. 501–520

Chapter 12

Design of baseband systems

12.1 Capacity and line rate

Digital line transmission systems have to justify their existence in competition with highly developed analogue services which use multichannel f.d.m. techniques. It is instructive to pay regard to the contrasting features of the two methods; in Table 12.1 we present several p.c.m. and f.d.m. transmission formats in terms of their channel capacity and either their top frequency limit or their line symbol rate, as appropriate. What is immediately apparent is the significantly higher frequency at which the line has to be worked in the case of the p.c.m. systems, even assuming that the p.c.m. system bandwidth is confined to half the information rate. Reference to the attenuation characteristics of twisted pair and coaxial lines given in Chapter 7 shows an alarming disproportion in the loss that the two types of system must anticipate for the same repeater section length. For example, both the 12 MHz f.d.m. system and the 120 Mbit/s digital system use a repeater spacing of 2 km on small-bore 4·4 mm coaxial cable; the loss of the repeater span at 12 MHz is typically 37 dB compared to 82 dB at half the p.c.m. system rate of 60 MHz. It is clear, therefore, that to compete effectively with f.d.m. systems of equivalent capacity, the digital systems require a much larger range of equalised amplification. This would, at first sight, seem to defeat one of the objectives of digital systems, namely their apparent simplicity; however, this superficial anomaly is largely resolved when it is realised that the repeater gain of the digital system, although significantly higher, may be allowed a significantly larger tolerance from nominal. Where in an f.d.m. system accuracies of 0·01 dB per repeater section may be required, an allowance of 1 dB may suffice for the digital system.

The transmission line as a channel
Any communication channel imposes its own peculiar constraints and characteristics. The transmission line has several features of both a theoretical and practical nature, which it is useful to identify at the outset.

(i) *Loss characteristic.* As was indicated in Chapter 6, the loss in logarithmic units

Table 12.1 Channel capacity of some f.d.m. and p.c.m. line systems

f.d.m.		p.c.m.	
number of channels	upper frequency	number of channels	information rate
	kHz		kbit/s
1[3]	3·4	1[3]	64
60[1]	240	30[5]	2048
960[1]	4028	120[5]	8448
2700[1]	12×10^3	1680[2]	120×10^3
10800[1]	60×10^3	7680[4]	565×10^3

Status in 1975:
1 — in service   2 — in development, UK
3 — single-channel requirement   4 — proposed
5 — specified for service

such as decibels rises predominantly as $\sqrt{f}$ and proportionately with distance.

(ii) *Low-frequency cut-off.* It has been standard practice to power sections of line containing a number of repeaters from power-feeding stations using the signal path itself for the direct current flow. This means that the actual signal has to be a.c. coupled, this being normally realised with transformers at the input and output of a repeater. In addition, other a.c. connections are likely within the repeater circuits. Practical considerations set a lower limit to the cut-off frequency of the a.c. couplings and this restricts the choice of line code.

(iii) *Crosstalk.* In cables made up of a large number of twisted-pair lines any one line will be subjected to crosstalk from its neighbours and this may contribute a dominant source of disturbance. Crosstalk characteristics were discussed in Chapter 6.

(iv) *Impulsive interference.* In some systems the transmitting terminal is situated in an environment rich in sources of impulsive interference generated by electro-mechanical equipment. Transmission terminals housed in telephone exchanges are typical examples; little predictable design can be or has been attempted in this context except to purposely shorten the end repeater-section spans.

(v) *Echoes.* Any line system is susceptible to discontinuities resulting in reflections or echoes at the sending end, as discussed in Chapter 7, and these must be included in the total noise and interference allocation in the system design.

(vi) *Compatibility.* Transmission lines have been used for many years as media for other, nondigital services. These have evolved historically into a reasonably standard format, particularly regarding spacing between manholes or other access points. Thus, where new digital systems are intended to work over existing cables there is a strong incentive to make the most use of the existing access points for repeater housing. In such systems, repeater spacing is a discretely rather than continuously variable design parameter.

12.2 Basic design procedure

The fundamental features of a digital line system are best illustrated by going through the steps involved in the design of such a system. Fig. 12.1 presents a simplified model of a digital line repeater section; the basic functions of regeneration and signal shaping have been treated in Chapter 9.

The fundamental objective of the design process is to compute the repeater output level for a given line symbol rate $1/T$ repeater section length and error rate. These four parameters are the ones which are of primary interest to the designer of the digital link, enabling him to assess the technical feasibility of any one set. They, in turn, enable computation of repeater power consumption, leading to an assessment of the economic feasibility.

With reference to Fig. 12.1, we can write several identities relating to signal spectra. The preamplifier/equaliser response $A(f)$, the cable loss $C(f)$ and the equalisation characteristic $R(f)$ are defined by:

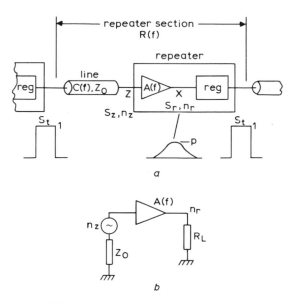

Fig. 12.1 Model of a digital line system
   a Functional model
   b Noise model

$$A(f) = \frac{S_r(f)}{S_z(f)}$$
(12.1)

$$C(f) = \frac{S_t(f)}{S_z(f)}$$
(12.2)

$$R(f) = \frac{S_r(f)}{S_t(f)}$$
(12.3)

The subscripts refer to the nodes of Fig. 12.1.
From which

$$A(f) = C(f)R(f)$$
(12.4)

Eqn. 12.4 simply states that the extent to which the cable loss is equalised by the preamplifier shaping depends on the equalisation characteristic $R(f)$. Complete equalisation would clearly be obtained with $R(f)$ constant, and in such a case the received signal $s_r(t)$, presented to the regenerator decision node, would be identical to the transmitted signal $s_t(t)$.

We proceed to consider the noise present at the node X. The basic noise is generated in the source resistor formed by the characteristic impedance Z_0 of the line. In fact, Z_0 is real only to a first approximation (see Chapter 6). Thus, the mean-square noise per hertz of bandwidth of the equivalent noise source, Fig. 12.1b, is

$$\sigma_z^2 = k\,T\,Z_0$$
(12.5)

where k and T are Boltzmann's constant and the absolute temperature, respectively. This noise is amplified and shaped by the preamplifier whose own circuit-noise performance is defined by the noise figure F, giving a mean-square noise level at its output of

$$\sigma_x^2 = k\,T\,Z_0 \int F\,A^2(f)\,df$$
(12.6)

The noise figure may be sufficiently frequency dependent to warrant inclusion within the integral, as shown.

The peak signal/r.m.s. noise ratio, s.n.r., at the decision node X determines the error probability of the regenerator as described in Chapter 9. Data on the relationship between s.n.r. and P_e may be found in that Chapter; we repeat here the approximation which is useful for hand computation

$$\text{s.n.r.} = 10 \cdot 65 + 11 \cdot 42 \log_{10} x, \text{ dB}$$
(12.7)

where $P_e = 10^{-x}$ for $4 < x < 15$.

The peak signal at X, V_x is therefore

$$V_x = \sigma_x \times \text{s.n.r.}$$
(12.8)

Finally, the peak output signal V_0 is given by

$$V_0 = V_x/p = (\sigma_x/p) \times \text{s.n.r.} \tag{12.9}$$

where p is the single-pulse equalisation loss as illustrated in Fig. 12.1a and depends on $R(f)$ and the manner in which $R(f)$ is specified. Usually, $R(f)$ is normalised to have a maximum value of unity causing p to be less than 1.

V_0 refers to the peak pulse amplitude for binary transmission, and may be interpreted in the general case as the difference between levels of a multilevel line code. The peak-to-peak output voltage is then $(L-1) V_0$, where L is the number of allowed code levels. At this stage, knowledge of the peak voltage requirements will provide the system designer with a necessary first criterion of feasibility for a given set of parameters.

Further examination of the above analysis affords a valuable insight into the behaviour of several design parameters. The noise at the decision node may be expressed in terms of the equalisation response $R(f)$ by substituting eqn. 12.4 into eqn. 12.6

$$\sigma_x^2 = k \, T Z_0 \int F \, C^2(f) R^2(f) \, df \tag{12.10}$$

This demonstrates formally that widening $R(f)$ to obtain a better formed received pulse $s_r(t)$ leads to an increase in noise level. The existence of a compromise solution or trade-off between an acceptable noise level and an acceptable received pulse shape was already mentioned in Chapter 9. We refrain here from using the word 'optimisation' because different users may have perfectly reasonable but different criteria for optimisation. A large volume of work has been published on various aspects of optimisation based on specific or tacit assumptions of varying practical relevance; we do not feel that the results obtained are of significant value.

In the context of line transmission, the trade-off becomes quite critical; in Chapter 6 it was shown that the line loss may be expressed as

$$\log C(f) = l(a_0 + a_1 \sqrt{f} + a_2 f) \tag{12.11}$$

where l is the line-length and a_0, a_1 and a_2 are constants. For the purpose of the present discussion we use the fact that a_1 is the dominant term and we express $C(f)$ by the less accurate expression

$$\log C(f) = l \, a \sqrt{f}$$

or

$$C(f) = e^{la\sqrt{f}} \tag{12.12}$$

where a is a constant differing slightly from a_1. Substituting in eqn. 12.10 we have

$$\sigma_x^2 = k \, T \, Z_0 \int F \, e^{2la\sqrt{f}} R^2(f) \, df \tag{12.13}$$

from which it is seen that the line loss contributes a $\sqrt{f}$ exponential weighting term which severely penalises any frequency extension of $R(f)$. In particular, it is imperative to ensure that $R(f)$ should decrease more rapidly than the increase of the line-loss term, beyond some chosen upper frequency, to prevent an unnecessary noise contribution from the top-frequency region. Because of the penalty incurred due to any extension in the frequency domain, it is obvious that the design of a digital line system necessarily reduces to the selection of an equalisation response $R(f)$ as restricted as possible and limited solely by the minimum acceptance standards placed on the resulting eye pattern (see Chapter 9).

It should be noted that the repeater output level computed above refers to theoretical conditions; allowance must be made for intersymbol interference and various imperfections in the realisation, so that an operating design margin is required; this will be discussed in a later Section.

Line encoding

A digital transmission system is normally required to transmit completely unrestricted digital information. As the line imposes certain restrictions on signals, it is clear that the information, normally binary, must be processed by means of a line encoder at the transmit terminal and the inverse decoder at the receive terminal. This may be regarded as the matching of an information source with a channel in the classical manner of information theory. The trend which this matching follows is not hard to trace when it is recalled that the line loss increases steeply with frequency, see eqn. 12.11. Consequently, any means of reducing the working rate over the line $1/T$ (the symbol rate) relative to the information rate B_{in} is highly desirable. This may be achieved (see Chapter 11) by resorting to a multilevel line code with adequate redundancy r to cover the code-design requirements such as error-monitoring capability. The relation between symbol rate, $1/T$ is then given by

$$1/T \log_2 L = (1 + r) B_{in} \tag{12.14}$$

which simply states that the information capacity of the L-level code, the left-hand side of eqn. 12.14, is made to exceed the input binary information capacity B_{in} bit/s by a fraction r, defined as the redundancy. As an illustration, taking a ternary code, if the information rate equals the symbol rate, then $r = 58\%$; with more sophisticated coding of the 4B3T type (see Chapter 11) $1/T = 0.75\, B_{in}$, signifying a useful rate reduction with $r = 19\%$. Clearly, increasing the number of levels offers even higher reductions which certainly satisfies the criteria stated earlier. In fact, the information rate of a transmission line has been analysed on an information theoretic basis by Pierce.[1] His deductions lead to an idealised optimum multilevel system using 52 levels, a forceful illustration of the trend. In practice, the number of levels used has to fall far short of the idealised optimum owing to practical difficulties in controlling intersymbol interference to within the extremely tight limits dictated by a large number of levels. For instance, for $L = 16$, using eqn. 9.6 we find that to achieve an eye opening of 70% the pulse spill-over at the next signalling instant must be no greater than 1%. In addition, aspects of tolerancing effectively preclude the use of more than about five levels in the presently conceived systems, giving rate reductions of not more than two to one, unless more complex control of the channel is resorted to by, for instance, adaptive equalisation.

12.3 Equalisation and transmitted pulse shape

In Chapter 9 we discussed general principles underlying the choice of the equalisation characteristic based on the criterion of an acceptable eye pattern. In Section 12.2 we also indicated that careless choice of $R(f)$ in terms of spectral occupancy can incur a severe noise penalty. We shall now examine this more closely, although without analytical rigour.

Our considerations will revolve around two received pulse waveforms $s_1(t)$ and $s_2(t)$. The first, $s_1(t)$, is a Gaussian function whose spectrum is given by

$$S_1(f) = \exp(-k_1 f T^2) \tag{12.15}$$

and has the gradual roll-off associated with such functions. An accelerated roll-off characteristic may be obtained whilst still preserving a near-Gaussian form at lower frequencies by modelling the spectrum by

$$S_2(f) = \exp(-k_2 f^2 T^2 - k_3 f^4 T^4) \tag{12.16}$$

Fig. 12.2 shows $S_1(f)$ and $S_2(f)$ for $k_1 = 3\cdot0$, $k_2 = 2\cdot5$, and $k_3 = 2\cdot5$. This corresponds to a pulse waveform $s_1(t)$ with just under 4% spill-over at the next signalling instants, see eqns. 9.11 and 9.12. The pulse waveforms $s_{1m}(t)$ and $s_{2m}(t)$ computed when $S_1(f)$ and $S_2(f)$ are associated with a minimum-phase network are shown in Fig. 12.2b. The minimum-phase, rather than linear-phase, network relationship is important in practice because it ensures that the specified function is physically realisable and avoids unnecessarily complex solutions.

Earlier, in Chapter 6, we showed that a transmission line may be taken as a minimum phase network in series with a delay network. Fig. 12.2c shows ternary eye patterns produced by $s_{1m}(t)$ and $s_{2m}(t)$ resulting in eye openings of $0\cdot9$ and $0\cdot8$, respectively.

We shall now concern ourselves with the noise level σ_x^2 produced at the decision node X in attaining these pulse waveforms. Referring to eqns. 12.3 and 12.13 we have, assuming the noise factor to be constant

$$\sigma_x^2 = c_1 \int e^{2al\sqrt{f}} \frac{S_r^2(f)}{S_t^2(f)} \, df \tag{12.17}$$

Expressing the line loss conventionally in terms of the loss a_w at the half symbol rate frequency $w = 1/(2T)$,

$$e^{al\sqrt{f}} = e^{al\sqrt{w}\sqrt{2fT}} = \exp\{a_w (2fT)^{\frac{1}{2}}\} \tag{12.18}$$

We next assume a rectangular transmitted pulse of unit amplitude and width kT; this implies a peak-limited system

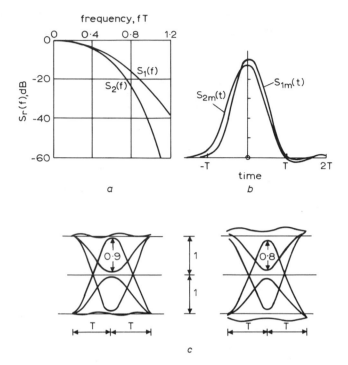

Fig. 12.2 Received pulse spectra, waveforms and eye patterns
  a Spectra
  b Waveforms
  c Ternary eye patterns

$$S_t(f) = (\sin \pi fkT)/(\pi f) \tag{12.19}$$

Combining eqns. 12.17, 12.18 and 12.19

$$\sigma_x^2 = \frac{\pi^2 c_1}{T^2} \int \frac{\exp(2a_w\sqrt{2fT})\, S_r^2(f)\, f^2 T^2}{\sin^2 \pi fkT}\, df \tag{12.20}$$

Modelling $S_r(f)$ by the Gaussian function $S_1(f)$ of eqn. 12.15 we get

$$\sigma_x^2 = \frac{\pi^2 c_1}{T^2} \int \frac{\exp 2(a_w\sqrt{2fT} - k_1 f^2 T^2)\, f^2 T^2}{\sin^2 \pi fkT}\, df$$

$$= n_1, \text{ say.} \tag{12.21}$$

For the accelerated Gaussian function $S_2(f)$ of eqn. 12.16 we get

$$\sigma_x^2 = \frac{\pi^2 c_1}{T^2} \int \frac{\exp 2(a_w\sqrt{2fT} - k_2 f^2 T^2 - k_3 f^4 T^4)\, f^2 T^2}{\sin^2 \pi fkT}\, df$$

$$= n_2, \text{ say.} \tag{12.22}$$

It is instructive to examine these seemingly formidable expressions for general trends; a more compact form is

$$\sigma_x^2 = n = \frac{c_1}{T^2} \int \frac{L\,(fT)\,df}{\sin^2 \pi fkT} \qquad (12.23)$$

The numerator of the integrand $L(fT)$ is the product of three functions; the line loss, the pulse spectrum and a weighting term $f^2 T^2$. From Fig. 12.2a it should be clear that this product has a maximum. This is shown in Fig. 12.3 for two values of line loss a_w of $9/\sqrt{2}$ and $11/\sqrt{2}$ nepers (55·2 and 67·6 dB).

Fig. 12.3 demonstrates a number of important features; first, for a given line loss the Gaussian received spectrum results in a higher maximum of $L(f)$ and hence in a higher noise level than that for the accelerated Gaussian spectrum. Reference to Fig. 12.2 indicates that the large difference is attributable predominantly to insufficiently steep roll-off of the Gaussian spectrum at the top end of the band beyond $f = 1/(2T)$. Secondly, the maxima of $L(f)$ occur at significantly different frequencies for the two spectral functions, but do not vary greatly for a given function over the range of line loss considered.

Examination of eqn. 12.23 in conjunction with Fig. 12.3 intuitively suggests that the noise will be minimised if the transmitted pulse width kT is chosen such as to coincide the frequencies at which the maxima of the numerator $L(f)$ and the denominator $\sin^2 \pi fkT$ occur. Denoting the frequency at which the maximum of $L(f)$ occurs by $(fT)_m$, then on the basis of this heuristic approach, the optimum value of k, k_0, which minimises the noise is approximately

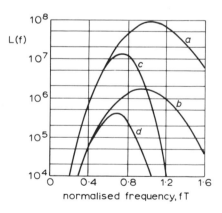

Fig. 12.3 $L(f)$ for shaping S_1 and S_2

curve	shaping	$\sqrt{2}a_w$
a	S_1	11
b	S_1	9
c	S_2	11
d	S_2	9

$$k_0 \simeq 1/2 \, (fT)_m \tag{12.24}$$

Numerical computation of eqn. 12.23 for the two spectra leads to Fig. 12.4 and confirms the above approximation. Fig. 12.4 shows the noise level for various pulse widths relative to the lowest value computed. It is again seen that the Gaussian received spectrum incurs a severe penalty, some 9·4 dB, relative to its steeper neighbour. This is only partly offset by the 1 dB improvement in eye opening shown in Fig. 12.2.

In the above, we have considered only line-section losses a_w near the current technical maximum. In some systems, the loss may be considerably smaller, as, for instance, in systems limited by crosstalk rather than noise. Fig. 12.5 shows the variation of optimum pulse width with section loss; it is seen that this increases with decreasing line-section loss, which is intuitively quite reasonable.

It should be noted that the above considerations relate to the case of unit transmitted pulse amplitude; this is relevant to systems with peak power limitation. In many cases, systems are mean-power limited implying unit transmitted pulse energy. Eqn. 12.19 is then replaced by

$$S_t(f) = (\sin \pi fkT)/(\pi fk) \tag{12.25}$$

and all eqns. up to 12.24 require due modification. The maximum of $(\sin^2 \pi \, fkT)/k$ occurs at $\pi fkT = 1\cdot165$ and hence eqn. 12.24 is replaced by

$$k_0 \simeq \frac{1\cdot165}{\pi(fT)_m} \tag{12.26}$$

It is seen that the relation between optimum transmitted pulse width for the cases of unit amplitude and unit energy ($\hat{k}_0$ and $\bar{k}_0$, respectively) is simply

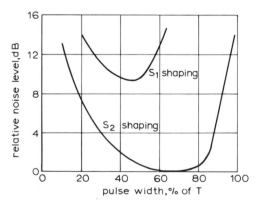

Fig. 12.4 Relative noise level against transmitted pulse width for peak-limited system
$\sqrt{2}a_w = 10$ Nepers

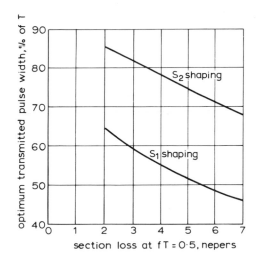

Fig. 12.5 Variation of optimum pulse width with section loss for peak-limited system

$$\bar{k}_0 \simeq 0\cdot808\ \hat{k}_0 \qquad (12.27)$$

It should be pointed out that hitherto most known digital systems used in practice, invariably employ transmitted pulses of 50% width ($k = 0\cdot5$) because of the ease of realisation. Fig. 12.9 shows the response characteristic $R_2(f)$ which, associated with a 50% width transmitted pulse, produces a received waveform $s_{2m}(t)$ with spectrum $S_2(f)$.

12.4 Crosstalk-limited systems

In Section 12.2 we described design procedures for systems where the dominant source of impairment is internally generated thermal noise. The tacit assumption was that the channel medium was of a form not allowing entry of external interference. As such, the channel model was appropriate to a fully shielded coaxial cable.

Any line which involves an open or semiopen type of construction must inevitably be subject to external interference; where a number of such lines are formed into a cable, simultaneous transmission causes crosstalk interference. Crosstalk characteristics, usually described by crosstalk attenuation, depend on the cable structure and were treated in Chapter 6 for the important case of twisted-pair lines. Fig. 12.6 illustrates possible crosstalk paths in a cable carrying both directions of transmission. Of interest is the crosstalk induced at the lowest-level signal point of a repeater section, namely the input to a repeater. The dominant contribution clearly occurs for crosstalk between lines carrying opposite directions of transmission and

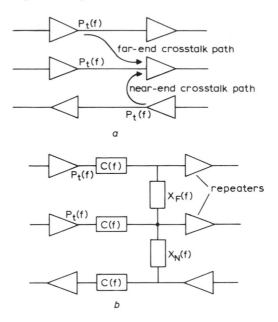

Fig. 12.6 Crosstalk paths and transfer functions
  a Crosstalk paths
  b Crosstalk model

is termed near-end crosstalk. For this reason, opposite directions of transmission are, where possible, allocated different units, preferably screened, of the cable. Even if near-end crosstalk is minimised by a cable-filling strategy, lines carrying the same direction of traffic are still subject to mutual interference, which is then termed far-end crosstalk.

The design still requires computation of the mean-square interference at the decision node X of the repeater, as described by eqn. 12.6 for the case of thermal noise alone. If, further, the probability distribution of the interference is known, then the probability of error can be computed. Referring to Fig. 12.6, let $P_t(f)$ = power spectral density of transmitted signal $X_F(f)$, $X_N(f)$ = far-end and near-end crosstalk transfer functions. Then for any two pairs near-end crosstalk power is given by

$$\sigma_x^2 = \int P_t(f)\, X_N^2(f)\, A^2(f)\, df \qquad (12.28)$$

and, for far-end crosstalk

$$\sigma_x^2 = \int P_t(f)\, \frac{X_F^2(f)}{C^2(f)}\, A^2(f)\, df \qquad (12.29)$$

In the latter case, the disturbing transmitted signal is coupled into the disturbed repeater via the line transfer function $C^{-1}(f)$ and the crosstalk function $X_F(f)$. It was shown in Chapter 6 that the crosstalk transfer functions may be expressed in the form

$$X_N^2(f) = k_N f^{3/2} \tag{12.30}$$

$$X_F^2(f) = k_F f^2 l \tag{12.31}$$

substituting and using eqn. 12.4 we get:

$$\sigma_x^2 = k_N \int P_t(f) f^{3/2} e^{2la\sqrt{f}} R^2(f) df \tag{12.32}$$

for near-end crosstalk, and

$$\sigma_x^2 = k_F \int P_t(f) f^2 \, l \, R^2(f) df \tag{12.33}$$

for far-end crosstalk.

An interesting feature of the above equations is the accumulation of crosstalk power along the length of the line. Comparison with eqn. 12.13 indicates that the near-end crosstalk accumulates in an identical manner to thermal noise. Far-end crosstalk power, however, varies linearly with the line length. A further obvious feature, already noted for the thermal-noise case, is the increase of crosstalk with the extension in the frequency domain of the equalisation response $R(f)$. Peculiar to crosstalk is the increase with extension of the transmitted-signal spectral density $P_t(f)$; this means that line codes with restricted spectra are desirable in crosstalk-limited systems. This feature is discussed in relation to bipolar and p.s.t. codes in Reference 16.

We now require to apply eqns. 12.32 and 12.33 to the case of a large number M of interfering systems. Measurements show[2,3] that the distribution of pair-to-pair crosstalk loss in a cable containing many pairs, expressed logarithmically, log-X, is normal. The total crosstalk is the power sum of individual components, and it may be shown[5,6] that this sum, in logarithmic units, namely $\log (X_1 + \cdots + X_M)$ is again normally distributed.

Not wishing to expand on this large topic we simply indicate that measurement data on a given type of multipair cable should furnish information about crosstalk constants referring to the power sum of M interferers. The information will be statistical by nature resulting in crosstalk constants $k_N(M,p)$ and $k_F(M,p)$ which depend on M and are specified with some probability p of not exceeding a given value. Eqns. 12.32 and 12.33 involving these constants now refer to the overall crosstalk $\sigma_x^2(M,p)$.

The total interference at the regenerator decision node is finally the sum of the mean-square crosstalk and thermal-noise contribution, σ_{th}^2

$$\sigma_{tot}^2 = \sigma_{th}^2 + \sigma_x^2(M,p) \tag{12.34}$$

If the crosstalk contribution is further assumed to have a near-Gaussian distribution, then σ_{tot}^2 can be treated as a Gaussian noise component allowing computation of error probability as described in Chapter 9.

Further relevant material may be found in Matsuda,[2] Aratani,[7] Mayo[8] and Aaron.[9]

12.5 Tolerancing and operating margins

In Section 9.6 we discussed basic forms of imperfection and the manner in which they affect the eye opening at the regenerator decision node. We now expand on this topic to take into account the primary causes of imperfection peculiar to baseband line transmission. The basic objective of the discussion is to deduce a factor by which the theoretical signal to noise ratio arrived at in Section 12.2 must be multiplied to guarantee the desired error rate in the presence of predicted impairments. We shall refer to this factor as the operating margin or the eye margin, because it is deduced using the eye opening of the decision-node eye pattern as a criterion. As already discussed (see Section 9.6 and Fig. 9.8), it is convenient to classify impairments according to those features of the eye pattern which they directly affect; we here identify two classes, those affecting the eye opening and those causing displacement or offset of the eye pattern. Synonymously, we associate eye-opening impairments with the shape of the eye pattern over a limited time scale of a few T, while we associate eye offset impairments with longer-term factors. This classification is somewhat arbitrary, and is intended to be helpful rather than pedantic. We can now enumerate the separate impairment factors which must be accounted for in determining the eye margin

  (i) tolerance on equalisation characteristic
 (ii) timing jitter
(iii) tolerance on transmitted eye pattern
 (iv) low-frequency wander
  (v) decision threshold uncertainty.

Table 12.2 classifies these impairments as described above, and we shall assess each in turn.

Tolerance on equalisation
The equalisation characteristic $R(f)$ specifies the overall frequency response from the output of one repeater to the decision node of the next. It is the product of line loss and equaliser gain and, in practice, will deviate from the nominal design response. The line loss varies with temperature and ageing; in addition, variations in repeater-section lengths around a nominal value must be automatically allowed for. For these reasons, the preamplifier/equaliser must have a variable-shaped gain characteristic controlled by an a.g.c. circuit; this cannot be made ideal economically. When, in addition, component tolerances are considered it is clear that the

Table 12.2

Impairment affecting, or due to	Affected parameter	
	eye opening	eye offset
equalisation characteristic	X	—
timing jitter	X	—
transmitted eye pattern	X	X
low-frequency wander	—	X
decision uncertainty	—	X

equalisation characteristic will be realised with a given tolerance. In turn, this tolerance in the frequency domain has to be translated into a tolerance in the pulse response; this, in general, requires numerical computation using the Fourier transform. However, an approximate analytical approach will be presented to expose significant trends. This is based on a method due to Davydov[4] used to study the effect of small arbitrary deviations in frequency response on the resulting pulse waveform.

Let $D_a(f)$ be the deviation of the response (real part) in nepers. Let $D_\phi(f)$ be the deviation of the response phase, in radians. Then, after some adaptation to the problem in hand, it may be shown[4] that the deviations in the received waveform $d_a(t)$ and $d_\phi(t)$, due to deviation in the attenuation and phase, are given by

$$d_a(t) = 2/f_0 \int_0^{f_0} R(f) D_a(f) \cos 2\pi ft \, df \tag{12.35}$$

$$d_\phi(t) = 2/f_0 \int_0^{f_0} R(f) D_\phi(f) \sin 2\pi ft \, df \tag{12.36}$$

for small values of $D_a(f)$ and $D_\phi(f)$. For an accuracy of about 10%, these may range up to about 0·4 Np or 3·5 dB, and 0·4 rad or 23°. The deviation of the time function is defined as the difference between distorted and undistorted waveforms. Of first interest is the effect on the waveform at its centre $t = 0$, and at the next signalling instant $t = T$. Concentrating on the effect of attenuation deviations and assuming further that $R(f)$ is zero beyond $f = 1/T$, thus defining the limits of integration, we have

$$d_a(t = 0) = 2T \int_0^{1/T} R(f) D_a(f) \, df \tag{12.37}$$

$$d_a(t = T) = 2T \int_0^{1/T} R(f) D_a(f) \cos 2\pi fT \, df \tag{12.38}$$

Before proceeding we note an important feature of eqns. 12.35–12.38, namely that

$R(f)$ appears as a weighting factor; from this we can immediately deduce that spectral deviations in regions where $R(f)$ is small, as for instance in the region approaching $1/T$, have little effect on the waveform.

To gain further insight we consider the effect of 'bump' deviations in several spectral regions on a nominal raised-cosine response $R(f)$, as illustrated in Fig. 12.7. The regions are A, B and C concentrated in the low, middle and high-frequency parts of the spectrum from 0 to $1/T$. Fig. 12.7b shows the areas corresponding to the integrals of eqn. 12.37, proportional to the distortion d_0 and d_T of the waveform at time $t = 0$ and $t = T$. The effect of these distortions on the eye pattern is interpreted in Fig. 12.7c. In case A, both d_0 and d_T are of like polarity and nearly equal, and hence the nett eye opening is nearly unchanged. The apparent offset will be eliminated by a combination of reactive coupling and the use of a d.c.-balanced line code. The conclusion here is that deviations of the amplitude frequency response occurring in a band, which is low compared with the symbol rate, are normally likely to have a small effect on the eye pattern.

In region B around the midband, d_0 and d_T are of opposite polarity and so cause expansion or contraction of the eye opening. However, in the presence of a.g.c. circuits acting on signal peaks, the peak value of $(1 + d_0)$ will be regulated to unity and the nett uncontrollable amount of additional intersymbol interference will be approximately d_T. Clearly, the illustration shows that midband deviation near the

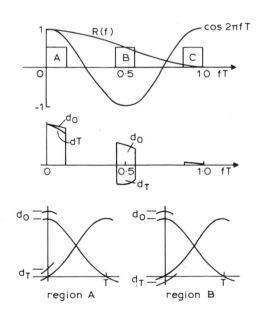

Fig. 12.7 Waveform distortion due to spectral deviations

half-symbol rate is a major contributor to eye degradation.

In region C, arguments similar to those used for region A apply, with the final values for d_0 and d_T reduced significantly by the weighting of $R(f)$. This demonstrates our earlier statement that top-band deviations produce negligible distortion. More detailed discussion and computed results will be found in Reference 17.

We can now give some quantitative results for a specific example which yields an analytic answer and is reasonably realistic. Let $R(f)$ be described by a raised-cosine function and let $D_a(f) = KfT$. This assumes the case where the attenuation distortion (in nepers or decibels) rises linearly with frequency, a not unreasonable proposition. Inserting this into eqn. 12.38 we have

$$d_a(T) = T^2 K \int_0^{1/T} f(1 + \cos \pi fT) \cos 2\pi fT \, df \qquad (12.39)$$

Trigonometric manipulation reduces this to

$$d_a(T) = T^2 K \int_0^{1/T} f\left(\frac{\cos \pi fT}{2} + \cos 2\pi fT + \frac{\cos 3\pi fT}{2}\right) df \qquad (12.40)$$

which has a solution given by

$$d_a(T) = \frac{10 K}{9\pi^2} = 0.126 K \qquad (12.41)$$

As an illustration, assuming the above, if the gain deviation is confined to ± 1 dB (0.115 N) at $f = 1/2T$ $(K = \pm 0.23)$, the waveform deviation at $t = T$, using eqn. 12.41 is 0.029: As the peak amplitude of the undistorted pulse is 0.5, the deviation at $t = T$ is nearly 6%.

The effect of phase deviations may similarly be assessed using eqn. 12.36; in broad terms deviations which tend to follow the $\sin 2\pi fT$ curve will have the most pronounced effect.

Timing jitter

As discussed in Section 9.6 we associate timing jitter with a direct reduction of eye opening, as shown in Fig. 9.8. The sources and characteristics of jitter are treated in Chapter 10, and as far as tolerancing is concerned all that is required is an indication that the jitter will not exceed a value $\pm h/2$ within given confidence limits.

Tolerance on transmitted eye pattern

The signal waveform at the output of a repeater would be, in a digital ideal world, a rectangular pulse characterised uniquely by its width t_w and its amplitude A. In practice, there will be a finite transition time t_r, and, in addition, these three parameters will each involve tolerance limits. We first consider the effect of variations in A, t_r and t_w on the received single pulse waveform at the regenerator decision node. The equalisation response $R(f)$ causes a loss in pulse amplitude p, illustrated in Fig. 12.1 and referred to in Section 12.2. Reduction of t_w from its nominal value causes a reduction in the received amplitude; however, an increase in t_w not only increases the received amplitude, but causes an increase in intersymbol interference. The pulse width must therefore be specified with an upper limit not greater than the nominal design value, the lower limit being dictated solely by acceptable signal loss. Fig. 12.8 shows the variation of the received pulse amplitude with transmitted pulse width, computed assuming $R(f)$ as in Fig. 12.9 designed for a nominal 50% pulse width, i.e. $t_w = T/2$. It is noticeable that the relation is linear, because it is the same case as demodulation of low-index p.w.m. with a low-pass filter.

Finite pulse-transition times have a much smaller effect on received amplitude; this feature is also shown in Fig. 12.8 where a raised-cosine transition is assumed, again in conjunction with the $R(f)$ of Fig. 12.9. This weak dependence clearly correlates with our earlier comments on the negligible effect of top-band spectral deviations, because the pulse transition contributes predominantly high-frequency energy.

In addition to the signal loss due to limits on the shape of a single transmitted pulse, a further effect associated with the transmitted eye pattern has to be considered. Taking ternary transmission as an example, the positive and negative pulses may not have the same shape. In the worst case, one polarity may have, within tolerance limits, minimum risetime, maximum width and maximum amplitude, while the opposite conditions may apply to the other polarity. This causes a difference in the eye amplitudes of opposite polarity resulting in a shift of the eye pattern relative to the nominal decision threshold settings. If the tolerances on transmitted pulse amplitude, width and risetime cause tolerances $\pm a$, $\pm w$, and $\pm r$, respectively, on the equalised pulse amplitude at the decision node, then it may be

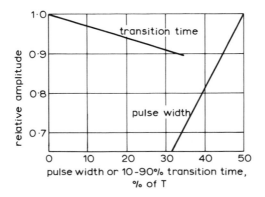

Fig. 12.8 Variation of equalised pulse amplitude with transmitted pulse width and transition time
Equalisation $R2$ of Fig. 12.9, 50% nominal pulse width

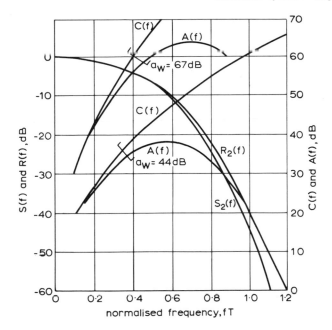

Fig. 12.9 Response $R_2(f)$, pulse spectrum $S_2(f)$ and associated gain $A(f)$ for two section-loss spectra $C(f)$

shown that the worst-case offset after a.c. coupling $d_{max} \simeq A(a + w + r)$ where A is the nominal received pulse amplitude. To illustrate the problem, assuming tolerances of 5% the worst offset is $0 \cdot 15\,A$ for a ternary eye opening of typically $0 \cdot 8\,A$ (see Fig. 12.2). It may be argued that specification of the tolerances on the pulse parameters separately is unrealistic; the alternative might involve an additional tolerance specification of the received amplitude with variation of the transmitted eye pattern as a result of the combined tolerances on the individual pulse parameters.

Low-frequency wander
A repeater section invariably incorporates a.c. coupling in the signal path contributed by input and output transformers and series capacitors within the preamplifier.

The effect of such d.c. removal and low-frequency attenuation may be illustrated by considering a single-pole network such as a series capacitor involving a single time constant. We simplify the discussion by assuming that the time constant τ is very long compared to the transmitted pulse width so that the exponential decay may be replaced by a linear droop, see Fig. 12.10.

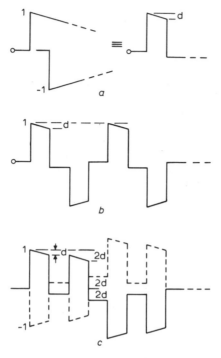

Fig. 12.10 Illustration of low-frequency wander
    a Single pulse
    b Code with DSV = 2
    c Code with DSV = 4

A single pulse thus exhibits a droop d and a tail of the same value; superposition of such pulses to form ternary codes with digital sum variations (d.s.v.) of 2 and 4 (see Chapter 11), is also shown in Fig. 12.10. It is noticeable that the pulse sequence acquires a wandering envelope, the effect being known as 'low-frequency wander'. The sequence of Fig. 12.10*c* acquires a negative peak wander of $2d$, and the inverse sequence would clearly acquire an equal positive value, shown dotted. The nett remaining eye opening is thus $(1 - 4d)$ or, in general, $(1 - d \times \text{d.s.v.})$. The initial rate of droop of a single time constant is $1/\tau$, and hence $d = kT/\tau$ where kT is the pulse width. The peak-to-peak wander $d \times$ d.s.v., contributes a wandering offset to the eye pattern and is the amount of eye closure introduced. Expressed in terms of the cut-off frequency f_0 the transmitted eye closure $= 2\pi kTf_0 \times$ d.s.v. The resulting eye closure at the regenerator decision node X, $(1 - E)$, may be deduced by noting the nearly direct correspondence between transmitted pulse width and received pulse amplitude (see above and also Section 12.6) giving

$$1 - E = 2\pi Tf_0 \times \text{d.s.v.} \tag{12.42}$$

relative to a unit pulse at X. For example, given bipolar coding (d.s.v. = 2), and low-frequency cut-off at $f_0 = 0{\cdot}01/T$, then an eye closure of $0{\cdot}126$ would result. To meet the value of $0{\cdot}024$ set in the design example at the end of this section, $f_0 T$ must not exceed $0{\cdot}0019$.

Our simple reasoning above is based on the assumption of a single-pole a.c. coupling. In practice, the number of poles may be at least three although it is unlikely that all will be coincident. The pulse response of a multipole high-pass filter can be damped oscillatory, and clearly our simple approximation does not then apply. The question of multipole networks has been considered by Saltzberg and Simon,[15] and their computations of error probability indicate that error performance gets worse with the number of poles. The above reference derives upper error bounds for the case of unrestricted format and partial-response coded random inputs, and hence has limited application. However, the general trend of the results confirms that eqn. 12.42 should be treated as a lower bound. Fukinuki and Sato[17] consider the effect of a multipole low-frequency cut-off on a single pulse, and show that the residual shift at the maximum of the equalised pulse is proportional to the number of poles; their result is identical in the limit to eqn. 12.42, and it is therefore reasonable to arrive at a heuristic measure for a system having m low-frequency poles, given by

$$1 - E = 2\pi T f_0 m \times \text{d.s.v.} \tag{12.43}$$

Threshold uncertainty
This form of imperfection was described in Section 9.6 and in Fig. 9.8. We assumed that a signal falling within the region of decision uncertainty d always results in an erroneous decision. In reality, the mechanism governing the behaviour of a decision-making device is quite complex and is dependent on the specific device. Thus, in practice, the response time between a decision output and an input stimulus depends on the level of the input relative to the threshold level. Hysteresis further complicates the issue and circuit tolerances add a probabilistic parameter. However, it is felt that our primitive model, although artificial, may be considered to be a simple reference equivalent to which the specification of more complex models[10] can be reduced.

Example
The quantitative implications of tolerancing and the relative contributions of the various toleranced parameters may best be demonstrated by means of a hypothetical example using practically realistic values. Even if, in a given system, the values may differ significantly, the example should indicate dominant parameters.

Let us take a ternary system with section response $R_2(f)$ and eye patterns as shown in Fig. 12.2 and 12.9. We can compile a 'balance sheet' relating to eye opening as shown in Table 12.3 and progress cumulatively through all the tolerancing items described. We start with item (a), the ideal rectangular pulse of unit amplitude and 50% width. Item (b) is the amplitude of a single received pulse

Table 12.3 Balance sheet of operating tolerances

Item		Amplitude
a	single transmitted pulse	1·000
b	single equalised pulse, nominal response	0·470
c	b, after tolerancing of item a: 0·81 b	0·380
d	equalised eye opening, nominal response: 0·80 c	0·304
e	d, after tolerancing of response: 0·68 c	0·258
f	e, after timing jitter : 0·9 e	0·232
g	offset after tolerancing of transmitted eye pattern	0·024
h	low-frequency wander	0·024
i	threshold uncertainty	0·024
j	total offset: g + h + i	0·072
k	nett eye opening: f − j	0·160

shaped by $R_2(f)$. Item (c) is the received pulse amplitude assuming a transmitted pulse of 45% width, transition time of $0.2T$ and an amplitude of 5% down on normal. Using Fig. 12.8 these lower limits introduce a factor of $0.91 \times 0.94 \times 0.95 = 0.81$. Item ($d$) is the eye opening with intersymbol interference produced by the nominal $R_2(f)$, see Fig. 12.2c, introducing a factor of 0.8 corresponding to $a = 5\%$ in eqn. 9.6. Item (e) gives the new eye opening allowing for deviations of $R_2(f)$ from nominal. Using our previous assumptions, and taking a peak-to-peak deviation of 1 dB at $f = 1/2T$, the extra intersymbol interference caused is 2·9% giving a total of 7·9% and leading to a loss factor, from eqn. 9.6, of 0·68. Item (f) is the effect of timing jitter and is taken as resulting in a loss factor of 0·9. At this stage, we have the total cumulative effect of all loss factors.

We take the tolerance on offset produced by differential polarity effects in the transmitted eye pattern, item (g), as 5% of the peak nominal received signal; threshold uncertainty, item (i), is allowed the same limit. We now estimate what low-frequency cut-off f_0 must be used to limit the low-frequency wander to the same proportion. Using our previous assessment and assuming a code d.s.v. = 2, we find that $f_0 = 0.0019/T$, which is a realistic requirement.

Subtracting the total offset, item (j), from the attenuated eye opening, item (f), leaves that portion of the eye which is guaranteed to remain open after allowing for all tolerances. The ratio b/k may be considered to be the margin by which the signal/noise ratio in the ideal case must be increased in practice. In the example, this is just short of 10 dB, indicating the importance of tolerance assessment in the system design.

12.6 Other design features

12.6.1 Empirical short cuts

It is useful to bring to light several empirical properties pertaining to digital line systems; some of these have already been implied in our treatment above. Consider the relation between the peak amplitude of the transmitted and that of the received equalised pulse at the regenerator decision node, V_0 and V_x, respectively, of Fig. 12.1 and eqns. 12.8 and 12.9. If the transmitted pulse is rectangular of width $T/2$, and the received pulse is a raised sinusoid of period $2T$, implying a dominant spectral term at $f = 1/2T$, then simple Fourier series analysis shows that V_x is just less than $V_0/2$, i.e. $p \simeq \frac{1}{2}$. Our earlier discussion indicates that this simplification is a reasonable first approximation, see Fig. 12.2. In practical designs, p will range from -6 to -8 dB (e.g. Reference 17). We note that this is also the likely range of the equalisation response $R(f)$ at $fT = \frac{1}{2}$. Hence we can argue that

$$p = V_x/V_0 \simeq R\left(f = \frac{1}{2T}\right)$$ (12.44)

Further, we examine eqns. 12.6 and 12.10, giving the decision noise level via the preamplifier gain $A(f)$. Reference to Fig. 12.2 and Fig. 12.9 indicates that $A(f)$ has a maximum at a frequency $f = \alpha/T$, say, it is found that α lies in the range 0·5–0·7 for most practical systems. The integrals of these expressions may be replaced by terms of equivalent area

$$\sigma_x^2 \simeq k\,T^0 Z_0 F C^2 \left(\frac{\alpha}{T}\right) R^2\left(\frac{\alpha}{T}\right) \frac{\beta}{T}$$ (12.45)

where β is the effective bandwidth expressed as a fraction of the symbol rate $1/T$; it can be taken as the 3 dB bandwidth and typically ranges from 0·2 to 0·3. Using the above equations in conjunction with eqn. 12.9 we have

$$V_0 \simeq \frac{\text{s.n.r.}}{R\left(\frac{0\cdot 5}{T}\right)} R\left(\frac{\alpha}{T}\right) C\left(\frac{\alpha}{T}\right) \left(k\,T^0 Z_0 F \frac{\beta}{T}\right)^{\frac{1}{2}}$$ (12.46)

The above is very useful where, in a specific system, particular solutions and sets of parameters are known, and it is required to extrapolate the solution for a different set of parameters. Alternatively, the approach may aid analytic demonstration of broad trends as exemplified by Aratani.[7] The latter author refers to α/T as the 'transmission equivalent frequency'.

Matsuda[2] defines this as the frequency at which the ratio of amplitudes of transmitted and received sinusoids equals the pulse ratio V_0/V_x under normal conditions. Similar reasoning applied to the case of crosstalk-limited systems leads to a 'crosstalk equivalent frequency' f_x defined as the frequency where

$$\frac{\text{received mean power of p.c.m. crosstalk}}{\text{received peak power of p.c.m. signal}}$$

$$= \frac{\text{received crosstalk power when sinusoid at } f_x \text{ is transmitted}}{\text{received signal power when sinusoid at } f_x \text{ is transmitted}}$$

Matsuda shows that for a wide range of conditions f_x for FEXT $\simeq 0.2/T$ and f_x for NEXT $\simeq 0.4/T$.

Eqn. 12.46 may be used to show the effect of altering the symbol rate $1/T$ by a factor $(1 + \delta)$; it can be shown that, for $\delta \ll 1$, the new output voltage V_{02} at a symbol rate $(1 + \delta)/T$, relative to the output voltage V_{01} at a rate $1/T$, is given by the following ratio in dB

$$\left[\frac{V_{02}}{V_{01}}\right] \mathrm{dB} \simeq \frac{\delta}{2} \left\{10 + C\left(\frac{\alpha}{T}\right)\right\} \tag{12.47}$$

where $C(\alpha/T)$ is the cable loss in dB at the critical frequency α/T. Taking for example $C(\alpha/T) = 60$ dB, a mediocre value for coaxial systems, we find that a 10% change in symbol rate increases the output level requirement by 3.5 dB, demonstrating the sharp increase in output level with an increase in symbol rate, which is a feature of digital line systems.

12.6.2 Hybrid digital/analogue systems

The normal rationale in designing a digital line link using fully regenerative digital repeaters is to minimise the total number of repeaters, or, in other words, to maximise the repeater spacing within technical and economic constraints on feasibility. However, it is possible to extend the objective to the minimisation of overall cost if devices other than digital repeaters are permitted. One such possibility is to use a mixture of digital and analogue repeaters. In such a hybrid arrangement, the spacing between digital repeaters is extended by interposing several analogue repeaters each comprising an equalising preamplifier and output stage. The logic is that, in saving the cost of timing extraction, the analogue repeater should cost significantly less than a digital one. Now, in practice, we feel that this saving is unlikely to exceed 30% calculated on the basis of total per unit cost including power separation filters, line build-out networks and a proportion of repeater housing cost. Assuming further that there must be a larger total of repeaters in the hybrid system, and, having regard to the high cost of siting repeaters, we find it difficult to envisage economical hybrid systems in the field.

The topic has been tackled from different optimisation points of view by several authors,[11-13] and the interested reader is referred to their lengthy analyses. References 12 and 13 include economic appraisals based on specific assumptions. Certain general conclusions may be seen to emerge; the ratio of analogue/digital repeaters should be high (20–200) and the number of code levels should be 4–16. We have reservations about the practical applicability of these results, although the trend towards the theoretical results obtained by Pierce[1] is interesting. The limiting case of such hybrid systems is a link with digital terminals and an analogue repeatered line which is very similar, in principle, to existing digital modems designed for use over f.d.m. channels.

Takasaki and Aoki[14] draw attention to the critical sensitivity of hybrid systems with regard to distortion produced by low-frequency and high-frequency cut-off; they quote that, with a low-frequency cut-off to symbol rate ratio of 0·001, only three analogue repeaters each with two time constants, produce 10% intersymbol interference.

The authors argue that proper phase equalisation in the cut-off regions is mandatory in practice. A further practical shortcoming is the need for linear output stages; because the power consumption of the output stage in a line repeater represents a significant fraction of the total repeater consumption, the use of linear output configurations with their attendant inefficiency is a serious drawback.

12.6.3 Quantised feedback

Earlier in Section 12.5 we considered low-frequency wander due to unavoidable l.f. cut-off in the equalisation response, and we concluded that it is a serious contributor to the reduction of operating margins. Hitherto, the standard method of controlling the effect has been by the use of line codes with small digital sum variation (d.s.v.). An alternative principle of long standing, which has found recent favour[18] is so-called quantised feedback, see Fig. 12.11.

The basic concept of quantised feedback is to allow a combination of code and l.f. cut-off which would, if unchecked, cause large baseline wander at the input of a

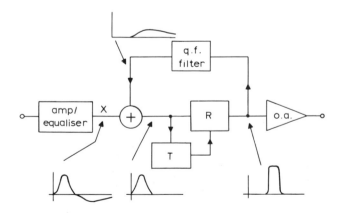

Fig. 12.11 Repeater with quantised feedback
    T = timing extractor
    R = regenerator
    o.a. = output amplifier
    q.f. = quantised feedback

regenerator, node X of Fig. 12.11; this, however, is prevented by adding an equal and opposite wander at the same node. The cancelling wander signal could be generated by passing the original code signal through a frequency shaping exactly complementary to the normal equalisation response with its low-frequency cut-off. In fact, the complementary low-pass filter derives the wander signal from the regenerator output itself, and applies it to the regenerator input X. On a single-pulse basis the effect is to cancel the long tail shown in Fig. 12.10 and 12.11.

At first sight, the need for complementary shaping of the feedback network might seem a disadvantage in the light of the uncertain l.f. cut-off of the networks involved; this is avoided (Reference 18) by inserting a single dominant network whose l.f. cut-off frequency is sufficiently higher than all the others involved to render their effects negligible. At the same time, this dominant l.f. crossover frequency must be sufficiently low compared to the half-symbol-rate frequency to avoid any fast transient effects around the regenerator quantised feedback loop. The exact choice of crossover frequency is not critical; 7% of the half symbol rate is the design value used in Reference 18.

Quantised feedback offers two main advantages. First, it minimises, and, in principle, eliminates, the need for careful balancing of the line code to minimise the d.s.v. This eases one of the severe restrictions under which code design has hitherto laboured and could enable the use of codes with marginal redundancy such as, possibly, 3B2T (three binary to two ternary digits). Reference 18, in fact describes a zero-redundancy binary system with timing provision by scrambling. Secondly, it permits the use of much higher l.f. cut-off frequencies with the attendant improvement in protection against induced lightning surges and other low-frequency interference.

12.7 References

1   PIERCE, J.R.: 'Information rate of a coaxial cable with various modulation systems', *Bell Syst. Tech. J.*, 1966, **45**, pp. 1197–1207
2   MATSUDA, S.: 'Equalisation and crosstalk in p.c.m. repeatered line using pair-type cable', *Electron. & Commun. Japan*, 1967, **50**, pp. 66–75
3   Bell Telephone Laboratories: 'Transmission systems for communications' (Western Electric Co., 1970) 4th edn.
4   DAVYDOV, G.B.: 'Deviation tolerances for phase, delay and attenuation frequency characteristics in pulse signal communication channels', *Telecommun. & Radio Eng.*, **20**, Pt.1, pp. 8–15
5   MARLOW, N.A.: 'A normal limit theorem for power sums of independent random variables', *Bell Syst. Tech. J.*, 1967, **46**, pp. 2081–2090
6   NASSELL, I.: 'Some properties of power sums of truncated normal random variables', *ibid.*, 1967, **46**, pp. 2091–2110
7   ARATANI, T.: 'Fundamental consideration of multilevel pulse code transmission in cables', *Electron. & Commun. Japan*, 1968, **51A**, pp. 26–33
8   MAYO, J.S.: 'A bipolar repeater for pulse code signals', *Bell Syst. Tech. J.*, 1962, **41**, pp. 25–98
9   AARON, M.R.: 'P.C.M. transmission in the exchange plant', *ibid.*, 1962, **41**, pp. 99–142

10 ARATANI, T.: 'Degradation in p.c.m. repeatered line due to incomplete pulse recognition', *Electron. & Commun. Japan*, **49**, 1966, p. 191

11 CHANG, R.W., and FREENY, S.L.: 'Hybrid digital transmission systems', Pt. 1: 'Joint optimisation of analogue and digital repeaters', Pt. 2. 'Information rate of hybrid coaxial cable systems', *Bell Syst. Tech. J.*, 1968, **47**, pp. 1665–1712

12 KAWASHIMA, M. *et al.*: 'Problems on digital repeatered lines', *Electron. & Commun. Japan*, 1970, **53A**, pp. 18–25

13 ERICSON, T., and JOHANSSON, U.: 'Digital transmission over coaxial cables', *Ericsson Technics*, 1971, **27**, pp. 191–272

14 TAKASAKI, Y., and AOKI, K.: 'A baseband hybrid digital transmission scheme', *IEEE Trans.*, 1973, **COM-21**, pp. 338–340

15 SALTZBERG, B.R., and SIMON, M.K.: 'Data transmission error probabilities in the presence of low-frequency removal and noise', *Bell Syst. Tech. J.*, 1969, **48**, pp. 255–274

16 IWAHASHI, E., and SAKURI, N.: 'Designing a p.c.m.-16M system repeatered line', *Rev. Electron. Commun. Lab.*, 1969, **17**, pp. 403–425

17 FUKINUKI, H., and SATO, Y.: 'Wave equalisation in p.c.m. repeatered line', *ibid.*, pp. 426–447

18 WALDHAUER, F.D.: 'Quantised feedback in an experimental 280-Mb/s digital repeater for coaxial transmission', *IEEE Trans.*, 1974, **COM-22**, pp. 1–5

Chapter 13

Digital modulated—carrier systems

13.1 Carrier-borne digital transmission

Many communication channels require further processing of the digital information before transmission. In many cases, the processing involves a translation of the digital signal into a suitable region of the frequency spectrum by means of modulation of a carrier signal. The need for carrier-borne transmission arises, in general, for one of two reasons; first, the transmission channel may be one which offers a definitively specified spectral window into which the information must be carefully fitted, such as an existing analogue f.d.m. channel. The second case occurs where the actual physical nature of the transmission channel calls for operation in a particular spectral range, waveguide and optical links being examples. In this Chapter, we shall examine the implications of digital transmission over carrier-borne systems.

As an interesting observation, reinforcing the view that very little in technology is genuinely novel, we shall recall that the earliest modulation systems of the spark-transmitter type were digital. The invention of the amplifying triode then shifted emphasis onto analogue-modulation systems for over half a century; our present interest in digital modulation might be regarded as a rebirth of the concept at a higher level of technology and expertise.

It is useful to note that modulation may be considered as an additional step in matching the source information to the communication channel, and hence as part of the channel coding process from the point of view of information theory. Identified as such, questions of the efficiency and capacity of any coding-modulation schemes properly arise, and also focus attention on the desirability of maintaining an overview of the digital system as a whole during design of the constituent parts. Thus, the main objectives of a digital carrier link remain identical to those of a baseband system, namely, the achievement of a specified error rate and jitter. Towards this aim a number of points arising from the nature of digital transmission, and again in common with baseband systems, require examination in the context of a carrier-based system. These are (i) the provision of timing information for synchronisation (ii) ensuring that the d.c. content, if any, of the source information is properly allowed for and (iii) provision of frame alignment, error monitoring and supervisory information.

The first two items must be regarded as *sine qua non*, the other may always be catered for with a varying degree of subtlety, depending on the system context. The above comment is reinforced in systems where repeatering involves demodulation to baseband, baseband regeneration and remodulation; since obviously any and all the baseband considerations described in Chapter 9 apply. In fact, the demodulation–remodulation method is hitherto universal wherever the carrier frequencies required are embarassingly high for existing circuit technology, so that controllable carrier amplification is not possible.

13.2 Optimal digital modulation and demodulation

A sinusoidal carrier is described by its amplitude, frequency and phase, and each of these parameters may be modulated by the digital source. If the digital source has M states or levels, each present for a duration T, then the modulated waveform corresponding to the ith state $s_i(t)$ may be described for the three types of modulation, as follows:

$$s_i(t) = A_i \cos \omega_0 t \tag{13.1}$$

indicating amplitude modulation or amplitude shift keying (a.s.k.).

$$s_i(t) = A \cos \omega_i t \tag{13.2}$$

indicating digital frequency modulation or frequency shift keying (f.s.k.),

$$s_i(t) = A \cos (\omega_0 t + 2\pi i/M) \tag{13.3}$$

indicating digital phase modulation or phase shift keying (p.s.k.). It should be noted that the above expressions define only the steady state of the modulated carrier under the modulating condition i, and they do not include a dynamic description of the modulating source. This simple form suffices for considerations of detection strategies and error rates in the presence of random noise. Simultaneous modulation of more than one of the three parameters is feasible and will be discussed later.

13.2.1 Representation of signals in signal space
One of the successful results of modern statistical communication theory is that we are now in a position to define optimum detection schemes for a large class of signal waveforms. As a start we note that a physically realisable waveform, may be expanded into a linear summation of D orthonormal waveforms $\phi_1(t)$, $\phi_2(t)$, ..., $\phi_D(t)$ (Lerner[1]):

$$s_i(t) = a_{i1} \phi_1(t) + a_{i2} \phi_2(t) + \ldots + a_{iD} \phi_D(t) \tag{13.4}$$

$$\int_0^T \phi_j(t)\,\phi_k(t)\,dt = 0 \quad \text{if } j \neq k$$

$$= 1 \quad \text{if } j = k \tag{13.5}$$

In particular cases where

$$\phi_k(t) = \frac{\sin \pi(2Wt - k)}{\pi(2Wt - k)}$$

we obtain the expansion in terms of a number $D = 2TW$ Nyquist samples, the coefficients a being then sample amplitudes. Another familiar case is when $\phi_k(t) = e^{jk\omega_0 t}$ giving the Fourier series expansion in exponential terms. Eqn. 13.4 therefore generalises a concept which is familiar in communications. It leads to the geometrical concept of a signal space (Shannon[2] Kotelnikov[3]) where each signal $s_i(t)$ is represented as a vector in a D-dimensional co-ordinate space, the projections of the signal vector on the axes being the coefficients $a_{i1}, a_{i2}, \ldots a_{iD}$.

13.2.2 Optimum detection

In transmission, noise is added to $s_i(t)$ so that at the receiver there arises a definite probability that the ith signal vector may be confused with its nearest neighbours. The optimum detection rule is one which minimises this probability of error. The representation of eqn. 13.4 and the orthonormality condition suggests a conceptual embodiment for such a detector. Let the set of describing waveforms $\phi(t)$ be locally available at the receiver, and let each of these be applied to D product integrators together with the received signal. Then, in the absence of noise, the outputs of the product integrators will be

$$\int_0^T s_i(t)\,\phi_k(t)\,dt = a_{ik} \tag{13.6}$$

namely the co-ordinates of the ith signal.

In the presence of noise, these co-ordinates are perturbed and the signal points in signal space become 'fuzzy'. A reasonable detection scheme to adopt is the so-called maximum-likelihood strategy, which for equiprobable transmitted signal states partitions the signal space into regions with boundaries equidistant from expected noise-free signal points. A noisy received signal falling into the ith region is then deemed to correspond to the ith transmitted signal. The above detection and decision strategy may be shown to be optimum for additive, zero-mean Gaussian noise and equiprobable source states (Harman[4]). Such a detector and decision stage are shown in Fig. 13.1a.

The scheme depicted in Fig. 13.1 may be recognised as implementing the so-called *correlation detection*. To justify this statement we note that the cross-correlation function of two signals $s_i(t)$ and $\phi_k(t)$ is defined by (see Lee[5])

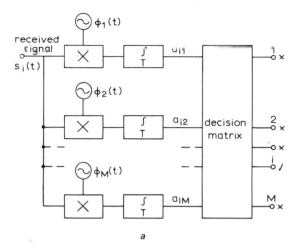

a

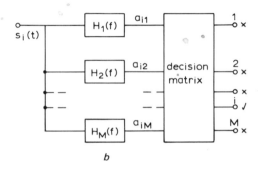

b

Fig. 13.1a Correlation receiver
       b Matched filter receiver

$$R_{ik}(\tau) = \frac{1}{T} \int_0^T \phi_k(t)\, s_i(t + \tau)\, dt$$

The optimum receiver thus emerges as a bank of correlators which form the correlation coefficients between the incoming signal and the local orthonormal waveforms.

It is also informative to note that the above scheme is synonymous with *matched filter* detection; the output $f_0(t)$ of a filter of impulse response $h(t)$, driven with an input $s_i(t)$, may be written as a convolution integral.[5]

$$f_0(t) = \int_0^T s_i(\tau)\, h(t - \tau)\, d\tau$$

hence

$$f_0(t = 0) = \int_0^T s_i(\tau) h(-\tau) d\tau$$

which should be compared with eqn. 13.6. If then we make $h_k(-t) = \phi_k(t)$, the kth filter output will be a_{ik}, as before. The bank of correlators may thus be replaced by a bank of filters whose impulse response time inverse is matched to the ortho-normal waveforms $\phi_k(t)$. The matched filter receiver is shown in Fig. 13.1b, and is exactly equivalent to the correlator form of Fig. 13.1a.

We can now apply these results to the cases of amplitude, phase and frequency shift keying in the knowledge that the resulting detector will be optimal.

It is first necessary to express the modulated waveforms given by eqns. 13.1, 13.2 and 13.3 by the orthonormal expansion defined by eqns. 13.4 and 13.5. The main difficulty is in the choice of the describing functions $\phi(t)$; mathematical procedures for this exist,[6] but the three cases above may be guessed by inspection.

For a.s.k., there is only one describing waveform, $\phi_1(t) = \sqrt{2/T} \cos \omega_0 t$ and the coefficient a_{i1}, at the output of the product integrator, reproduces the original amplitude

$$a_{i1} = \sqrt{\frac{2}{T}} \int_0^T A_i \cos^2 \omega_0 t \, dt = \frac{A_i}{\sqrt{2}} \tag{13.7}$$

(We assume that $\omega_0 = 2\pi m_0/T$, m_0 being some integer). The optimum detector for a.s.k. is shown in Fig. 13.2a. The decision stage is a threshold device with equidistant thresholds defining the permitted amplitude regions.

For p.s.k. we find that two describing functions

$$\phi_1(t) = \sqrt{\frac{2}{T}} \cos \omega_0 t$$

and

$$\phi_2(t) = \sqrt{\frac{2}{T}} \sin \omega_0 t$$

are needed, resulting in a 2-dimensional signal space. The two output coefficients are then

$$a_{i_1} = \frac{A}{\sqrt{2}} \cos \frac{2\pi i}{M} \tag{13.8}$$

and

$$a_{i2} = -\frac{A}{\sqrt{2}} \sin \frac{2\pi i}{M} \tag{13.9}$$

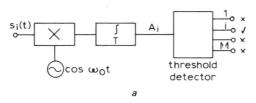

a

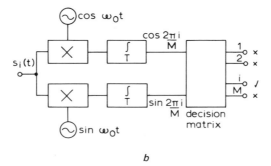

b

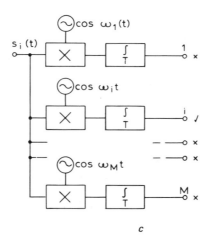

c

Fig. 13.2a Optimum receiver for a.s.k.
    b Optimum receiver for p.s.k.
    c Optimum receiver for f.s.k.

The optimum detector for p.s.k. thus requires two local quadrature carriers as shown in Fig. 13.2b.

For the case of 4-level p.s.k., $M = 4$, the decision stage here comprises two threshold devices and a logic matrix to implement the following truth table

i	1	2	3	4
$\dfrac{\sqrt{2}}{A} a_{i1}$	0	-1	0	$+1$
$\dfrac{\sqrt{2}}{A} a_{i2}$	-1	0	$+1$	0

For f.s.k. the number of describing functions D is found to coincide with the number of states M which, in this case, are the discrete modulated frequencies

$$\phi_k(t) = \sqrt{\frac{2}{T}}\ \cos \omega_k t \qquad (13.10)$$

and the detected outputs are

$$a_{ik} = \frac{A}{\sqrt{2}} \quad \text{if } i = k$$

$$= 0 \quad \text{if } i \neq k$$

The optimum detector for f.s.k. is shown in Fig. 13.2c, and is a bank of local sources with frequencies coincident with those transmitted, feeding the product integrators. Only one output at a time is active, this being taken as the correct state by the decision stage.

Detector signal spaces for the cases of 4-level a.s.k. and p.s.k. and 3-level f.s.k. are shown in Fig. 13.3.

13.2.3 Incoherent detection
It should be noted that the optimum detectors described above all assume that at the demodulator local carrier waveforms in exact phase with the transmitted carriers are available. Such demodulation is referred to as coherent. In many practical systems coherent detection involves excessive receiver complexity, and hence it is of importance to consider the implications of incoherent detection.

The situation defining incoherence is described by a transmitted carrier $\cos \omega_0 t$ and a local receiver signal $\cos (\omega_0 t + \alpha)$, where α is a random phase whose variations are assumed to be slow relative to the signalling interval T. In all cases, the outputs of the product integrators (eqn. 13.6) considered thus far would then contain functions of the unwanted angle α. For the cases of a.s.k. and f.s.k., the situation is remedied by resorting to a set of two carriers, quadrature related, for each one used previously; thus, for a.s.k. we use two local functions

$$\phi_1(t) = \sqrt{\frac{2}{T}}\ \cos \omega_0 t$$

and

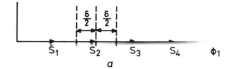

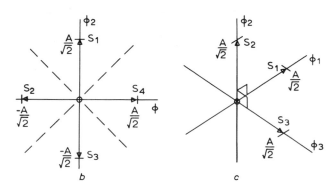

Fig. 13.3 Detector signal spaces for a.s.k.4, p.s.k.4 and f.s.k.3

    a a.s.k., *M* = 4
    b p.s.k., *M* = 4
    c f.s.k., *M* = 3

$$\phi_{1q}(t) = \sqrt{\frac{2}{T}} \sin \omega_0 t$$

while for f.s.k. we supplement each

$$\phi_k(t) = \sqrt{\frac{2}{T}} \cos \omega_k t$$

with

$$\phi_{kq}(t) = \sqrt{\frac{2}{T}} \sin \omega_k t$$

This is rather obvious from eqns. 13.1 and 13.2 which may be expanded into quadrature components cos ωt and sin ωt. The outputs of the pairs of product integrators then become

$$a_{i1} = \frac{A_i}{\sqrt{2}} \cos \alpha \tag{13.11}$$

$$b_{i1} = -\frac{A_i}{\sqrt{2}} \sin \alpha \tag{13.12}$$

The original information in both cases is thus contained in the modulus $(a^2 + b^2)^{\frac{1}{2}}$, and therefore, the detection process for incoherent a.s.k. and f.s.k., at least conceptually, differs little from the coherent case. This is recognised as envelope detection and, in fact, is not quite an optimal strategy for incoherent a.s.k. (Reference 6).

In the case of incoherent p.s.k., the situation alters more drastically, because the unwanted phase α is inextricably added to the wanted phase $2\pi i/M$. The solution to this potential ambiguity lies in modifying the significance of the modulating process. Instead of the ith state being associated with an absolute phase $2\pi i/M$, we associate it with *a change* in phase of $2\pi i/M$ relative to the previous transmitted phase. This strategy is known as differential phase shift-keying, or d.p.s.k. As both the ith signal vector and the $(i-1)$th are required for demodulation then, because noise affects both equally, it is clear that the signal/noise ratio must suffer. A reasonable guess at the amount would be 3 dB, on the basis of doubled noise power, and this will be found to be asymptotically true for large M.

13.3 Probability of error in additive noise

The geometrical model of a signal space provides a powerful basis of attack on problems associated with error performance in the presence of additive random noise. The classic works of Shannon,[7] on channel capacity and of Kotelnikov,[3] on optimum noise immunity rely on the signal-space concept. In essence, the formulation of the problem is quite simple; for a given modulation scheme, what is required is the probability that, given the transmitted message i noise added to the detected signal vector s_i will perturb the vector to just beyond the boundary with its nearest neighbour. This condition defines an error, and assuming the usual zero-mean Gaussian noise model, it is in most cases possible to define upper and lower limits on the probability of error in integral form. A thorough tutorial treatment of the techniques involved is to be found in Arthurs and Dym.[6] We confine the following treatment to the essentials needed in presenting and using the concepts.

Consider the signal space at the detector output described by the coefficients a_i of the previous Section; taking the simplest case of a signal space containing two signal points, the nearest neighbour distance δ is then the distance separating the two points, and the probability of error P_e, is equal to the probability that the noise component n_{12}, along the line connecting the two signal points exceeds $\delta/2$. This case is identical to that treated in Chapter 9 for baseband transmission. Hence

$$[P_e]_{M=2} = \text{prob}\left(n_{12} > \frac{\delta}{2}\right)$$

$$= \frac{1}{\sqrt{2\pi}} \int_{\frac{\delta}{2\sigma}}^{\infty} e^{-x^2/2} \, dx = \bar{\Phi}\left(\frac{\delta}{2\sigma}\right), \text{ say} \qquad (13.13)$$

where σ^2 is the variance or mean square value of the noise. Note that eqn. 13.13 is *quite general for any equiprobable binary signal space*; different modulation schemes simply require different values for the nearest neighbour distance δ. For nonequiprobable states, although eqn. 13.13 is correct, the partitioning of the signal space at $\delta/2$ between signal points, no longer represents an optimum detection strategy inasmuch as it does not then minimise probability of error.

13.3.1 P_e for optimal binary modulation

δ may be easily computed from eqns. 13.7–13.10, as follows:
For binary a.s.k. with one state zero, from eqn. 13.7 it is seen that $\delta = A/\sqrt{2}$. For binary p.s.k., from eqns. 13.8 and 13.9, $\delta = a_1 + a_2 = 2A/\sqrt{2}$, since the two signal points have phases $0°$ and $180°$. For binary f.s.k. eqn. 13.10 indicates that the two signal output coefficients are both $A/\sqrt{2}$ on perpendicular co-ordinates, and hence $\delta = A$. These values may be inserted directly in eqn. 13.13 to obtain the error probability with signal amplitude A as parameter. Such a case is of interest where the transmission system is peak-level limited. Alternatively, A may be expressed in terms of the average signal power S (mean square value in a unit load); assuming equiprobable binary states, the power S, for the binary a.s.k., p.s.k. and f.s.k. is $A^2/4$, $A^2/2$ and $A^2/2$, respectively. Hence, P_e may in turn be expressed in terms of S. We present the above results for binary a.s.k., p.s.k., and f.s.k. in Table 13.1.

Table 13.1 Optimal probability of error in binary modulation, P_e

Modulation	Detector nearest-neighbour distance	Mean signal power, S	P_e as function of A	P_e as function of S	Relative peak signal, dB	Relative mean signal, dB
a.s.k.	$\dfrac{A}{\sqrt{2}}$	$\dfrac{A^2}{4}$	$\overline{\Phi}\left(\dfrac{A}{2\sqrt{2}\sigma}\right)$	$\Phi\left(\dfrac{\sqrt{S}}{\sqrt{2}\sigma}\right)$	+6	+3
p.s.k.	$\dfrac{2A}{\sqrt{2}}$	$\dfrac{A^2}{2}$	$\overline{\Phi}\left(\dfrac{A}{\sqrt{2}\sigma}\right)$	$\Phi\left(\dfrac{\sqrt{S}}{\sigma}\right)$	0	0
f.s.k.	A	$\dfrac{A^2}{2}$	$\overline{\Phi}\left(\dfrac{A}{2\sigma}\right)$	$\Phi\left(\dfrac{\sqrt{S}}{\sqrt{2}\sigma}\right)$	+3	+3

A = carrier amplitude
S = mean-square value of signal in unit load
$$\overline{\Phi}\left(\frac{h}{\sigma}\right) = \frac{1}{\sqrt{2\pi}} \int_{h/\sigma}^{\infty} e^{-x^2/2} \, dx = \tfrac{1}{2} \operatorname{erfc}\left(\frac{h}{\sigma}\right)$$
The last two columns indicate relative peak and mean-signal levels required to maintain equal P_e for given σ.

13.3.2 P_e for optimal multilevel modulation

For multilevel systems with M states the nearest neighbour argument of the binary case (eqn. 13.13) becomes more involved due to the multidimensional geometry relating signal and noise vectors. However, even for the general M-ary coherent modulation system, simple limits for P_e have been found by Arthurs and Dym[6]

$$\overline{\Phi}\left(\frac{\overline{\delta}}{2\sigma}\right) \leqslant P_e \leqslant (M-1)\,\overline{\Phi}\left(\frac{\delta_{min}}{2\sigma}\right) \tag{13.14}$$

where $\overline{\Phi}$ is the function defined by eqn. 13.13, and $\overline{\delta}$ is the average neighbour distance

$$\overline{\delta} = \frac{1}{M} \sum_{i=1}^{M} \delta_i \tag{13.15}$$

and δ_{min} is the nearest neighbour distance.

In logical terms eqn. 13.14 states that P_e must be greater than the probability of error based on average neighbour distance $\overline{\delta}$, since some neighbours are closer than $\overline{\delta}$, and that P_e must be less than the probability of error based on the nearest neighbour distance δ_{min} since some neighbours are further away than δ_{min}.

Eqn. 13.14 has the elegance of total generality and extreme simplicity, coupled with the practical advantage that, for most cases, the relevant distances $\overline{\delta}$ and δ_{min} may be found by inspection. Its one drawback is that for many cases it is possible to find tighter bounds by direct consideration of the case in hand. In the following, we list expressions for the P_e of digital modulation systems largely based on the work of Arthurs and Dym.[6] We have restricted the list to expressions containing single error functions and exponentials, for which standard tables are available, even where tighter bounds are possible with considerably more complex integral expressions. In the latter case the simpler expression is presented together with the alternative reference. We justify this by noting that, since, in general, P_e is expressed in terms of the error function, $\overline{\Phi}(k \times \text{s.n.r.})$, where k is a factor depending on the modulation system, a very small change in the limit, i.e. s.n.r., is required to cause a large change in P_e. Hence, relatively loose bounds on P_e are covered by a very small range of s.n.r. For instance, for 4-level f.s.k. the listed upper and lower bounds on P_e are in the ratio of three to one. This, however, corresponds to a range of about 0·2 dB at $P_e = 10^{-8}$.

13.3.3 Interpretation of P_e

The interpretation of the probability of error in a multilevel system requires some caution. First, the errors refer to symbol errors and, since each symbol has an information capacity of $\log_2 M$ binary digits, an error in a given symbol must affect at least one binary digit in $\log_2 M$ and at most every binary digit. Hence, the bit error probability lies between $P_e/\log_2 M$ and P_e, the exact relation depending purely on the coding scheme used. In this respect, the Gray code calls for attention because of its property that wherever the multilevel signal changes to the next adjacent symbol, the corresponding binary word changes by only one digit. As errors due to noise, in systems having high s.n.r., are predominantly caused by confusion of nearest neighbour signals, the Gray code offers a real advantage and obtains the lower limit quoted above.

As a further implication, since a multilevel system carries an information rate $\log_2 M$ times greater than that carried by a binary system, the bandwidth required for transmission may be scaled down by the same ratio. It follows that, for the same average signal energy the noise energy is also scaled down by $\log_2 M$, assuming a white noise spectral density, and that the s.n.r. requirement for the same error probability is reduced by $\log_2 M$. Hence, comparisons based on equal information rates will exhibit different trends.

As an illustration of the comparative behaviour of the different modulation systems the symbol probability of error is shown in Fig. 13.4 for binary and 4-level systems for both coherent and incoherent detection. The range of error probabilities covered, 10^{-5} to 10^{-12}, is the normal range of interest in the design of practical transmission systems. The s.n.r. is defined as $S/2\sigma^2$ for reasons which will become clear in Section 13.4.

We wish to point out that the geometrical signal space approach is not the only method available in calculating error probabilities. In the case of p.s.k. and f.s.k. an alternative approach exists based on the known probability distribution of the phase of a sine wave with added noise.[12] This approach is well illustrated by the work of Bennett[8] for the general case, of Cahn,[9] Prabhu[10] for p.s.k., and of Mazo and Salz[11] for f.s.k.

13.3.4 Formulas for P_e for optimum systems

S = mean signal power

σ = r.m.s. noise

The r.f. s.n.r. is defined as $S/2\sigma^2$ (see Section 13.4).

a.s.k., coherent

$$P_e = \frac{2(M-1)}{M} \overline{\Phi}\left(\frac{\delta}{2\sigma}\right) \tag{13.16}$$

a.s.k., incoherent (envelope detection, see Section 13.8)

$$P_e > \frac{1}{M} \exp\left(-\delta^2/8\sigma^2\right)$$

$$\leqslant \quad \exp\left(-\delta^2/8\sigma^2\right) \tag{13.17}$$

in the above δ the amplitude separation between adjacent levels, is given by

$$\delta = \{6S/(M-1)(2M-1)\}^{\frac{1}{2}} \tag{13.18}$$

or, in terms of the peak amplitude, A_p

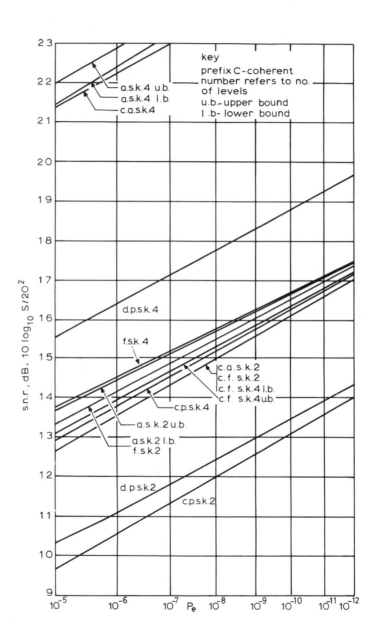

Fig. 13.4 S.N.R. against probability of error for optimal systems

$$\delta = A_p/(M-1) \tag{13.19}$$

p.s.k., coherent
for $M > 2$

$$\overline{\Phi}\!\left(\frac{\sqrt{S}}{\sigma}\,\sin \pi/M\right) < P_e < 2\overline{\Phi}\!\left(\frac{\sqrt{S}}{\sigma}\,\sin \pi/M\right) \tag{13.20}$$

The above upper bound is the same expression as the approximate equality derived by Cahn.[37]
for $M = 4$

$$P_e = 2\overline{\Phi}\!\left(\frac{\sqrt{S}}{\sqrt{2}\sigma}\right) - \overline{\Phi}^2\!\left(\frac{\sqrt{S}}{\sqrt{2}\sigma}\right) \simeq 2\overline{\Phi}\!\left(\frac{\sqrt{S}}{\sqrt{2}\sigma}\right) \tag{13.21}$$

p.s.k., incoherent
for $M = 2$

$$P_e = \tfrac{1}{2}\exp\left(-\frac{S}{2\sigma^2}\right) \tag{13.22}$$

for $M \geqslant 4$

$$P_e \simeq 2\overline{\Phi}\!\left(\frac{\sqrt{S}}{\sigma}\,\sin\frac{\pi}{\sqrt{2}M}\right) \tag{13.23}$$

f.s.k., coherent

$$\overline{\Phi}\!\left(\frac{\sqrt{S}}{\sigma\sqrt{2}}\right) \leqslant P_e \leqslant (M-1)\,\overline{\Phi}\!\left(\frac{\sqrt{S}}{\sigma\sqrt{2}}\right) \tag{13.24}$$

Eqn. 13.24 is obtained directly from eqn. 13.13.
f.s.k., incoherent

$$P_e = \frac{1}{M}\exp\left(-\frac{S}{4\sigma^2}\right)\sum_{q=2}^{M}{}^{M}C_q\,(-1)^q\,\exp\left(\frac{S(2-q)}{4q\sigma^2}\right) \tag{13.25}$$

for $M = 2$ this reduces to

$$P_e = \tfrac{1}{2}\exp\left(-\frac{S}{4\sigma^2}\right) \tag{13.26}$$

For graphical results encompassing a more comprehensive range of parameters, the reader is referred to Prabhu[10] who offers large-scale plots for c.p.s.k. and d.p.s.k. for $M = 2, 4, 8$ and 16.

We leave this Section on a cautionary note prompted by the two underlying propositions which govern the above results—optimum detection and Gaussian-noise statistics. In practice, optimum criteria sometimes cannot be applied because of technical and economic constraints, and alternatives, such as envelope detection for a.s.k., or discriminator detection for f.s.k., may be preferred; we shall discuss these in Section 13.8. In some systems, such as radio links, the Gaussian noise model must be regarded as merely a convenient basis for comparison of different systems, and significant deviations from the theoretical results so obtained, particularly at high s.n.r. values, should be anticipated.

13.4 Spectra of digital modulation

Of immediate concern in the choice of any one modulation system is its bandwidth occupancy. Analysis of modulated carrier spectra has received much attention with significant contributions dating back over the last half-century.[13,14] The methods used have reflected the direction of application and the mathematical modelling of the period. We thus find the early treatments based on sinusoidal modulating functions and harmonic analysis. With the birth of statistical communication theory, the modulating function became a continuous random variable leading to an analysis leaning heavily on statistical techniques.

In digital modulation, the modulating function is a digital sequence which may, with caution, be modelled by a discrete random variable. The proviso is inserted to alert the unwary against the possibility of certain unique deterministic rather than probabilistic features, in the presence of which the statistical model is not fully descriptive and may lead to misrepresentation. An obvious example might be the requirement for transmission of a unique digital pattern, in which case, spectral analysis based on a statistical regime would be completely misleading.

13.4.1 Amplitude modulation

Amplitude modulation is the most familiar modulation process, the modulated signal being described by

$$s(t) = x(t) \cos (\omega_0 t + \theta) \tag{13.27}$$

The above equation differs from eqn. 13.1 in that amplitude is now allowed to be a function of time. In fact, $x(t)$ is now assumed to be a digital sequence. The analysis of the spectrum of such an a.m. process might seem to be a matter of unnecessary formalism in the light of the well-known a.m. spectrum for the case of a modulating sinusoid. However, the few steps involved are useful in drawing attention to basic differences between harmonic and statistical analysis, and introducing the basic method of spectral computation which is applied to the case of more complex modulating processes, such as angle modulation.

We therefore note that if and only if, $x(t)$ is known as a specific time sequence, the spectrum may be obtained directly by simple harmonic analysis, because the spectrum of $x(t)$, $S_x(f)$, then exists and may be described as either an amplitude spectrum with Fourier harmonic coefficients if $x(t)$ is repetitive, or as an energy spectrum using the Fourier transform, if $x(t)$ is aperiodic. In either case the modulated spectrum is then given by the Fourier transform of eqn. 13.27

$$S(f) = S_x(f) * \tfrac{1}{2}\{\delta(f - f_0) + \delta(f + f_0)\}$$
$$= \tfrac{1}{2}S_x(f - f_0) + \tfrac{1}{2}S_x(f + f_0) \tag{13.28}$$

In the above we invoke Duhamel's theorem[5] that multiplication in the time domain is equivalent to convolution (denoted by *) in the frequency domain. If, however, $x(t)$ is a random sequence then it can only be described in statistical terms involving averaged functions such as the autocorrelation function of $x(t)$, $R_x(\tau)$. The general method of attack is thus to obtain the autocorrelation function of the modulated signal $R_s(\tau)$ and hence the power density spectrum $W_s(f)$ which is related to $R_s(\tau)$ by a Fourier transform (Wiener-Khinchin theorem[5]).

It may be shown that

$$R_s(\tau) = \tfrac{1}{2}R_x(\tau) \cos \omega_0 \tau \tag{13.29}$$

Again, applying Duhamel's theorem and the Wiener-Khinchin theorem, we have

$$W_s(f) = \tfrac{1}{2}W_x(f) * \tfrac{1}{2}\{\delta(f - f_0) + \delta(f + f_0)\}$$
$$= \tfrac{1}{4}W_x(f - f_0) + \tfrac{1}{4}W_x(f + f_0) \tag{13.30}$$

It is seen from the above that, whether the signal $x(t)$ is deterministic or random, the a.m. signal spectrum consists of two sidebands symmetrically disposed around f_0, the carrier frequency.

The process described by eqn. 13.27 refers to a suppressed carrier a.m. system, the only one considered here on the grounds of efficiency. Sideband manipulation is easily described by consideration of eqns. 13.28 and 13.30. Single sideband a.m. (s.s.b.) may obviously be obtained by band-pass filtering of the double-sideband spectrum. Where $x(t)$ contains d.c. or near-zero-frequency components separation of sidebands with real filters becomes impractical. In such a case, a modified technique known as vestigial sideband modulation (v.s.b.) must be adopted, in which a carefully controlled portion of the unwanted sideband is left. It is known[15,16] that if the amplitude spectrum is made skew-symmetric about the carrier, and if the phase spectrum is linear, then, with the use of coherent product demodulation, distortionless detection of the modulating signal is possible.

S.S.B. and v.s.b. are special cases of asymmetric sideband signals; in this context we draw attention to the existence of an alternative method of analysis based on the representation of the modulated waveform in analytic signal terms. The basic reason for such a formulation is that in Fourier transform analysis a spectrum extends in both positive and negative frequencies; this leads to a simple representation involving only real terms. However, it is sometimes convenient to formulate positive frequency spectra and accept the consequent complex signal representation. In such a formulation $x(t)$ is expressed as

$$x(t) = y(t) + j\,z(t) \tag{31.31}$$

the condition for a positive frequency spectrum being satisfied if y and z are Hilbert transform related. This formulation and its application are described in the work of Voelcker.[17] It is shown therein, that sidebands can be manipulated without affecting the envelope or bandwidth of a signal, leading to a 'common-envelope' set of signals, and also that under certain conditions, asymmetric sideband waveforms may be demodulated by envelope detection.

13.4.2 Definition of s.n.r.

We here justify our earlier convention regarding s.n.r. Let the modulating signal be confined to a bandwidth of W hertz. Let n_0 be the white-noise power spectral density extending double-sided around zero frequency. Then the baseband noise power, or variance in a unit load, is $\sigma^2 = 2n_0 W$. After modulation we assume that an r.f. bandwidth of $2W$ is used, resulting in a noise power of $2\sigma^2$. If the average modulated signal power is S, then the s.n.r., defined in terms of average powers is $S/2\sigma^2$.

13.4.3 On the spectra of angle-modulated processes

The analysis of the spectra of phase- or frequency-modulated processes is one of the most challenging problems in communication theory, and has received much effort. An early milestone was the paper by Carson in 1922[13] indicating, contrary to current belief, that frequency modulation need not necessarily result in a narrower bandwidth than amplitude modulation. The best known approach is the classical method of analysis[16,18] based on the assumption of a sinusoidal modulating signal. However, this becomes unmanageable for more complicated modulating waveforms. The application of statistical techniques led to the characterisation of the modulating process as a Gaussian random variable. The autocorrelation function of a signal, angle modulated by such a process is quite simple in form,[19] but the Fourier transform necessary to obtain the spectrum resulted in a maze of approximation terms of varying complexity. Major contributions are to be found in the work of Middleton,[19] Woodward,[20] Blachman,[21] and Medhurst.[23]

Conclusions about two limiting cases emerge from the Gaussian model treatment which have significance in the digital-modulation context. For the case where the resulting frequency deviation is large compared to the bandwidth of the modulating signal, the power-density spectrum approaches the 1st-order (simple) probability-density function of the modulating signal. This case may be described as slow, strong angle modulation, and the above statement is sometimes called Woodward's theorem.[22] It may also be labelled as rather obvious if the power-density spectrum over a given band is regarded as the proportion of time which the modulated signal spends in that frequency band.

The second case arises where the resulting mean frequency deviation is small compared to the frequencies present in the modulating wave, the modulated spectrum then approaching that obtained by amplitude modulation.[19,21]

Apart from the above limiting cases work on Gaussian angle modulation exhibited significant lack of generality; the situation was altered by an important paper by Abramson[24] in which he domonstrated an approximate solution to the spectrum applicable to all cases, based on a power series expansion of the modu lated function. Abramson noted certain strikingly simplifying features in the expansion, hitherto unnoticed, leading to an elegant solution.

The work on the spectra of angle modulation by Gaussian processes is not directly applicable to modulation by a random digital signal. Spectral analysis of digital angle modulation has resulted in a new body of work, the results of which are unfortunately complicated and only suited to machine computation. It is our intention to present the underlying philosophy of the methods which have been evolved.

The general expression for the angle-modulated waveform may be described by

$$s(t) = A \cos (\omega_0 t + \theta(t) + \alpha) \tag{13.32}$$

where $\theta(t)$ is the angle modulation, in radians, and is a random phase angle. Denoting the modulating message by $x(t)$ we have, for phase modulation

$$\theta(t) = x(t)$$

and for frequency modulation

$$\frac{d\theta}{dt} = \omega_d x(t)$$

or,

$$\theta(t) = \omega_d \int_0^t x(t') dt' \tag{13.33}$$

ω_d is a constant representing the angular frequency deviation per unit of the modulating signal. If the modulating message is digital with signalling intervals of duration T, then

$$x(t) = \sum_{k=-\infty}^{\infty} x_k(t - kT)$$

$$= \sum_{k=-\infty}^{\infty} a_k g(t - kT) \tag{13.34}$$

the latter description separating the shape of the waveform $g(t)$ from the value of the digital weighting coefficient a_k, pertaining to the kth interval. The waveforms whose spectra are desired may therefore be described by

$$s(t) = A \cos \left\{ \omega_0 t + \sum_{-\infty}^{\infty} a_k g(t - kT) + \alpha \right\} \tag{13.35}$$

for p.s.k., and

$$s(t) = A \, \cos \left\{ \omega_0 t + \omega_d \sum_{-\infty}^{\infty} a_k \int_0^t g(t' - kT) dt' + \alpha \right\} \tag{13.36}$$

for f.s.k.

The a_k are assumed to be discrete random variables. Two major distinct methods for the analysis of the spectra of the signals described by eqns. 13.35 and 13.36 have been reported, and represent major contributions. The first is the method due to Anderson and Salz[25] and the second that due to Shimbo.[26]

The first is best described by quoting the authors themselves. 'Our method of attack on the problem is direct. We calculate the segmented Fourier transform (of eqn. 13.36), obtain its magnitude squared, average over all random variables, divide by the length of the segment, and then evaluate the limit as the length increases without bound. After considerable amount of bookkeeping, we obtain general formulas.' The segmented Fourier transform referred to is the Fourier transform of a portion of signal of duration N signalling intervals. The Anderson and Salz analysis assumes independent modulating symbols a_k and a fixed modulating waveform shape $g(t)$. The final expression is complicated, but suitable for numerical calculation by a digital computer.

The second method, due to Shimbo[26] is a specific application of a very general method of spectrum evaluation developed by the same author.[27] In this formulation the digitally modulated signal is described by

$$Q(t) = \sum_{n=-\infty}^{\infty} s(t - nT; w_n) \tag{13.37}$$

where w_n denotes a modulated parameter such as amplitude, width or shape of a pulse, or phase of a sinusoid. The autocorrelation function of $Q(t)$ is shown to be

$$R_Q(\tau) = \sum_{n=-\infty}^{\infty} R_n(\tau - kT) \tag{13.38}$$

where

$$R_0(\tau) = \mathop{E}_{w} \left\{ \frac{1}{T} \int_{-\infty}^{\infty} s(t; w) \, s(t + \tau; w) \, dt \right\} \tag{13.39}$$

and

$$R_n(\tau) = \mathop{E}_{w_0 \times w_n} \left\{ \frac{1}{T} \int_{-\infty}^{\infty} s(t; w_0) \, s(t + \tau; w_n) \, dt \right\} \tag{13.40}$$

$\mathop{E}_{w}\{\ \}$ signifies the average taken over the ensemble of all possible modulated parameters taken singly. $\mathop{E}_{w_0 \times w_n}\{\ \}$ signifies the average taken over the ensemble of all possible modulated parameters taken in pairs spaced by n signalling intervals. Hence, $R_0(\tau)$ is the ensemble average of the autocorrelation functions, and $R_n(\tau)$ is the ensemble average of the crosscorrelation functions with a time difference of nT.

The power density spectrum is then expressed as the Fourier transform of eqn. 13.38.

$$W_Q(f) = W_0(f) + 2 \sum_{n=1}^{\infty} W_n(f) \cos 2\pi fT \qquad (13.41)$$

where

$$W_n(f) = \int_{-\infty}^{\infty} R_n(\tau) e^{j2\pi fT} d\tau \qquad (13.42)$$

To apply the formulas to the analysis of the spectrum of p.s.k. or f.s.k., $s(t)$, as described by eqns. 13.35 or 13.36, is inserted into eqn. 13.37. The procedure for an f.s.k. signal is described in Reference 26; that for p.s.k. is to be found in Reference 28.

13.4.4 Spectra of p.s.k.

Applying the above analysis it may be shown that the spectrum of p.s.k. is given by

$$W(f) = T \left(\frac{\sin \pi fT}{\pi fT} \right)^2 (1 - b^2) + b^2 \delta(f) \qquad (13.43)$$

where

$$b = E \{\exp (j\theta_k)\}$$

As $\exp(j\theta_k)$ is the signal vector with modulated phase θ_k, b is simply the average of the modulated vectors. The above formula refers to the spectrum centered on a fictional zero-frequency carrier. Successive modulating symbols are assumed to be independent.

Two important properties immediately stand out. First, the spectrum *is independent of the number of modulated phases, M.* Secondly, *if the average of the modulated vectors is zero, then no discrete line component at the carrier frequency is present.* The latter condition pertains when all symbols are equiprobable and when the phases are equally disposed within 2π. When this is so, the spectrum reduces to

$$W(f) = T \left(\frac{\sin \pi fT}{\pi fT} \right)^2 \qquad (13.44)$$

This is the familiar $\text{sinc}^2 x$ function and is shown in Fig. 13.5.

The method of analysis developed by Shimbo enables consideration of modulating messages whose symbols are not independent; in this case, eqn. 13.43 is

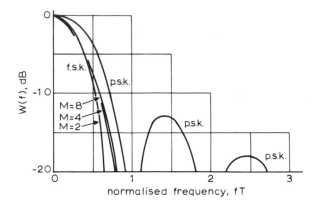

Fig. 13.5 Power-density spectra of p.s.k. and f.s.k.
For f.s.k. $2f_dT = 1/M$

modified by the addition of a term dependent on the correlative properties of the message. This term has, as might be expected, the effect of compressing the spectrum, and hence, as far as spectral occupancy is concerned, the assumption of independent symbols leans towards a worst-case situation.

In the above, $g(t)$ the modulating function has been assumed to be rectangular of duration T. Other cases where $g(t)$ has nonzero transition times have been considered; Marshall[28] derives spectra for linear transitions, and Glance[34] those produced by a raised-cosine modulating function. The lobes of the $(\text{sinc}^2 x)$ spectrum are in both cases somewhat modified in detail, but the differences are not considered to be significant in practice.

13.4.5 Spectra of f.s.k.

In considering the spectral properties of f.s.k. it is necessary to distinguish two disparate general cases, which depend on the behaviour of the phase angle α in the expression for the modulated signal, eqn. 13.36. If α is assumed to be random and uniformly distributed within 2π, then it bears no correlation to the modulation and may, at the signalling transitions, take any random value. This leads to a possible phase discontinuity, as shown in Fig. 13.6. Phase continuity may be achieved by forcing α to have a definite correlation with the modulating message.[26,29] In practice, both phase-continuous and phase-discontinuous f.s.k. may be achieved; the latter by frequency-keying separate, unlocked oscillators, the former by frequency deviating a single oscillator.

Unfortunately, the expression for the spectrum of f.s.k.[25,26] takes the form of an unwieldy series, and we do not feel that its inclusion here would add any insight.

Anderson and Salz[25] present graphs of the spectra of f.s.k. for $M = 2$, 4 and 8 for different values of the frequency deviation parameter ω_d. The case treated is phase-discontinuous. It should be noted from eqn. 13.36 that the frequency

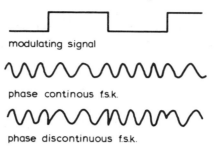

Fig. 13.6 F.S.K. waveforms

deviation between adjacent levels or states is $2\omega_d$, and that the peak-to-peak deviation is $2\omega_d(M-1)$. The particular case where $2f_d T = 1/M$ is shown in Fig. 13.5.

13.5 Effects of filtering, distortion and interference

In Sections 13.2 and 13.3 we considered digital modulation systems with optimal properties; in particular, each modulated waveform corresponding to a given modulating message was assumed to 'stand alone' unperturbed by the presence of neighbouring waveforms. The probability of error so obtained pertained to an optimal quasistatic situation. The real situation involves a transmission channel whose frequency response is limited and must be chosen by design within the specified limits. A model of such a modulation system is shown in Fig. 13.7, the receiver filter now forming a new element in the design. In practice, it is usual to find a transmitter filter and possibly a baseband filter before the modulator. We again encounter the trade-off generic to digital systems, an example of which arose in Chapter 12 in connection with baseband systems. If the effective filter bandwidth is reduced, the noise energy is reduced, but the signal intersymbol interference effects are increased; if the filter bandwidth is increased, the noise energy increases, but intersymbol interference is reduced. The design problem is to choose a filter characteristic such as to minimise the probability of error for a given signal/noise ratio.

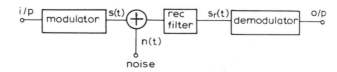

Fig. 13.7 Model of modulation system with filtering

We begin by observing that for the case of a.s.k. all the baseband filtering considerations described in Chapter 9 are relevant, since a baseband signal is synonymous with a zero-frequency carrier amplitude-modulated signal. We therefore here confine discussion to the p.s.k. and f.s.k. case.

Referring to Fig. 13.7 if, as before (eqn. 13.32), we let

$$s(t) = A \cos \{\omega_0 t + \theta(t)\} \tag{13.45}$$

where $\theta(t)$ is the angle modulation, the output $s_r(t)$ of the receive filter may be written as

$$s_r(t) = B(t) \cos \{\omega_0 t + \theta(t) + \beta(t)\} + n(t) \tag{13.46}$$

The distorting action of the filter is described by the envelope function $B(t)$, which is no longer constant, and by $\beta(t)$, a phase deviation. Both depend on the filter and the modulating message. The noise vector may be decomposed so that $s_r(t)$ may be rewritten as

$$s_r(t) = V(t) \cos \{\omega_0 t + \theta(t) + \beta(t) + \alpha(t)\} \tag{13.47}$$

where $\alpha(t)$ is the phase contribution from the noise. The problem then becomes one of finding the probability distribution of $\alpha(t)$ in the presence of the distortion $\beta(t)$, and hence the probability of error. A complete analysis of p.s.k. and d.p.s.k. for $M = 2$ and $M = 4$ is given in Calandrino *et al*[30] and Marshall.[31] Tjhung and Wittke[32] cover the case of binary f.s.k.

At the outset, in choosing a filter design for any angle-modulated system, we would expect, intuitively, filters with linear phase (constant group delay) to perform better than those having unspecified phase characteristics. This, in fact, is demonstrated by Marshall[31] to be the case. Of the six types of realisable filters considered by him, the Butterworth, which is characterised by a maximally flat amplitude-frequency response, is worst, while the delay-specified filters such as the Bessel, equal-ripple delay and Scanlan[33] types are best. The Gaussian frequency-response filter is found to lie between the Butterworth and Bessel types in terms of performance.

Of great interest is the choice of filter bandwidth for optimum noise immunity. In this context, the filter bandwidth is defined as the equivalent noise bandwidth, B

$$B = \tfrac{1}{2} \int_{-\infty}^{\infty} |H(\omega)|^2 \, d\omega/2\pi \tag{13.48}$$

The noise variance at the filter output is

$$\sigma^2 = 2n_0 B$$

where n_0 is the white noise double-sided power spectral density.

In Fig. 13.8 we reproduce Marshall's results for binary d.p.s.k. giving the signal/noise ratio at the receiver input (see Section 13.4 for definition) required to maintain a $P_e = 10^{-6}$, against normalised filter bandwidth BT, for Gaussian filters. It

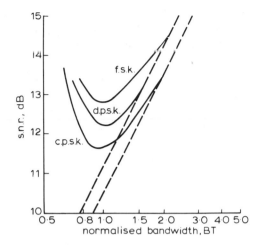

Fig. 13.8 S.N.R. against filter bandwidth for binary c.p.s.k., d.p.s.k., and f.s.k. for $P_e = 10^{-6}$, Gaussian filtering

is seen that s.n.r. is minimised at a value of BT very close to one. This is also found to hold for quaternary d.p.s.k. The straight line corresponds to the optimal quasistatic modulation considered in Sections 13.2 and 13.3, in which we assumed $BT = 1$.

Tjhung and Wittke[32] consider the case of binary incoherent f.s.k. with Gaussian and rectangular filtering. Their results indicate that, for Gaussian filtering, a choice of the frequency deviation index $2T f_d = 0.7$ minimises the s.n.r. required to achieve a given P_e. We show this case in Fig. 13.8 using data extracted from the above reference. We can, therefore, compare the effects of Gaussian filtering on binary f.s.k. and binary d.p.s.k. It is seen that the minimum s.n.r. for binary d.p.s.k. is about 0·6 dB lower than that for binary f.s.k.

It is interesting to note that the minimum s.n.r. for the f.s.k. system also occurs at $BT = 1$. The Scanlan-filtered d.p.s.k. outperforms Gaussian filtered d.p.s.k. and is about 0·5 dB worse than the quasistatic system at $BT = 1$.

Calandrino *et al*,[30] in addition, present results for coherent p.s.k. ($M = 2$ and 4) with Gaussian filtering and their data for the binary case is also plotted in Fig. 13.8. It is clear that coherent p.s.k. outperforms all rivals; again its s.n.r. minimum occurs at $BT = 1$, and is about 1 dB worse than for the optimal p.s.k. system.

We therefore conclude that the comparative trends in modulation-system performance apply to both the optimal quasistatic regime of Fig. 13.4 and to the spectrally distorted regime of Fig. 13.8.

Tjhung and Wittke[32] present a comparative treatment of binary f.s.k. and coherent p.s.k. with rectangular (truncation) filtering, in the course of which they draw attention to an interesting feature. Given a free choice of the bandwidth BT, p.s.k. requires ₁ smaller s.n.r. than f.s.k.; if, however, the bandwidth is restricted to less than about 1·2 then the reverse is true and f.s.k. outperforms p.s.k. This phenomenon is no more than a demonstration of the relative bandwidth occupancy of p.s.k. and f.s.k. as indicated by Fig. 13.5. Clearly, at $BT = 1$ the p.s.k. spectrum is severely truncated, much more so than the f.s.k. spectrum.

Extraneous interference

Certain types of modulated systems suffer from the potential presence of interference from other systems. In radio point-to-point networks there is a risk of interference from channels sharing the same spectral window, referred to as cochannel interference, and from channels allocated to neighbouring spectral windows, referred to as interchannel interference. In concept, the problem is similar to crosstalk in line systems. The main body of work on the subject is exclusively devoted to p.s.k. systems, reflecting the widespread application of such systems to point-to-point digital radio networks. The problem involves multiple parameters; the type of transmit and receive filters used, the channel frequency separation; the attenuation of the interfering channels. It is therefore unreasonable to expect general results, and problems have to be dealt with on their merits.

Benedetto *et al*[35] derive an error analysis for multilevel coherent p.s.k. for interchannel and cochannel interference. They present comprehensive graphical results for the s.n.r. required to maintain $P_e = 10^{-6}$ using Butterworth receiver filters. It appears that the optimum receiver filter bandwidth is 1·1, a value not significantly different from the optimum obtained above (Fig. 13.8), under somewhat different conditions. The results presented should facilitate the choice of system parameters in any given situation.

The case of binary p.s.k. subject to multiple cochannel interferences has been analysed by Rosenbaum[36] who also presents exhaustive graphical results.

Marcatili[50] studies a system consisting of separate frequency division multiplexed channels each carrying a.s.k. He considers the simultaneous choice of within-channel intersymbol interference and channel frequency separation f_s to minimise the product $T f_s$. Again it is difficult to draw general conclusions from a large body of graphical data based on specific assumptions.

13.6 Combined modulation systems

In previous Sections, we assumed that modulation of a carrier was achieved by working on one single parameter, either amplitude, phase or frequency. A natural extension is to consider modulation of more than one of these parameters at once. The motivations for the study of such schemes may be quite diverse; in data transmission, for instance, the pursuit of efficient signal formats suffices. In radio links, the s.n.r. under no-fade conditions is very excessive, and the attractive prospect emerges of sending lower-priority information by combined modulation with the main signal, thus utilising the channel at a higher capacity.

The possibility of combined modulation has been recognised for some time, an example being quadrature amplitude modulation[43] in which two independent modulating messages are amplitude modulated onto two sinusoids of quadrature phase and the two modulated quadrature components are subsequently summed. Using the nomenclature of Section 13.2, the quadrature shift-keyed (q.s.k.) signal, corresponding to the ith message of one modulating signal and the kth message of the other, may be expressed as

$$s_{ik}(t) = A_i \cos \omega_0 t + B_k \sin \omega_0 t \qquad (13.49)$$

Because the two message sets are assumed to be independent the total number of possible modulated states is $M = M_a M_b$ where M_a and M_b are the individual message alphabets. For instance, if four amplitudes are used then $M = 16$. The q.s.k. signal space is shown in Fig. 13.9b. Quadrature modulation was born some time before the concept of signal space and, while its functional form as described by eqn. 13.49 is simple, and leads to a simple implementation, it is clear that the signal space so obtained is just one of a countless number which may be designed by more

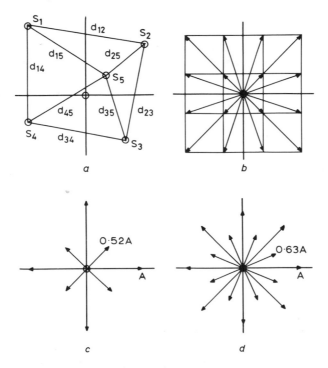

Fig. 13.9 Combined modulation transmitter signal spaces
  a 5-point signal space
  b 4-level q.s.k., $M = 16$
  c Combined a.s.k. − p.s.k., $M = 8$
  d Combined a.s.k. − p.s.k., $M = 16$

sophisticated coding and modulation. In particular, the implications of 2-dimensional manipulation by combined amplitude and phase modulation have been appraised by several workers.[37-39]

Again, the principal property of interest in assessing the efficiency of any one scheme is its noise immunity, or, in other words, the relation between the s.n.r. and P_e. We can draw directly on our earlier treatment (Section 13.3) to gain an insight into the problem in the present context; in particular, the nearest-neighbour concept will be seen to be the key to the solution. As an illustration, consider a signal space containing five signal points, $S_1 - S_5$, as shown in Fig. 13.9a. Under conditions of high s.n.r. with Gaussian noise, the probability of confusion between distant signal points, such as S_1 and S_3, becomes insignificant in comparison with that pertaining to immediate neighbours, such as S_1 and S_5. The probability of an error, given that S_1 was transmitted, $p(e|S_1)$ may therefore be equated to the probability that the noise of variance σ^2, will transgress the halfway distances, $d_{12}/2, d_{14}/2$ and $d_{15}/2$ to the immediate neighbours

$$p(e|S_1) = \bar{\Phi}\left(\frac{d_{12}}{2\sigma}\right) + \bar{\Phi}\left(\frac{d_{14}}{2\sigma}\right) + \bar{\Phi}\left(\frac{d_{15}}{2\sigma}\right) \tag{13.50}$$

The above assumes that all signals are equiprobable. The overall average probability of error P_e is then the average of the individual probabilities above, over the complete set of five signals.

We generalise the above example by calling the set of local immediate neighbour signal points around a given point S_i, D_i, and we have:

$$p(e|S_i) = \sum_{D_i} \bar{\Phi}\left(\frac{d_{ij}}{2\sigma}\right) \tag{13.51}$$

and

$$P_e = \frac{1}{M} \sum_{i=1}^{M} p(e|S_i) \tag{13.52}$$

It is clear that in designing a combined modulation scheme the immediate-neighbour distances should be made as large as possible. This can obviously be done by making the signal distances from the origin large, at the expense, however, of increased signal energy. The question arises whether, for a given total number of states M, it is possible to design combined modulation schemes whose noise immunity is better than that achievable with simple a.s.k., p.s.k. or f.s.k. Hancock and Lucky[38,39] deduce that this is so. Their findings, pertaining to Gaussian noise for $M = 2$, 4, 8 and 16 may be briefly summarised as follows: when the signal is peak power-limited p.s.k. alone is optimum up to and including $M = 8$, while combined p.s.k.–a.s.k. becomes optimum for $M = 16$; when the signal is average power-limited, p.s.k. alone is optimum up to and including $M = 4$, combined p.s.k.–a.s.k. taking over for $M = 8$ and $M = 16$. The optimum p.s.k.–a.s.k. schemes proposed by Hancock and Lucky are shown in Fig. 13.9c and d.

Recent investigations[40,41] confirm these results in experimental systems. Foschini *et al.*[42] extend their treatment to include the effects of random phase jitter and frequency offset. They derive a number of signal spaces based on a concentric ring structure identified by the number of signal points on each ring; 1–6–9, for instance, describes a scheme having one point at the origin, six in the first ring and nine in the outer ring. Their results indicate that, for $M = 16$, with no phase jitter, q.s.k. is marginally best, while with a jitter of $3°$ r.m.s., the 1–6–9 and 1–5–10 signal spaces consistently outperform other schemes.

Yet another recent variation on this theme[48] bases the signal space design on a structure comprising concentric hexagons. This results in a 'honeycomb' signal space which is claimed to be an optimum geometry based on the tightest sphere-packing approach, leading to a near optimum noise immunity.

13.6.1 Modulation of an incoherent carrier
Hitherto, we have assumed that the carrier is a sinusoidal signal of determinate amplitude, phase and frequency. In some systems, some or all of these parameters have random components, which lead in the limit to noise-like carrier properties. Certain optical, infrared and microwave devices generate such carriers, and hence the feasibility of detection of modulated incoherent carriers arises. It is possible to make some rather obvious eliminations from the list of candidate modulation schemes. Because noise may be regarded as a summation of individual sinusoids of equal energy and random phase, it is unreasonable to attempt to convey a message by phase modulation; this eliminates p.s.k. If the noise carrier is a narrow-band process whose centre frequency can be shifted by the modulation, then f.s.k. is a candidate. However, in most cases of practical interest, the noise carrier bandwidth is wide compared to the realisable frequency deviation, which clearly precludes f.s.k. Of the schemes considered we are therefore left with a.s.k. Although we cannot use optimum detection design because our previous signal representation ideas are not applicable, it is easy to propose a practical detection scheme. Because the modulated signal has unknown phase, the detector must operate on a power rather than on an amplitude basis; as such an r.m.s. form of detector could thus demodulate the message.

Such a system, operating with a square-law detector, is analysed by Ducot[49] who shows that the detected signal to noise ratio is proportional to the ratio of noise-carrier bandwidth/message bandwidth. The model used assumes a detector of unlimited bandwidth followed by a filter of message bandwidth. In practice, the noise-carrier bandwidth will be limited by the demodulator, and hence the carrier bandwidth in the above ratio should be replaced by the demodulator bandwidth. The general result seems reasonable if the message bandwidth is viewed as the reciprocal of the time allowed for measurement of a given symbol.

13.7 Comparison of digital modulation systems

To round off and summarise the work described in this Chapter we shall endeavour to assess the comparative performance of the different modulation systems. This is less easy than might be expected, since it is difficult to propose a unique criterion of excellence.

A physically meaningful measure of the efficiency of a system has been proposed by Sanders[46] in connection with satellite communication, who defines an index of efficiency β as;

$$\beta = (S/N)\,(B/R) \tag{13.53}$$

where S, N = signal and noise mean power, respectively

B = channel bandwidth required, Hz

R = information rate, bit/s

Using our previous notation, $S/N = S/2\sigma^2$, the signal/noise ratio. It is clear that a desirable system is one which requires a small s.n.r. to achieve a high information rate R in a small bandwidth, B; β, therefore, should be as small as possible. β may alternatively be interpreted as an effective signal/noise ratio required to transmit 1 bit/s using 1 Hz of bandwidth.

Using Fig. 13.4 we have extracted results for 2- and 4-level systems for $P_e = 10^{-6}$. These are presented in Table 13.2. In the case of the amplitude-

Table 13.2 Comparison of modulation systems for
$M = 2, M = 4$ at $P_e = 10^{-6}$

System	s.n.r. $\dfrac{S}{2\sigma^2}$	R/B	Effective s.n.r., β
	dB		dB
a.s.k.2	14·4	2	11·4
c.a.s.k.2	13·6	2	10·6
d.p.s.k.2	11·1	1	11·1
c.p.s.k.2	10·6	1	10·6
f.s.k.2	14·2	1	14·2
c.f.s.k.2	13·6	1	13·6
a.s.k.4	22·9	4	16·9
c.a.s.k.4	22·2	4	16·2
d.p.s.k.4	16·4	2	13·4
c.p.s.k.4	13·8	2	10·8
f.s.k.4	14·5	2	11·5
c.f.s.k.4	13·9	2	10·9

modulated systems, we have assumed that s.s.b. transmission is used; this, in practice, would require a small pilot carrier for the case of coherent reception, but for simplicity, this is not allowed for in the s.n.r. It is at once possible to note certain peculiar differences between the systems.

(i) *Coherent binary versus incoherent binary*
   For all systems the difference is no more than 0·8 dB.

(ii) *Binary systems comparison*
   A.S.K. and p.s.k. are equally good, while f.s.k. is some 3 dB worse.

(iii) *Coherent multilevel versus incoherent multilevel (M = 4)*
   For a.s.k. and f.s.k. the difference is no more than 0·7 dB. d.p.s.k., however, is 2·6 dB worse than c.p.s.k. This value reaches 2·3 dB asymptotically[44] for very high s.n.r.

(iv) *Multilevel systems comparison (M = 4)*
   F.S.K. outperforms all others with exception of c.p.s.k., whose performance it nearly equals with coherent reception. In particular incoherent f.s.k. is nearly 2 dB better than d.p.s.k. A.S.K. slips to the bottom of the league table.

We feel that the above comparison, although restricted in the number of levels, serves adequately to identify the above principal trends. Comparisons for larger M are to be found in References 44, 45, 39 and 40. We wish to remind the reader that for $M = 8$ the combined systems described in the previous section must be considered.[40]

While the criterion of system efficiency as described above certainly represents a meaningful measure of goodness, there are occasions when the system parameters have associated cost criteria which are heavily biased. It may be, in one instance, that restriction of spectral occupancy is of first importance, the necessary signal power being no problem, while, in another instance, restriction of signal power might outweigh spectral considerations. An example of the first is the transmission of visual telephony over an analogue f.d.m. link, while transmission from a space station illustrates the second case.

Where spectral occupancy is a limitation and other parameters are unimportant, a.s.k. systems have the distinct advantage of being directly amenable to single or vestigial sideband working, and must therefore be rated first. f.s.k. comes next, because its spectrum may be to some extent tailored by suitable choice of the frequency deviation, which is an additional independent parameter. A measure of the spectral occupancy of f.s.k. is given by a rule accredited to Carson, which is found to apply quite well to digital modulation:[47]

$$B = 2(B_m + f_{dmax}) \qquad (13.54)$$

where B_m is the modulating signal bandwidth and f_{dmax} is the maximum frequency deviation. For rectangular modulation $B_m = 1/2T$ and hence,

$$B = \frac{1}{T} + (M - 1)f_d \qquad (13.55)$$

For p.s.k., if we assume that it is necessary and sufficient to preserve the first lobe of the sinc x function then we arrive at a value $B = 2/T$. This is probably unrealistic, because we have seen that for Gaussian rather than abrupt filtering B may be constrained to $1/T$. However, since Carson's rule is, in the digital f.s.k. context, somewhat generous, the above proviso should not be read as an invitation to equate the spectral occupancy of p.s.k. and f.s.k.; examination of Fig. 13.5 shows that f.s.k. can be made to exhibit a more tightly bounded spectrum.

Where spectral occupancy is not a major problem and noise immunity is important then the choice rests between p.s.k. and f.s.k. For binary transmission, p.s.k., either coherent or differentially detected, clearly outperforms f.s.k. In 4-level transmission, however, f.s.k., either coherent or incoherent, is superior to d.p.s.k., while the coherent systems perform nearly equally.

13.8 Some nonoptimal detection schemes

Hitherto, we have considered modulation systems with optimum detection, the only concession to practical limitations being the discussion of the effects of filtering and interference. However, nonoptimal detection schemes, well-tried in analogue modulation systems, can be applied to digital information; particularly well-known are envelope detection of amplitude modulation and discriminator detection of frequency modulation.

Envelope detection of a.s.k.
In this scheme, the modulus of the received waveform is measured, and we therefore note that this is identical to the detection rule described for incoherent a.s.k. in Section 13.2. In fact, this form of detection is not quite optimum for incoherent a.s.k., but is very nearly so[6] for very high s.n.r. (greater than 20 dB). Kotelnikov[3] proves that under such conditions envelope detection is equivalent to optimum detection.

Discriminator detection of f.s.k.
Here the receiver consists of a filter followed by a discriminator whose output is sampled by a decision circuit at the signalling rate, to establish whether the input is above or below a threshold value. Shaft[51] analyses the case for binary f.s.k. and presents plots of P_e against s.n.r. from which the conclusion can be drawn that discriminator detection of f.s.k. 2 is less than 1 dB worse than optimum c.f.s.k. 2 for $P_e = 10^{-4}$. The deduction is also made that the best value of frequency separation is $0 \cdot 796/T$ with a filter noise bandwidth $BT = 1$.

13.9 References

1   LERNER, R.M.: 'Representation of signals' *in* 'Lectures on communication system theory', BAGHDADY, E. (Ed.) (McGraw-Hill, 1961)
2   SHANNON, C.E.: 'Communication in the presence of noise', *Proc. IRE*, 1949, **37**, pp. 10−21
3   KOTELNIKOV, V.A.: 'Theory of optimum noise immunity', (McGraw-Hill, 1959) English edn.
4   HARMAN, W.W.: 'Principles of the statistical theory of communication' (McGraw-Hill, 1963)
5   LEE, Y.W.: 'Statistical theory of communication' (McGraw-Hill, 1960)
6   ARTHURS, E., and DYM, H.: 'On the optimum detection of digital signals in the presence of white Gaussian noise − a geometric interpretation and a study of three basic data transmission systems', *IRE Trans.*, 1962, **CS 10**, pp. 336−373
7   SHANNON, C.E.: 'A mathematical theory of communication', *Bell Syst. Tech. J.*, 1948, **27**, pp. 379−423 and 623−656 Reprinted in book form as SHANNON, C.E., and WEAVER, W. (University of Illinois Press, Urbana, 1949)
8   BENNETT, W.R.: 'Methods of solving noise problems', *Proc. IRE*, 1956, **44**, pp. 609−638
9   CAHN, C.R.: 'Performance of digital phase-modulation communication systems', *IRE Trans.*, 1959, **CS-7**, pp. 3−6
10 PRABHU, V.K.: 'Error-rate considerations for digital phase modulation systems', *IEEE Trans.*, 1969, **COM 17**, pp. 33−42
11 MAZO, J.E., and SALZ, J.: 'Theory of error rates for digital f.m.', *Bell Syst. Tech. J.*, 1966, **45**, pp. 1511−1535
12 RICE, S.O.: 'Mathematical analysis of random noise', *ibid.*, 1944, **23**, pp. 282−332 and 1945, **24**, pp. 46−157
13 CARSON, J.R.: 'Notes on the theory of modulation', *Proc. IRE*, 1922, **10**, pp. 57−64
14 HARTLEY, R.V.L.: 'Relations of carrier and sidebands in radio transmission', *Bell Syst. Tech. J.*, 1923, **2**, pp. 90−112
15 WHEELER, H.A.: 'The solution of unsymmetrical sideband problems', *Proc. IRE*, 1941, **29**, p. 446
16 BLACK, H.: 'Modulation theory' (Van Nostrand, 1953)
17 VOELCKER, H.: 'Toward a unified theory of modulation', *Proc. IEEE*, 1966, **54**, pp. 340−354 and 735−756
18 CUCCIA, C.L.: 'Harmonics, sidebands and transients in communication engineering' (McGraw-Hill, 1952)
19 MIDDLETON, D.: 'Introduction to statistical communication theory' (McGraw-Hill, 1960)
20 BLACHMAN, N.M.: 'A generalisation of Woodward's theorem on f.m. spectra', *Inf. & Control*, 1962, **5**, pp. 55−63
21 BLACHMAN, N.M.: 'Limiting frequency modulation spectra', *Inf. & Control*, 1957, **1**, pp. 26−37
22 BLACHMAN, N.M.: 'Noise and its effects on communication' (McGraw-Hill, 1966)
23 MEDHURST, R.G.: 'R.F. spectra and interferring carrier distortion in f.m. trunk radio systems with low modulation ratios', *IRE Trans.*, 1961, **CS-9**, pp. 107−115
24 ABRAMSON, N.: 'Bandwidth and spectra of phase-and-frequency-modulated waves', *IEEE Trans.*, 1963, **CS-11**, pp. 407−414
25 ANDERSON, R.R., and SALZ, J.: 'Spectra of digital f.m.', *Bell Syst. Tech. J.*, 1965, **44**, pp. 1165−1189
26 SHIMBO, O.: 'General formula for power spectra of digital f.m. signals', *Proc. IEE*, 1966, **113**, pp. 1783−1789
27 SHIMBO, O., OHIRA, T., and NITADORI, K.: 'A general formula on power spectra of pulse-modulated signals and its application to p.c.m. transmission problems', *J. Inst. Electr. Commun. Eng. Japan*, 1963, **46**, pp. 9−17
28 MARSHALL, G.J.: 'Power spectra of digital p.m. signals', *Proc. IEE*, 1970, **117**, (10), pp. 1909−1914

29   BENNETT, W.R., and RICE, S.O.: 'Spectral density and autocorrelation function associated with binary frequency-shift keying', *Bell Syst. Tech. J.*, 1963, **42**, pp. 2355–2385

30   CALANDRINO, L., CRIPPA, G., and IMMOVILLI, G.: 'Intersymbol interference in binary and quaternary p.s.k. and d.c.p.s.k. systems – Pt.1 and Pt.2', *Alta Freq.*, May and Aug. 1969, **38**, pp. 337–344 and pp. 562–567 (in English)

31   MARSHALL, G.J.: 'Problems of receiver filter design for digitally modulated carrier system', *J. Sci. & Tech.*, 1971, **38**, pp. 174–182

32   TJHUNG, T.T., and WITTKE, P.H.: 'Carrier transmission of binary data in a restricted band', *IEEE Trans.*, 1970, **COM-18**, pp. 295–304

33   SCANLAN, J.O.: 'Transfer functions with elliptic distribution of poles at equal frequency spacings', *IEEE Trans.*, 1965, **CT-12**, pp. 260–266

34   GLANCE, B.: 'Power spectra of multilevel digital phase modulated signals', *Bell Syst. Tech. J.*, 1971, **50**, pp. 2857–2879

35   BENEDETTO, S., BIGLIERI, E., and CASTELLANI, V.: 'Combined effects of intersymbol, Interchannel and cochannel interferences in *M*-ary c.p.s.k. systems', *IEEE Trans.*, 1973, **COM-21**, pp. 997–1008

36   ROSENBAUM, A.S.: 'Binary p.s.k. error probabilities with multiple cochannel interferences', *ibid.*, 1970, **COM-18**, pp. 241–253

37   CAHN, C.R.: 'Combined digital phase-and-amplitude modulation communication systems', *IRE Trans.*, 1960, **CS-8**, pp. 150–154

38   HANCOCK, J.C., and LUCKY, R.W.: 'Performance of combined amplitude and phase-modulated communication systems', *ibid.*, 1960, **CS-8**, pp. 232–237

39   LUCKY, R.W., and HANCOCK, J.C.: 'On the optimum performance of *N*-ary systems having two degrees of freedom', *ibid.*, 1962, **CS-10**, pp. 185–193

40   SALZ, J., SHEEHAN, J.R., and PARIS, D.J.: 'Data transmission by combined a.m. and p.m.', *Bell Syst. Tech. J.*, 1971, **50**, pp. 2399–2419

41   DAVEY, J.R.: 'Modems', *Proc. IEEE*, 1972, **60**, pp. 1284–1292

42   FOSCHINI, G.J., GITLIN, R.D., and WEINSTEIN, S.B.: 'On the selection of a two-dimensional signal constellation in the presence of phase jitter and Gaussian noise', *Bell Syst. Tech. J.*, 1973, **52**, pp. 927–965

43   DAY, A.V.T.: US Patent 1885010, October 25, 1932

44   BENNETT, W.R., and DAVEY, J.R.: 'Data transmission' (McGraw-Hill, 1965)

45   SALZ, J.: 'Communications efficiency of certain digital modulation systems', *IEEE Trans,*, 1970, **COM-18**, pp. 97–102

46   SANDERS, R.W.: 'Communication efficiency comparison of several communication systems', *Proc. IRE*, 1960, **48**, pp. 575–588

47   MAZO, J.E., ROWE, H.E., and SALZ, J.: 'Rate optimisation for digital frequency modulation', *ibid.*, 1969, **48**, pp. 3021–3030

48   SIMON, M.K., and SMITH, J.G.: 'Hexagonal multiple phase-and -amplitude-shift-keyed signal sets', *IEEE Trans.*, 1973, **COM-21**, pp. 1108–1115

49   DUCOT, C.: 'Detection d'un signal vehicule par une porteuse incoherente', *L'onde Electr.*, April 1963, pp. 452–455

50   MARCATILI, E.A.: 'Errors in detection of r.f. pulses embedded in time crosstalk, frequency crosstalk, and noise', *Bell Syst. Tech. J.*, 1961, **40**, pp. 921–950

51   SHAFT, P.D.: 'Error rate of p.c.m.-f.m. using discriminator detection', *IEEE Trans.*, 1963, **SET-9**, pp. 131–137

Chapter 14

Aspects of realisation

14.1 Baseband systems

The usual arrangement of a baseband-repeatered line link takes the form shown in Fig. 14.1. At the terminal stations the digital information to be transmitted may be generated by a single source or may be derived from many sources via a multiplexer. It is encoded into a code format suitable for transmission over the line using one or other of the coding techniques described in Chapter 11. Finally, the individual pulses are transmitted onto the line in a standard signal format with specified pulse parameters (amplitude, width and transition times). Repeaters placed at regular intervals along the line have as their main objective reconstruction of the degraded signals arriving at their inputs into standard signal format at their outputs; in the case of digital repeaters this process involves regeneration, as described in Chapter 9. Repeater stations are usually below-surface chambers with manhole access, housing cable ends and repeater cases each of which accommodates a number of bothway repeaters.

It is normal practice to power intermediate repeaters by feeding direct current along the line from designated power-feeding repeater stations, and these, because of the bulky equipment required, are normally larger, above-ground buildings; because of the high cost of these installations compared with standard or dependent repeater stations, the minimisation of their numbers is normally a major design aim.

Types of digital repeaters
Digital repeatering involves regeneration. However, it may or may not involve amplification/equalisation. Furthermore, regeneration may or may not involve timing, and timing may be provided in more than one way. The above statements are contrived to demonstrate that there are a number of repeater configurations which may be called digital; these are shown in Fig. 14.2 progressing in complexity and performance. (*a*) shows the simplest arrangement of an amplitude regenerator such as might be used in a logic interconnection system, the regenerator being then a logic element. This system allows modest line loss so as to work within the

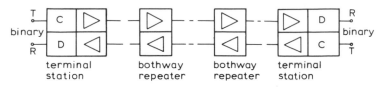

Fig. 14.1 Repeatered link
  C = coder   *T* = transmit
  D = decoder   *R* = receive

noise-immunity tolerance of the logic family. In (*b*), transmission range is extended by providing amplification and, more significantly, equalisation of the line loss. In (*c*), timing information is conveyed over a separate timing channel; this system there-fore incorporates all the basic operations which can help in recovering the original signal. However, the provision of a separate channel devoted to timing information is a luxury difficult to justify in a long transmission system and, moreover, may encounter some difficulty due to variable delay inequalities in the information and timing channels. These snags are avoided in scheme (*d*), incorporating timing extrac-tion from the information channel; this form of repeater is known as self-timed, forward-acting (see Section 14.3), and is currently establishing itself as a standard configuration.

To put more detailed discussion of subsequent Sections in perspective, Fig. 14.3 shows a typical block schematic of such a repeater[1-3,6-8] for a binary line code.

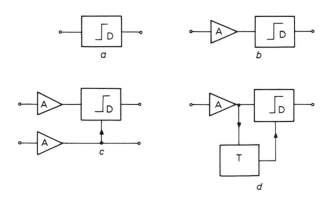

Fig. 14.2 Regenerative repeater forms   Key: A = amplifier/equaliser
  a Untimed   D = decision circuit
  b Untimed, preamplified   T = timing circuit
  c Separately timed, preamplified
  d Self-timed, preamplified

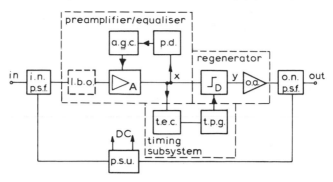

Fig. 14.3 Typical baseband line repeater
    i.n. = input network
    o.n. = output network
    p.s.f. = power separation filter
    p.s.u. = power separation unit
    l.b.o. = line build-out network
    t.e.c. = timing-extraction circuit
    t.p.g. = timing pulse generator
    D = decision circuit
    o.a. = output amplifier
    A = equaliser/amplifier
    p.d. = peak detector

Four main subsystems are immediately recognised

  (i) the equaliser/amplifier
 (ii) the timing subsystem
(iii) the regenerator
(iv) the power subsystem.

In addition, some designs incorporate supervisory circuits which are required to operate while traffic is being carried; in these a supervisory subsystem of some complexity has to be included. In what follows, although the timing and regenerator subsystems are described in a baseband context in which digital practice is well established, they are in most cases directly applicable to carrier systems at the detected baseband port.

14.2 Baseband amplifiers and equalisation

In the majority of systems, the regenerator proper is preceded by some form of input amplifier. The main function of this is not to supply gain, because any required gain can, in principle, be achieved by increasing the sensitivity of the threshold detector section of the regenerator. The function is rather to reshape and stabilise the amplitude of the signal before regeneration. It is usual to employ a linear amplifier to perform these functions, although this is not essential; nonlinear arrangements can be used. Two situations arise, first, those in which the loss in the transmission path, although variable with time, is independent of frequency. Conditions of this sort arise in the case of microwave radio systems, in which the frequency dependence of the transmission path is due mainly to the finite bandwidth of the band separation filters and the i.f. amplifier, while instability in the transmission path appears as variations in the loss. The baseband amplifiers required in this situation are similar to those employed in other pulse applications, and the reader is referred to the extensive literature on the design of pulse amplifiers for further information.[20–23] Variations in the level of the input signal may be dealt with by a simple automatic gain-control system, using the peak output level of the amplifier as a reference.

A second more complex situation arises in the case of systems in which the intervening transmission path includes a significant length of cable. In this case, the basic transmission-path response is a function of frequency, and both the frequency characteristics and the absolute magnitude may be time dependent. The amplifiers used in these cases almost always employ negative feedback. The choice of this approach stems from the fact that frequency-selective feedback provides a convenient method of controlling the frequency response of the amplifiers, while a useful improvement in the stability of the amplifier characteristics is obtained. The associated improvement in linearity, which is of vital importance in the case of f.d.m. systems, is not a significant feature in digital transmission. Feedback may either be applied over a number of stages[2,23] or may be restricted to local feedback on each stage separately.[6,8] This latter approach is, of course, simpler in very high speed systems.

The gain and response variation needed to cover variations in cable temperature, and small variations in cable length, may be provided by automatic gain-control systems. The networks employed are of course rather more complex than those required for simple frequency independent gain adjustment. Larger variations in cable length may be covered by employing preset cable simulator networks to build out the cable to the equivalent of a section of standard length. All of these problems parallel closely the corresponding problems in the design of the linear repeaters employed in f.d.m. systems, and much useful information can be obtained from papers dealing with the design of such systems.[24,25] The main differences are that in the case of digital systems, high linearity is not required, but rather greater attention has to be paid to the transient response. Another factor which requires attention is the possibility of low-frequency overload. As we have seen in previous Chapters, the energy of digital signals can be high at low frequencies. This is the region in which the cable losses are small, and as a result the low-frequency components can appear at the input to the next amplifier at an unpleasantly high level. In some cases, it is even worth deliberately increasing the transmission path loss at low frequencies by the provision of a high-pass network at the output of the repeater; this affords the added benefit of protecting the output port against low-frequency surges such as lightning effects.

14.3 Timing subsystem

As already stated, the most common timing arrangement is one where the timing wave is extracted from the information, such a repeater being then classified as 'self-timed'. The first task of the timing subsystem is therefore to provide a means of timing extraction, and the second is to use the timing wave to best effect in conjunction with the decision circuit to achieve regeneration. The method of timing extraction and timing-insertion (retiming) jointly define the timing subsystem. In principle, a number of methods have been proposed for each, and the most significant are summarised in Table 14.1. The crossed entries signify incompatible combinations.

Table 14.1 Timing-subsystem options

Retiming methods	Timing-extraction methods	Self timed			
		forward acting		backward acting	
		linear	nonlinear	linear	nonlinear
Partial (ADD)		√	√	√	√
Permissive (AND)		√	√	X	X
Complete (sample/reset)		√	√	X	X

√ = permissible   X = not permissible

Timing extraction

In a self-timed system, timing extraction may be either 'forward-acting', or 'backward-acting', depending on whether the point of extraction is from the input side or the output side of the regenerator respectively. In Fig. 14.3 we show forward-acting extraction; for backward-acting extraction the block TEX would be connected to point Y. The difference between the two modes is quite basic because backward extraction incorporates a feedback loop which includes the decision circuit, while forward extraction is an open-loop system. Consequently, there are certain forbidden feedback-loop configurations which preclude backward extraction and, as we shall see, the choice between the two methods depends on the method of timing reinsertion or retiming.

The second major factor which must be considered is whether the timing extraction should involve linear or nonlinear processing. Linear extraction involves filtering out the discrete signalling or clock component by means of a narrow-band filter, and clearly depends on the presence of this component in the spectrum of the equalised pulse sequence. An analysis of this approach in relation to noise and other degradations is given in Reference 5 and 9. However, in most line systems the equalisation design is such as to almost suppress any discrete timing wave component at the signalling frequency (see Chapter 12), a situation which generally pertains to pulse shapes having a duty factor of near unity (Reference 9, Fig. 4). The *existence of a random component in the spectral region of interest is no guarantee* of a linearly extracted timing component; it simply means that there exists a long-term average power component with unspecified phase relations, including some which would have the effect of quenching the tuned filter, as, for example, identical patterns with a phase difference of π, shown in Fig. 14.4.

For this reason, linear timing extraction has not found favour in baseband line systems. The unfortunate phase relationships can, however, be eliminated by nonlinear processing involving full-wave rectification, or clipping, or both.[9] This is illustrated in Fig. 14.4 where the full-wave rectification forces a timing component at $f = 1/T$ by eliminating the damaging phase relations present in the original pattern at $f = 1/2T$. Takasaki[11] examines the choice of nonlinearity and concludes that full-wave rectification followed by downward clipping (allowing only inputs larger than a threshold set at half-way to peak value to pass) is an effective process

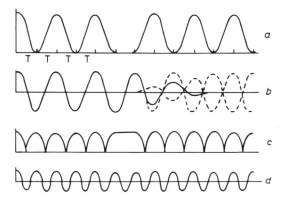

Fig. 14.4 Timing-extraction waveforms
    a Input pattern
    b Circuit tuned at $f_0 = 1/2T$, driven by *a*
    c Pattern shifted and rectified
    d Circuit tuned at $f_0 = 1/T$, driven by *c*

for binary or ternary codes; he deduces more complex laws for multilevel codes.

Retiming

Given that a sinusoidal timing wave is available, the question arises of utilising it in the regenerator so as to define the transitions of the regenerated pulse sequence. Three basic methods are available, classified by Aaron[4] as partial, permissive and complete retiming. In partial retiming the timing sinusoid is clamped (to ground, say) and added to the equalised signal at the input of a threshold circuit; it therefore bears close resemblance to the simplest AND gate arrangement. The regenerated pulsewidth obviously depends on the amplitude of the timing wave; the method is only of interest inasmuch as it is compatible with backward-acting timing extraction (Table 14.1), and because of its inherent simplicity of realisation, a feature which was important when the method was first considered in the mid-fifties. Sunde,[10] in an exhaustive treatment, shows that partial retiming is inferior in jitter-accumulation properties to complete retiming. In permissive retiming the equalised input is connected to a decision circuit via a linear gate which opens during one halfcycle of the timing sinusoid; the opposite halfcycle resets the regenerator. The whole process is therefore akin to an AND operation because a regenerated '1' occurs when the input exceeds the threshold AND the timing wave is positive.

Finally, in complete retiming the decision device is interrogated at a predetermined optimal instant under the control of the timing wave. As such it can be described by various circuit models, the most common of which is shown in Fig. 14.5. The input is sampled with the set pulses and again applied to a decision circuit which is returned to a 'low' state by the action of the reset pulses. This

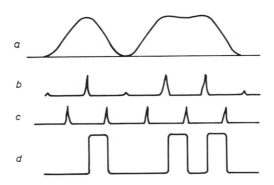

Fig. 14.5 Regeneration with complete retiming of sample and reset type
    a Input pattern
    b Sampled input
    c Reset pulses
    d Regenerator output

method provides complete control of regenerated pulsewidth and is less prone to jitter because the decision instant is well defined. However, as pointed out by Sunde,[10] it cannot be used with backward-acting timing extraction because of the consequent instability of the timing loop. That this is so is intuitively predictable because the input sequence then has no control over the transitions of the output.[5]

Realisation

The realisation of timing extraction and retiming circuits has evolved into a nearly standard form; the diagram of Fig. 14.6 would describe with minor variations the systems described in References 1, 2, 3, 6, and 8, ranging in symbol rate from 1·5 Mbit/s to over 200 Mbit/s. It refers to a self-timed, forward-acting system with complete retiming and nonlinear extraction. The full-wave rectifier and threshold circuit (downward clipper) perform the nonlinear extraction process, which forces a timing component of varying level depending on the input pattern. This is selected by means of a narrowly tuned tank circuit, the tuning, if any, of the tank buffer amplifiers being weak so as not to contribute variations to the overall phase. To prevent amplitude/phase conversion in the timing pulse generator, it is desirable that it be driven from a nearly constant-level input. This is the function of the limiter. The timing-pulse generator may be any one of several circuits; a zero-crossing switch followed by a differentiator, or a step recovery diode circuit.

Possible refinements are the provision of an automatic level control (a.l.c.) loop to further compress the range of the tank output[6,26] in which the latter is peak detected and a d.c. signal proportional to the peak is fed back to alter the threshold of the threshold circuit.

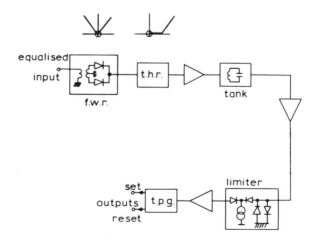

Fig. 14.6 Baseband repeater timing chain
　　f.w.r. = full-wave rectifier
　　t.h.r. = threshold rectifier
　　t.p.g. = timing-pulse generator

Design features
In designing the timing chain three dependent parameters must be considered

(i) the Q factor of the tank circuit
(ii) the overall timing chain gain
(iii) the dynamic range of the timing wave.

In an ideal situation, random signal-dependent jitter varies inversely with Q (Reference 9, p. 1522) and so the highest possible Q would be a desirable aim. However, more realistically, the tank circuit of any one repeater will be to some extent mistuned from the nominal symbol rate component f_0. For a steady signal consisting of impulses at a rate f_0 exciting a tank mistuned to $f_0 - \Delta f$, the phase shift ψ may be shown to be given by[10]

$$\tan \psi \simeq 2Q \frac{\Delta f}{f_0} \tag{14.1}$$

When these impulses comprise an independent information-carrying signal, Bennett[9] shows that the r.m.s. phase error ψ_{rms} is

$$\psi_{rms} \simeq \frac{\Delta f}{f_0} (\pi Q)^{\frac{1}{2}} \tag{14.2}$$

Thus, the choice of a high Q must be reconciled with a proportionally high tuning stability. Most practical systems have hitherto used a value for Q of around 100.

The overall timing-chain gain and dynamic range are determined by considering the range of level of the timing wave component at the timing extractor input. This is totally dependent on the characteristics of the transmission code and statistics of the information. Certain rules of thumb have, however, evolved for certain well known cases such as the bipolar code.[1] Consider a bipolar input of $\pm V$; after rectification it ranges from 0 to V, and after passing over a threshold of $V/2$ it ranges from $V/2$ to V. The clipping removes predominantly low-frequency energy and so we may treat this signal as a rectified sinusoid of amplitude $V/2$, whose timing (fundamental) amplitude is

$$\simeq \left(\frac{4}{3\pi}\right)\left(\frac{V}{2}\right) = \frac{2V}{3\pi}$$

This refers to the maximum timing component in the presence of continuous marks. The minimum component may be approximately deduced from this on the basis of the minimum frequency of marks[1,2] $1/N$ simply as $2V/3\pi N$. Further allowance should be made for the damping loss of the tank circuit after M intervening spaces; this is given by $\exp -\pi M/Q$. The minimum timing amplitude at the tank-circuit input is therefore

$$\frac{2V}{3\pi N} \exp \frac{-\pi M}{Q}$$

and the dynamic range which must be accommodated in the limiter is

$$20 \log_{10} N - 25 \cdot 7 \frac{M}{Q}, \text{dB} \qquad (14.3)$$

In early 24-channel p.c.m. designs,[1,2] this range, using simple bipolar coding, added to the equalised input variation, amounted to more than 20 dB; current designs using filled bipolar codes forbidding M to exceed 3 (in the HDB3 code) should be 10 dB or so better.

Finally, we should point out that recent developments suggest that crystal tank circuits of sufficient stability to warrant Q values of 500 or more may be used.[12] An alternative in achieving high Q is the phase-locked loop (p. 1.1., Reference 19). This has not been used to any extent, except in experimental systems, because of two factors – first, problems in tuning stability and acquisition range, and secondly, the presence of the unlocked oscillation in the event of failure of the input.

14.4 Regenerator

In practical systems, the regenerator unit provides amplitude quantisation by means of a decision circuit and, in a system with complete retiming, a means of time quantisation. In some cases, the regenerator circuit is also the output stage of the repeater, and for this reason we shall also consider output arrangements in this Section, although conceptually they need not be part of the regenerator. Most designs use the basic arrangement of Fig. 14.7 in which the input signal is sampled by means of an interrogating or 'set' pulse, the result is compared by the decision circuit with a threshold, and the decision state is held until the circuit is reset to a given state on arrival of the reset pulse after a predetermined interval of time. The decision output is, at least in principle, a 'hard' logic state, and it is used to drive the final output stage which has to deliver the specified signal level and waveform to the line. The sampling gate may use either diodes[1,2] or transistors[8] in a current steering mode, as shown in Fig. 14.7. This arrangement may be recognised as the sampling regenerator described in Section 9.8; it is not the only method of obtaining complete retiming. An alternative is to reverse the sampling and decision functions in much the same manner as described in Chapter 2 in the general context of quantisation; we have no firm conclusions as to the comparative efficacy of this scheme.

The choice of decision circuit or quantiser is dependent on operating speed, and is therefore rather wide. Basically, all forms fall into one of two classes depending on whether positive feedback is used or not. Generically, the positive feedback quantiser features a simple circuit and potentially high speed, but suffers from hysteresis and low stability. The nonpositive feedback quantiser is usually a limiting amplifier deriving high gain from many stages, and by its nature is free from hysteresis.

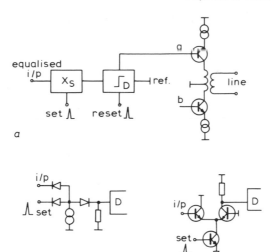

Fig. 14.7 Typical baseband regenerator
a General arrangement
b Diode sampling gate
c Current steering sampling gate

Key: S = sampling gate
D = decision circuit

Of the more common positive-feedback decision circuits, early designs for 1·5 Mbit/s operation invariably used blocking oscillators with a reset facility,[1,2] yielding two important advantages— high efficiency, and the capability of driving the transmission line directly via a transformer which also provided the necessary feedback. However, the poor definition of the decision threshold and of the output waveform, and, in addition, modest operating speed, set limits on the applicability of this circuit.

The classic bistable flip-flop is unsuitable on counts of inefficiency and poor threshold definition; however, several variants have been produced with much improved performance,[6,8] and, with the continuing advances in semiconductor technology, it is expected that variants of increasing complexity in terms of the number of active elements will evolve.

A further decision-circuit candidate is any device with inherently bistable characteristics, such as the tunnel (Esaki) diode.[3] The action of such a tunnel diode as a decision element is based on quantum mechanical phenomena rather than circuit feedback, and, as such, it is nearly an ideal decision circuit. However, its inherently low output level, low efficiency and allegedly doubtful life expectancy have all contributed to its current low popularity.

The practical decision circuit suffers from three types of imperfection: a finite-slope rather than a discontinuous characteristic, decision-threshold uncertainty and hysteresis. The finite-slope characteristic implies that signals falling in that range will be simply amplified to levels some way short of the permitted outputs, and may propagate to some extent down the repeater chain.[13] The threshold uncertainty has the effect of reducing the operating margin of the system (Reference 13 and Chapter 12). Hysteresis might at first be thought to be of some benefit in causing a reluctance of the system to changes due to added noise; however, the corollary is also true inasmuch as legitimate signal transitions encounter a reluctant response and may be ignored due to added noise. In fact, it may be shown (Chapter 9) that, for a random signal, the probability of error is almost the probability that random noise exceeds $(V - h)/2$, where h is the hysteresis.

Finally, the basic feature required from the output stage is high efficiency pulse generation. The blocking oscillator mentioned earlier offers high drive capability, efficiency and regeneration in one circuit. However, because the incentive in saving active devices has currently almost disappeared and, because of the poor waveform produced by the blocking oscillator, this configuration has been abandoned in most designs.

A typical output arrangement in lieu of a blocking oscillator is shown in Fig. 14.7, consisting of a push-pull current drive circuit; with a ternary signal, a and b would be the two separate decision outputs. Line coupling by means of a transformer is still widely preferred in view of its good isolation properties for high-level signals induced on the line, particularly of the common-mode type.

In some applications, particularly in high-speed systems, the design of the output transformer presents some difficulty; for instance in a system operating at 100 Mbaud, the transformer response must be maintained up to, say, 200 MHz, and may be required to descend to perhaps 100 kHz to minimise low-frequency wander. This implies a ratio of upper to lower cutoff frequencies of some 2000 : 1!

14.5 Power feeding over lines

The standard practice in line systems is to power the line repeaters using the line itself for d.c. power transmission. In systems using paired lines the usual arrangement is to derive a phantom circuit via centre-taps on the input and output transformers, and to feed direct current over this circuit, typically as shown in Fig. 14.8. The power-feed loop is driven at constant current I from a power-feeding station and each repeater converts this to the required number of voltage rails in its power-separation unit, p.s.u. This may consist simply of a string of Zener diodes. As the p.s.u. forms a direct connection between the output and input of a repeater which may be at levels differing by some 60–80 dB, adverse feedback effects must be guarded against by very careful filtering. This feature is less critical with the balanced system shown, as the centre-tap affords a large amount of attenuation. However, in coaxial-line systems separate power-separation filters must be provided on either side of the p.s.u., as indicated in Fig. 14.3, affording up to 100 dB loss.[6]

The power supply at the power-feeding station delivers a constant current, typically 50 mA, at a specified maximum voltage of $\pm V$ (typically ± 250 V); for a

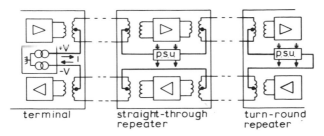

terminal   straight-through   turn-round
               repeater   repeater

Fig. 14.8 Power-feeding loop on paired lines

given repeater power consumption and loop resistance, V determines the maximum number of repeater sections in the power loop.

14.6 Synchronisation in carrier systems

The range of application and diversity of environments of carrier systems are so wide as to render impossible a comprehensive discussion of their form of realisation in one Section. As an illustration, we find data modems for telephone lines with information capacities of a few hundred bit/s and carrier frequencies within the voiceband; we have p.c.m. systems with capacities from a few to hundreds of Mbit/s operating over microwave line-of-sight carriers at frequencies of a few gigahertz to tens of gigahertz; finally, there are optical systems in the experimental phase. We therefore intend to dissociate the discussion from the modulation-demodulation hardware and concentrate on aspects of realisation exclusive to digital working.

The transmission of digital information by modulated carrier calls for the inter-working of two disciplines which have initially evolved on separate paths — digital baseband techniques and modulation techniques. Digital transmission is associated with certain constraints which the overall system must satisfy; these require the provision of what may be termed 'house-keeping information', the most important being provision of digit timing or synchronisation. Our earlier discussion of timing extraction in baseband systems emphasised this feature and its dependence on the line code used. If the latter is visualised as a form of modulation, then the modulation process may be seen as either imposing specific constraints, or offering added degrees of freedom in the overall channel coding process. In which particular light the situation is viewed depends on the degree to which the system can be designed with an integrated philosophy. We shall therefore limit this Section to a discussion of available methods for the provision of timing information or synchronisation.

Synchronisation by combined modulation
Given a sufficiently high s.n.r. it is quite feasible (see Chapter 13) to convey the timing wave or some convenient function derived from it as an auxiliary modula-

tion of the main modulation system. Thus, given a p.s.k. or f.s.k. main system the timing wave may be amplitude modulated onto the same carrier with a certain loss of noise immunity. It has been reported[14] that, for an a.m. index of 0·1, the degradation in s.n.r. suffered by a binary f.s.k. main system is about 0·9 dB relative to an undisturbed f.s.k. system.

Synchronisation from filtering distortion
An ideal p.s.k. or f.s.k. signal has a constant carrier envelope. However, band-limiting at carrier and i.f. frequencies introduces a disturbance of the envelope due to the finite transition times allowed by the filter. Envelope detection followed by baseband timing extraction techniques as described earlier, can thus provide a timing content.[15]

Synchronisation from coherent detection
It is sometimes possible to design systems in which the carrier frequency bears an integral or simple rational fraction relationship to the symbol rate. The latter can then be derived by division from, or phase-locking to, the coherently recovered carrier. Elements of this technique in different circumstances can be found in References 16 and 17.

Synchronisation in p.s.k. systems
4-phase p.s.k., and in particular d.p.s.k., is finding wide acceptance as an efficient modulation format. The simplest phasor space in such systems is one of $0, 90°, 180°, -90°$; however, this implies that one signal state, corresponding to the $0°$ phase, causes no transition of the carrier phase and long runs of this state will give no timing content. The situation can be alleviated by the use of scramblers and descramblers. An alternative and widely preferred approach,[18] is to shift the phasor space by $45°$ resulting in phases of $\pm45°$ and $\pm135°$. This means that, in a d.p.s.k. system, a constant input signal state will cause continuous modulated phase changes of at least $45°$, at the symbol rate. This, in conjunction with finite filter bandwidth, causes sufficient envelope disturbances at the symbol rate to guarantee a timing content, as described earlier.

14.7 References

1   MAYO, J.S.: 'A bipolar repeater for pulse code signals', *Bell Syst. Tech. J.*, 1962, **41**, pp. 25–98
2   INOUE, N., SAKURAI, N., and IWAHASHI, E.: 'A repeater for short-haul p.c.m. system', *Rev. Electr. Commun. Lab. Japan*, 1965, **13**, pp. 972–1004
3   DORROS, I., SIPRESS, J.M., and WALDHAUER, F.D.: 'An experimental 224 Mbit/s digital repeatered line', *Bell Syst. Tech. J.*, 1966, **45**, pp. 993–1044
4   AARON, M.R.: 'P.C.M. transmission in the exchange plant', *ibid.*, 1962, **41**, pp. 99–142
5   ROWE, H.E.: 'Timing in a long chain of regenerative binary repeaters', *ibid.*, 1958, **37**, pp. 1543–1598

6 ARATANI, T.: 'An experimental 200 Mbit/s p.c.m. repeater', *Rev. Electr. Commun. Lab. Japan*, 1969, **17**, pp. 22–48

7 FUKINUKI, H., KASAI, H., SENMOTO, S., and MATSUSHITA, M.: 'Study on PCM–100M repeatered line', *ibid.*, 1973, pp. 254–275

8 OHTSUKI, M., FUDEMOTO, I., and YOSHIBAYASHI, T.: 'A new high-speed p.c.m. repeater utilising hybrid integrated circuit technology', *Fujitsu Sci. & Tech. J.*, 1971, 7, pp. 109–128

9 BENNETT, W.R.: 'Statistics of regenerative digital transmission', *Bell Syst. Tech. J.*, 1958, **38**, pp. 1501–1542

10 SUNDE, E.D.: 'Self-timing regenerative repeaters', *ibid.*, 1957, 37, pp. 879–938

11 TAKASAKI, Y.: 'Timing extraction in baseband pulse transmission', *IEEE Trans.*, 1972, **COM-20**, pp. 877–884

12 CUNNINGHAM, P.B., *et al.*: 'A new T-1 regenerative repeater', *Bell Lab. Rec.*, 1971, pp. 109–114

13 ARATANI, T.: 'Degradation of error rate in p.c.m. repeatered line caused by incomplete pulse recognition', *Electron. & Commun. in Japan*, 1966, 49, p. 191

14 SINGH, H., and TJHUNG, T.T.: 'f.s.k. bit synchronisation by combined a.m.', *IEEE. Trans.*, 1973, **COM-21**

15 NAKAMURA, S., and FUKETA, H.: 'Design of p.c.m. repeatering equipment in the 20 GHz band', *Rev. Electr. Commun. Lab. Japan*, 1969, 17, pp. 223–240

16 VAN GERWEN, P.J., and VAN DER WULF, P.: 'Data modems with integrated digital filters and modulators', *IEEE Trans.*, 1970, **COM-18**, pp. 214–222

17 HUBBARD, W.M.: 'The effect of intersymbol interference on error rate in binary d.p.s.k. systems', *Bell Syst. Tech. J.*, 1967, 46, pp. 1149–1192

18 DAVEY, J.R.: 'Modems' *Proc. IEEE*, 1972, 60, pp. 1284–1292

19 BALL, J.R.: 'A 120 Mbit/s digital regenerator employing a phase locked loop for timing recovery', British Post Office Research Dept. Report 413, Dec. 1973

20 KALMANN, H.C., SPENCER, R.E., and SINGER, C.P.: 'Transient response', *Proc. IRE*, 1945, 33, pp. 169–195

21 FRAME, R.G.M.: 'Transistor amplifiers with nanosecond rise times', Ph.D thesis, Heriot-Watt University, 1971 (see *Science Abstract* 9713B/1971)

22 RICHARDS, J.C.S.: 'Very wideband amplifiers using transistor feedback pairs', *Proc. IEE*, 1970, 117, (10), pp. 1949–1952

23 CHERRY, E.M., and HOOPER, D.E.: 'The design of wideband transistor feedback amplifiers', *ibid.*, 1963, 110, pp. 375–389

24 BLECHER, F.H. *et al.*: 'The L4 coaxial system', *Bell Syst. Tech. J.*, 1969, **48**, pp. 821–1093

25 ELMENDORF, C.H. *et al.*: 'The L3 coaxial system', *ibid.*, 1953, 32, pp. 779–1005

26 BERRY, A.D., LARNER, D.S., and MUMFORD, H.: 'Long distance coaxial cable system for 120 Mbit/s digital traffic', *in* 'Telecommunication transmission', *IEE Conf. Publ.*, 131, 1975

Chapter 15
Supervisory systems and fault location

15.1 Introduction

The ultimate test of the performance of any system must be based on some measure of its success in meeting the original design requirements, and because, in practice, we are always dealing with systems having limited reliability, and subjected to external disturbances, our assessment of performance must include allowance for these effects. In the case of communication systems two alternative situations may arise. In the first of these, the system is only used rarely, but when it is used, it is essential that it operates successfully. Examples of this type of requirement arise in military weapons systems, remote control of machinery, alarm systems, and such like. These are usually cases in which the need for correct functioning far outways any consideration of costs, and it is then normal to provide duplicated or triplicated transmission channels to ensure that at least one operational channel is available at any time.

At the other extreme, we have the case exemplified by commercial tele-communication systems, in which the objective is to provide an almost continuous service in a way that maximises its earning capacity. In this case, the economic penalty of providing an almost perfect performance is unacceptable, and short breaks in the service, or some transmission delays, are permissible. The earning capacity of the system will be increased by making these breaks as short as possible, but this must be balanced against the costs of the added equipment needed for this purpose.

In both cases, the role of the supervisory and fault-location systems is crucial. In the military type of system, there is the certainty that sooner or later one of the parallel channels will fail, and when this happens the protection provided will be reduced or will vanish. The sooner this faulty channel can be repaired or replaced, the greater the overall system reliability. This introduces a need to monitor system performance, and to detect and locate faults as rapidly as possible. In the case of commercial systems, similar considerations arise, although here the speed of fault location must be balanced against the cost of provision of this facility. However, a fast system usually represents a saving in labour time for the operating staff, and this is often an important factor in assessing the true economic costs.

15.2 Required facilities

The supervisory arrangements for a transmission system must therefore provide a number of facilities. These include

(a) *Detection of transmission errors.* In almost all cases there is a need to monitor the transmission performance and detect the onset of any degradation in this. In the case of digital systems, such degradation appears as digital errors in the received signal, and as discussed in Chapter 11, monitoring of transmission impairment may take place at several levels of sophistication. At the highest level we may attempt to detect and correct the majority of errors, below this we may merely detect errors, and at the lowest level we only detect that errors are occurring, and get a rough estimate of their numbers through some statistical sample.

(b) *Detection of other system abnormalities.* Apart from direct detection of the error rate, identification of other deviations from normal operating conditions, for example, changes in power supply voltages or currents, temperature rises etc., may provide a useful indication of the onset of fault conditions. In the case of both (a) and (b) it may be possible to detect abnormalities before these cause any significant transmission impairment, and thereby introduce preventative maintenance measures.

(c) *Fault location.* The location of the point in the system where the abnormality first becomes apparent is usually an essential adjunct to its detection. The extreme example of this arises in cable transmission systems, where we may employ several hundred regenerators, in unattended locations along the transmission route. To achieve speedy repair of a fault the maintenance staff must know the point on the route where the malfunction becomes apparent.

(d) *Supervisory transmission facilities.* Supervision and control of the system must usually take place at one selected location, usually one of the system terminals. All the supervisory information must therefore be transmitted to this terminal, and appropriate control information must be transmitted from the terminal to other points in the system. It is frequently necessary to provide auxilliary communication channels, for example, telephony, or teleprinter channels, for the use of the maintenance staff.

15.3 System principles

The design of any supervisory system involves a number of principles, whose existence and significance should be appreciated by the designer. The first, and most important, of these is the degree of isolation between the supervisory system and the main transmission system. In the extreme case, the supervisory system may employ completely independent equipment, separately powered, and using a separate transmission path, thus making its operation almost completely independent of the main channel. This, however, is not entirely advantageous.

The provision of two independent transmission channels which must remain associated not only increases costs, but also leads to operational complications; for example, the complications which arise in dealing with two channels when re-routing or reorganising a transmission link. Because of this it is frequently decided that the main and supervisory systems should employ the same cable or transmission path for both sets of equipment.

A second distinction lies in the difference between in-service and out-of-service monitoring operations. In the former, checking of the equipment may be carried out while it is in use for normal information transmission, while the latter is performed by taking the system out of service and applying some special test signals or conditions.

Another distinction arises between systems where supervisory data is automatically transmitted from remote stations to the control centre whenever any abnormality is detected, and systems which only transmit data when interrogated by the terminal. These differences are associated with the question of station identification. In the former case, each station must have some unique identifier which is included in its transmission, whereas, in the second, it is possible to avoid this. For example, each station may transmit its data, and then relay the interrogation signal to the next station, thus leading to a set sequence for the arrival of the data, which allows station identification.

Finally, there is the question of active and passive fault-monitoring equipment. Some form of passive equipment is usually included to cover the case of fault location when the ultimate catastrophy occurs and electrical power is lost at the intermediate stations.

15.4 Implementations

We conclude by reviewing some examples of the way in which these principles have been applied in practice.

(a) *Stand-by systems.* While, outside the military field, it is not usually possible to afford complete duplication of equipment, it if often found that the complete transmission capacity between two points consists of several transmission systems in parallel, and it is then possible to provide a single stand-by system that may be substituted for any one of the working systems if a fault develops. Examples arise in coaxial cable transmission, in which a 12 tube cable is commonly employed to provide five bidirectional working systems and one stand-by system. In microwave radio links it is common practice to include a stand-by radio channel on each physical route.

The switching operations involved in this scheme can become quite complex. Typically, the operating sequence runs along the following lines. First, the decision to switch to the spare channel is made when some parameter of an operating system, for example, its error rate or noise level, deviates outside some defined limit. At the transmitting end of the route the signal is then applied to both the old working channel, and the stand-by channel in parallel, and an appropriate control message is transmitted to the receiving terminal over the supervisory channel to indicate what is being done. The receiving terminal then monitors the signal coming over the standby channel, and providing this is superior to that received on the working channel, switches to it. If this is not the case, the attempt to change over is abandoned. In either case, appropriate messages are returned to the transmitting terminal.

A further complication arises if some further working channel later becomes unsatisfactory. It is normal to assign some hierarchy to the working channels, and if this second channel is higher in ranking than the previous channel it will then seize the standby channel, leaving the previous user to revert to the faulty channel.

Care must be taken to minimise the disturbances to transmission during the actual switching operations. In particular, it is difficult to ensure identical propagation delays through the two channels, so, in the case of digital transmission, some timing disturbance, and possibly some spurious digits, may arise at this time.

(*b*) *Error-detecting and correcting codes.* These have already been dealt with in Chapter 11. We merely note that in the case of an error-correcting system it will be necessary to monitor the rate at which corrections are being made to assess the transmission quality.

(*c*) *Use of signal structure.* In the case of random binary data, no direct check on the error rate can be made, but as discussed in Chapter 5 some structure is usually imposed on the signal by previous multiplexing operations. The multiplex frame-alignment words then provide a systematic feature which may be used to monitor the transmission error rate. Two difficulties arise in this case. First, the location and checking of the frame-alignment words is complex and is not practical at intermediate repeater locations. Secondly, the framing content is small, typically 5%, and as a result it takes an appreciable time to obtain a significant measure of low error rates. For example, with a system operating at 2 MBit/s it would take several seconds to determine the magnitude of an error rate in the region of 1 in 10^6.

(*d*) *Ternary systems.* The use of redundant ternary codes in cable-transmission systems has been dealt with in Chapter 11, and the ways in which this redundancy can be used for error monitoring have been discussed. All of these, in fact, depend on keeping a check on the running digital sum variation of the code, and in the case of bipolar and c.h.d.b. codes the implementation is almost trivial. An outline of the implementation in the general case is shown in Fig. 15.1. These schemes are sufficiently simple to permit implementation in the intermediate regenerators in line systems.

(*e*) *Transmission of supervisory information.* So far this has only received extensive attention in the case of cable systems. Two techniques have been considered. First, the use of an additional physical conductor pair within the same cable sheath, over which the supervisory data may be transmitted either as a set of audio tones, or as a low-speed digital signal. Secondly, the transmission of the data over the same electrical circuit as is used for power feeding the intermediate regenerators.

(*f*) *Out-of-service monitoring.* For binary signals, the standard technique is to replace the input by a pseudorandom sequence and check for errors at other points in the system. The p.r. sequence is assumed to be a fair representation of a random signal, although in some cases this may not be an entirely valid assumption, and, in any case, fairly long sequences (e.g. $2^{15} - 1$) should be used. The orthodox error-detection scheme is to arrange, by some means, to generate an identical sequence at the receiving point. The main difficulty in this is in achieving pattern synchronism between the two sequences. An elegant solution to this problem has been proposed by R.J. Westcott of the British Post Office.[1] This involves using a similar feedback shift register as a detector, but with the feedback loop broken, as shown in Fig. 15.2. It will be apparent that in this case we are comparing the incoming signal with the expected next digit,

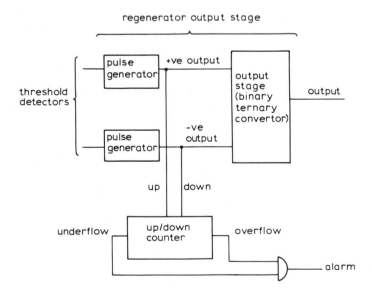

Fig. 15.1 Arrangements for error monitoring through a digital sum variation check

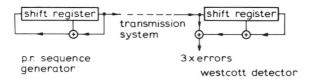

Fig. 15.2 Use of pseudorandom sequences for system tests

based on the feedback logic being used. Any discrepancy then represents an error. It can be seen that this completely avoids the synchronisation problem, but it should be noted that every isolated transmission error now results in a number of discrepancies in the receiving detector. It will also be apparent that some multiple-error patterns will not be correctly identified. These effects are however not important under normal test conditions.

A second example of out-of-service montoring is the trio scheme used by the British Post Office for checking 24 channel p.c.m. systems.[2,3] This is a particularly ingeneous scheme. The normal bipolar signal is replaced by a pattern of the form shown in Fig. 15.3. This consists of repeated triplets of positive and negative going pulses, with periodic reversal of the sign of the pattern. As shown, both the triplet density, and the frequency of sign reversals, may be varied. The effect of this is to introduce a very low-frequency square-wave component into the transmitted signal. Now at the first regenerator, due to the low-frequency cutoff of the input transformer and preamplifier, this component is not transmitted through to the threshold detector, which has the result of producing an offset in the signal at that point, as shown in Fig. 15.4. As the triplet density increases, the amount of this offset also increases, with the result that if the operation of the regenerator is marginal, failure to correctly regenerate the pattern will occur at some triplet density. Up to this point, correct regeneration will take place, and so a low-frequency square wave will reappear in the output. A low-frequency narrow bandpass filter is connected to the regenerator output which picks off the fundamental of this square

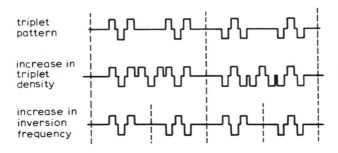

Fig. 15.3 Use of triplet patterns in testing ternary regenerators

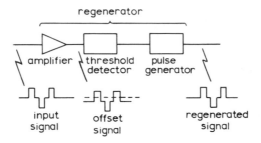

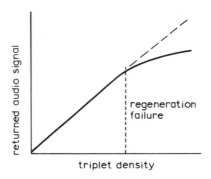

Fig. 15.4 Response of regenerator to a triplet pattern

Fig. 15.5 Detection of regenerator failure threshold

wave, and feeds it back to the terminal over an auxilliary pair in the cable. Thus, the returned signal may be compared with the transmitted triplet density, and the point where regeneration failure takes place may be detected, as shown in Fig. 15.5. A different reversal frequency and associated filter are used with each regenerator, thus making it possible to check each regenerator separately.

15.5 References

1   DIECKMAN, D.J.: 'Pseudo random sequence binary digit generators and error detectors', *Post Off. Electr. Eng. J.*, 1972, **64**, p. 245
2   BENNETT, G.H.: 'Testing techniques for 24 channel p.c.m. systems', *ibid.*, 1972, **65**, Pt.3, pp. 182–190
3   BENNETT, G.H.: 'Measurement techniques associated with p.c.m. systems', *in* 'Digital methods of measurement', IEE Conference Proceedings, July 1969, pp. 133–142

Present systems and future developments

Chapter 16

16.1 Current scene, worldwide

In this final Chapter we hope to present a concise summary of worldwide activity in digital transmission, and to indicate developments in fields with which it will become increasingly related. The growing trend towards integration of such hitherto diverse fields as transmission and switching is a particularly important case. The information presented should not be taken as in any way comprehensive; rather, it is intended as a sketch serving as perspective to the more serious content of previous Chapters. In this Section we give pride of place to systems which have seen service. Tables 16.1 and 16.2 present concise data on selected systems and are based on published sources in our list of References.

The greatest direct spur towards the application of digital techniques in communications was, it is generally agreed, the introduction in 1962 of the T1 24-channel p.c.m. carrier system by the Bell System in the USA. This was originally intended as an interexchange junction facility to provide more circuits over existing cables in urban congested areas; its economic advantages were soon appreciated and T1 links are now specified for operation over up to fifty miles of twisted pair. At January 1975 some 40 million circuit miles had been installed in the USA.[1] The T1 system leads naturally to a review of North American activity.

16.1.1 North American developments

Since the pioneering T1 system a complete hierarchy (see Chapter 5) of multiplex levels to interconnect various services with various transmission facilities has evolved; although the system offered by Bell is well defined, the presence of an increasing number of manufacturers and operators is creating a somewhat confused situation; Reference 2 implies at least eleven different multiplex levels. However, concentrating on the submissions made to the CCITT, Fig. 16.1 and Table 16.1, based on Reference 3, illustrate the scope of the proposed digital system. It is seen that the 1·5 (1·544) Mbit/s level is a general-purpose flexibility point offering transmission over the T1 paired line, or as a d.u.v. (digits under voice) system over

Table 16.1 Multiplex hierarchies

System	First		Second		Third		Fourth
N. American	1·544 (24)	X4	6·312 (96)	X7	44·736 (672)	X6	274·176 (4032)
Japanese	1·544 (24)	X4	6·312 (96)	X5			
		X5	7·876 (120)	X4	32·064 (480)	X3	97·728 (1440)
European (CEPT)	2·048 (30)	X4	8·448 (120)	X4	34·368 (480)	X4	139·268 (1920)

Values refer to information rates in Mbit/s
Values in brackets show number of speech channels

analogue microwave radio. The latter[4] fits the 1·5 Mbit/s information into the 'empty' 564 kHz band below the 600-channel f.d.m. system, by a process of rate reduction and partial-response coding with seven received levels.

Particularly noteworthy is the presence of the No. 4 ESS digital switching system,[5] this being the first large-scale example of the integration of switching and transmission, long-promised by digital t.d.m. techniques. This switch, with a capacity of over 100 000 trunks is accessed either by direct speech-interface units each of 120 channels, or from the 1·5 Mbit/s level via an ancillary multiplex termed a 'digroup terminal'.

Strict control of crosstalk by attention to cable design and repeater housings[28] has enabled development of the T1C system capable of operating at double the T1 rate over the same route facilities. The T2 digital link[6] at 6·3 Mbit/s uses special low capacitance cable and is intended for intercity links up to 500 miles long. The 6·3 (6·312) Mbit/s level can also be transmitted via a modem over analogue f.d.m. carrier facilities. The modem fits two 6·3 Mbit/s services into a 2·5 MHz band vacated by a 600-channel f.d.m. assembly, by resorting to 15-level partial response coding and adaptive equalisation. The facility offers transmission of visual telephone services.

The third multiplex level at about 45 Mbit/s does not have a transmission outlet but provides access for coded 600-channel f.d.m. mastergroup (American) and could, in due course accept a data-compressed television signal. The latter requires about 90 Mbit/s with no compression. The fourth level, at 274 Mbit/s can either be transmitted over the T4M coaxial line system intended as a high-capacity feeder in metropolitan areas, or over the DR 18 radio system operating at 18 GHz. T4M uses specially designed 3/8" coaxial tube and, in the Bell System, abandons the

Table 16.2 Some examples of digital transmission systems

System designation	Country or administration	Information rate Symbol rate Mbit/s Mbaud	Channel	Repeater spacing	Code or modulation	Repeater consumption	Application area	Status
T1	Bell, USA	1·544	tw. pair	6000 ft	AMI	0·6 W	exchange junctions 50 m	service
T1C	Bell, USA	3·152	tw. pair	6000 ft			exchange junctions	development
	CEPT	2·048	tw. pair	2000 m	HDB3	0·6 W	exchange junctions	service
T2	Bell, USA	6·312	low cap. pair	14800 ft max.	BZS[1]		intercity 500 miles	service
PCM — 16M PCM — 120	Nippon TT	16 8	tw. pair[3] tw. pair[4]	1·4 km on 0·65 mm or 1·9 km on 0·9 mm	AMI			service
	Italian PTT	34·368 25·776	microcoax. 0·65/2·8 mm	2100 m max.	4B3T[2]	1 W		experimental
PCM — 100M	Nippon TT	97·728	special pair cable	1500 m	AMI with scrambler	3 W		service
	UKPO	120·000 90·000	coaxial 1·2/4·4 mm	2100 m max.	4B3T[2]	2 W		experimental

Table 16.2 (continued)

System designation	Country or administration	Information rate Symbol rate Mbit/s Mbaud	Channel	Repeater spacing	Code or modulation	Repeater consumption	Application area	Status
T4M	Bell, USA	274·176	coaxial 0·375 inch	5700 ft max.	binary[6]		short feeder	service
LD–4	Canada	274·176	coaxial 2·6/9·5 mm	1900 m	B3ZS	14 W	long-haul	service
PCM – 400M	Nippon TT	400·352	coaxial 2·6/9·5 mm	1600 m max.	AMI with scrambler	6·6 W		experimental
DR–18	Bell, USA	274·176	radio 18 GHz	1–5 miles[5]	PSK 4		short-haul pole-mounted	development
WT–4	Bell, USA	274·176	circular waveguide	25 miles	PSK 2		long-haul	experimental

1 = general class of bipolar with zero substitution
2 = General 4B3T class
3 = Go and return in separate cables
4 = Go and return in same cable
5 = Depends on local weather
6 = With quantised feedback and scrambling

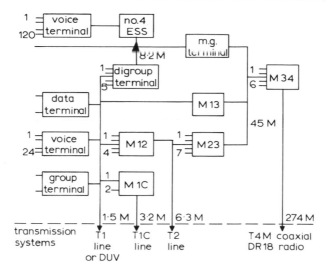

Fig. 16.1 Digital family of the Bell System (based on Reference 3)
M after number = Mbit/s
m.g. = mastergroup

traditional correlation of high capacity with long-distance working; its ten tubes offer 40320 circuits. An interesting difference between the American T4M system and the Canadian LD4 system exists in the choice of line code, see Table 16.2. The former uses scrambled binary with quantised feedback, the latter favours bipolar with three-zero substitution (B3ZS). DR18 is an 18 GHz microwave system with pole-mounted aerials. It uses 4-phase p.s.k., with seven radio channels plus one for standby, resulting in a total capacity of 28 224 circuits. In due course, a 2·5″ diameter circular waveguide system WT4 will enter service with 60 channels each of 274 Mbit/s, in the band 40—110 GHz, offering some 230 000 circuits, to fulfil the long-haul role. A field-trial of a 20 mile section in N. Jersey is due in 1975. It is possible that, after consolidation of the T4 level, a further level T5 at some 564 Mbit/s may be developed for similar transmission media.

16.1.2 Japanese developments

The Japanese 24-channel p.c.m. system went into service in 1965; since then digital-system developments in Japan have been strongly pursued and have evolved along lines somewhat different from those of N. America and Europe.[21] The multiplex hierarchy, see Table 16.1, shows the coexistence of two separate second levels. The 7·9 Mbit/s level is transmitted either over the PCM — 16M[7] system at about 16 Mbit/s or over the PCM — 120[8] system at 7·9 Mbit/s using specially designed pair cable. In the double-rate system, go and return paths are conveyed over separate cables, whereas the lower-rate system can use a single cable. Further details are given in Table 16.2. The 2nd-order level can also be transmitted over 2 GHz microwave radio.[11]

F.D.M. group, supergroup and mastergroup assemblies can gain access to the digital system at the first, second and third multiplex levels, respectively. The second level at 7·9 Mbit/s provides access for visual telephone, while colour television enters at the fourth level.

The fourth level at about 100 Mbit/s may be transmitted over either special pair or coaxial cable via the PCM − 100M system.[9] Higher rate experimental multiplexers have been described[10] operating at about 200, 400 and 800 Mbit/s; these are intended for coaxial cable and 20 GHz microwave radio links.

16.1.3 European developments

Early work in Europe initially followed the pattern set by the 24-channel p.c.m. system and several countries, among them the UK, placed large numbers of such systems in service from 1965 to 1974. However, in the UK version, detailed format was altered; in particular the 193rd framing digit was abandoned in favour of a multiframe structure, see Chapter 5, resulting in a rate of 1·536 Mbit/s. In due course, experience with the 24-channel system indicated that its capacity was somewhat lower than could be reliably achieved, and several countries, notably France, proposed a 32 time slot system at 2·048 Mbit/s. This was adopted by the CEPT and CCITT as the primary level, see Table 16.1. It has several advanced features, the most important being the use of common-channel signalling (c.c.s), see Chapter 5, which interworks naturally with stored−program controlled digital switching. The number of speech channels is 30. Another unsatisfactory feature of the early primary systems, namely the inadequate timing content of the bipolar code, was removed by the adoption of a filled bipolar code in this case HDB3.

The multiplex hierarchy is currently fully specified in CCITT recommendations up to the second level, see Table 16.1; there is, however, wide agreement on a uniform hierarchy based on powers of four. Partly because of the delay in adopting standards, transmission systems beyond the primary level have not yet seen service. However, experimental systems in the region of 120 and 140 Mbit/s over coaxial cable are under test;[22−24] an interesting feature here of European design is the bias towards rate-reducing codes of the 4B-3T type, for instance, to 'stretch' the cable capacity, see Table 16.2.

Transmission at the second and third levels is also receiving attention; Reference 12 indicates the feasibility of transmitting 8·448 Mbit/s over shielded-unit pair cable with repeater spacings of up to 4·3 km. An Italian contribution to the CCITT[13] describes transmission of the third level, 34·368 Mbit/s over 0·65/2·8 mm microcoaxial cable with 2 km spacing. Reference 12 also suggests that this is practicable over shielded-unit pair cable.

16.2 Related activities

The success of digital transmission has not gone unnoticed in other tele communication disciplines, and, now that technology is beginning to provide economically attractive means of realisation, it is possible that these disciplines are on the verge of an explosive evolution.

16.2.1 Digital switching
The assets and investments in the worldwide telecommunication network are dominated by the switching equipment and the local telephony distribution network. Present switching technology is based on electromechanical devices, be they crossbar, step-by-step or reed-relay switches. Such devices, by their nature, require different forms of signal for the information to be switched (speech) and for their control; this in turn requires awkward signalling interfaces between the switching machine and the transmission channel. Once, however, the information and the control data are in digital form, the extraction of the control data, its processing, and the routing through the digital switching machine can be performed in much the same way as in a digital computer, with the added constraint of real-time operation. The overall result is a switching machine with significantly lower cost; the application of digital switching to the trunk network is, therefore, being developed in almost every technically advanced country (Reference 14 gives an excellent review), the American No. 4 ESS System being a prime example.[5] The expected rapid evolution of such systems will greatly affect digital transmission – first, it will stimulate even more strongly the trend towards it, and secondly, it will with time improve the overall transmission quality through the network due to the progressive reduction of connections with a large number of coding-decoding operations (see Section 4.2). The last point implies that at some future date, say around 2000 AD, our current specifications for speech coding will be altogether too stringent; we have the ironic situation of requiring stringent performance from technology which is primitive by the standards of 2000 AD, by which date the performance may be no longer required.

16.2.2 Digital satellite systems
Commercial communication satellite systems have up to the present provided fixed point-to-point transmission using f.d.m. channels frequency modulated onto microwave carriers. Each ground station transmits and receives in a specific band, the satellite acting as a relay to other ground stations within its earth-coverage aerial illumination. Any two ground stations are thus linked with a preassigned capacity; this is satisfactory for pairs of destinations with a high traffic intensity, as for instance, between the USA and France, but tends to be very uneconomical for pairs with little common traffic. Clearly, the ideal system would be one offering a given overall channel capacity to a network of ground stations, each one of which could

demand a variable fraction of the overall capacity to be routed to variable destinations. Although the f.d.m. systems can provide multiple access simply by providing at any one ground station receiving equipment for the f.d.m. assemblies emitted by multiple stations, the facility of variable destination and variable capacity would involve f.d.m. multiplexing–demultiplexing equipment on board the satellite. This would be unacceptable in terms of size and weight. In addition, there are advantages in departing from the earth-coverage aerial in favour of n spot-beam aerials directed at individual ground stations; basically this offers an effective usable bandwidth n times larger than the spectral allocation. However, in f.d.m. satellite systems level control is achieved by monitoring the return signals from the satellite — but this is no longer available with spot-beam operation. For these and other reasons the desirable variable capacity, variable destination satellite system with spot beams cannot be realised by an f.d.m. system without severe restrictions.

On the other hand, a time division multiple access (t.d.m.a.) system is well-suited to the task, Reference 15. The most advanced concepts envisage a multibeam satellite with an onboard digital switch to provide interconnection and capacity on demand. In an experimental system tested by Intelsat[16] a 50 Mbit/s channel is provided over a p.s.k. microwave link offering a nominal capacity of 782 64 kbit/s speech channels. Each station uses a 125 μs frame into which it inserts sequentially bursts of digital information for any destination in the network of ten stations. The bursts can be of variable length depending on the capacity required. Because transmission is effected in a burst regime, an essential design feature is to ensure reliable synchronisation and alignment. This is achieved by means of a 'preamble' of 57 bits which precedes actual channel information, and is made up as follows: 5 bits are allowed as guard time to preclude any overlapping, 10 bits for carrier recovery for coherent demodulation, 10 bits for synchronisation; 20 bits for sending station address code, 4 bits (part of a 32-bit multiframed word) for channel allocation and routing data, and finally 8 bits provide an order-wire channel. The station address code also defines the exact start of the burst by correlation detection. The 57-bit preamble displaces nearly eight speech channels per burst or a maximum of 80 of the normal 782 channels, leaving a utilisation efficiency of about 90%. Reference 17 describes an interesting development on this general approach; the overall capacity is extended to 90 Mbit/s by resorting to 8-phase p.s.k. for the information channels, with an available bandwidth of 30 MHz. However, the crucial preamble data is transmitted using binary p.s.k.

T.D.M.A. techniques are now regarded as beyond the experimental stage and should find service in future generations of communications satellites. They are particularly interesting on two counts; first, in using burst communication, and secondly, as offering truly integrated digital transmission and switching.

16.2.3 Special systems
The emphasis of our work has been directed towards commercial telephony which is by far the largest communication system made by man. However, there are a

number of areas which are sufficiently isolated from the public network as to permit the use of special techniques; military systems form the most important class. In such systems, the criteria for speech quality described in Chapter 4 do not apply. The emphasis is on adequate intelligibility at the lowest possible information rates. In such a context, the general class of delta modulation[18] systems is very suitable, providing satisfactory performance at about 19 kbit/s; a good example in the context of a military network is described in Reference 19. Each terminal station codes speech into companded delta modulation and is multiplexed with 14 others and a framing channel. The coding rate is selectable as either 32 kbit/s or 16 kbit/s offering a balance between capacity and quality as required. The multiplex rates are then 512 or 256 kbit/s which can then be readily fitted into the commercial network primary p.c.m. level at 2048 kbit/s, if required. Alternative coding rates of 19·2 and 38·4 kbit/s which are military preferred values can also be used.

The above example illustrates the possibility of accommodating 60 or even 120 channels at the 2·048 Mbit/s level instead of the normal thirty p.c.m. channels, if a degradation of quality is allowed; this might offer possibilities of economic leased digital line working to large communications users.

16·3 Future trends

Future activities in digital transmission are likely to fall within two classes; one of evolutionary improvements, the other of a more basic nature. Systems which offer ever increasing capacity we would identify as evolutionary; developments which offer increased penetration of digital working into areas hitherto held by analogue techniques may be loosely described as revolutionary.

16.3.1 Higher-capacity systems
The evolutionary trend toward higher capacity was already implied in considering current developments. Referring to Table 16.1, systems operating at rates beyond the fourth multiplex level are under consideration. Brief data on a 400 Mbit/s experimental system is given in Table 16.2; a prototype 800 Mbit/s system using a feedback-balanced 5-level line code is described in Reference 20. In a similar fashion, the capacity of the 274 Mbit/s system may be doubled in the future by resorting to multilevel coding. In Europe, transmission of a yet incompletely specified fifth level multiplex at around 565 Mbit/s is also a likely development[25]

16.3.2 Wider penetration
As we have seen, the penetration of digital techniques in transmission and switching into the trunk portions of the telephone network is an established fact. However, the local area of the network which serves individual subscribers and which accounts for possibly 50% of the total value of the telephony network remains

analogue. We feel that there are three strong currents which in due course may sweep digital transmission into the local area, possibly even so far as to locate the digital conversion process in the subscriber's instrument. First, the cost of the complex circuitry required will continue to decline with continuing advances in solid-state technology, secondly, local exchanges are likely to adopt digital switching, and thirdly, multimode optical-fibre cables requiring digital operation will be developed. Reference 26 surveys some possibilities.

In connection with optical-fibre transmission[27] initial efforts concentrated on the application of single-mode fibres to high-capacity digital systems; an increasing awareness that the value of optical fibres lies in their compactness in making up multipair cables has diverted emphasis to lower-capacity systems. Time may show, therefore, that multimode fibre cables operating at very modest rates may be the most economical solution; their principal application would then be in the local and junction networks.

16.4 References

1   Bell Laboratories Record., Jan. 1975
2   FISH, R.H., and WALKER, A.C.: 'Digital transmission from d.c. to light', International Conference on Communication, ICC 74, Paper 42-B
3   THOMPSON, M.J.: 'New developments in digital transmission planning for the Bell system', ICC 74, Paper 24-A
4   PRIME, R.C., and SHEETS, L.L.: 'The 1A radio digital system makes "data under voice" a reality', *Bell Lab. Rec.* 1973, **51**, pp. 335–339
5   JOHNSON, G.D.: 'No. 4 ESS – long-distance switching for the future', *ibid.*, 1973, **51**, pp. 226–232
6   BOBSIN, J.H., and FORMAN, L.E.: 'The T2 digital line – extending the digital network', *ibid.*, pp. 239–243
7   KUMAGAI, D. *et al.*: 'P.C.M.-16M system', *Rev. Electr. Commun. Lab. Japan* 1969, **17**, pp. 349–361
8   SAKASHITA, T., and MIYAKE, N.: 'A 120-channel p.c.m. system on symmetrical pairs – the PCM – 120 System', *Japan Telecommun. Rev.*, 1969, **11**, p. 143
9   KUMAGAI, D. *et al.*: 'P.C.M.-100M system', *ibid.*, 1973, **21**, pp. 247–253
10 KAWASHIMA, M. *et al.*: 'High-speed up to 400/800 Mb/s digital multiplexers', ICC 72, Philadelphia, June 1972
11 NTT: '2 GHz p.c.m. radio system', *Rev. Electr. Commun. Lab. Japan* Mar/April, 1969
12 POPISCHIL, R., and DOMER, J.: '8-448 Mbit/s p.c.m. transmission over twisted – pair cables with shield unit', *Nachrichtentech. Z.*, 1974, **27**, pp. 297–301 (in German)
13 CCITT: Special Study Group D Cont. No. 101, 1974
14 LUCAS, P.: 'Progress of electronic switching throughout the world', *Commutat. & Electron.*, 1974, **44**, pp. 45 (in English)
15 SCHMIDT, W.G.: 'An onboard switched multiple-access system for millimeter wave satellites', International Conference on 'Digital satellite communication' IEE Conf. Publ., 59 1969, pp. 339–407
16 SCHMIDT, W.G. *et al.*: 'Mat-1: INTELSAT's experimental 700-channel t.d.m.a./d.a. system', IEE Conf. Publ. 59, pp. 428–440
17 OGAWA, A. *et al.*: 'A t.d.m.a. terminal for the spot-beam satellite and its field test results', IEEE International Communication Conference, 1974, ICC 74, Paper 36E
18 SCHINDLER, H.R.: 'Linear, nonlinear, and adaptive delta modulation', *IEEE Trans.*, 1974, **COM-22**, pp. 1807–1823

19   ZIEKMAN, C., and ZWAAL, P.: 'DELTAMUX: a design element for military communication networks', *Philips Telecommun. Rev.*, 1974, **32**, pp. 78–92

20   KUMAGAI, D.: 'Cable communications in Japan', *IEEE Trans.*, 1972, **COM-20**, pp. 707 717

21   KONDO, S.: 'P.C.M. multiplex frame structure and higher-order hierarchy planning', *Electron. & Commun.*, 1974, **57-A**, pp. 11–19

22   BERRY, A.D., LARNER, D.S., and MUMFORD, H.: 'Long distance coaxial cable system for 120 Mbit/s digital traffic', Conference on 'Telecommunication transmission', IEE Conf. Publ. 131, 1975, pp. 115–118

23   BETTS, M.C., NORMAN, P., and WATERS, D.B.: 'Factory and field trial experience of the 120 Mbit/s digital line system', *ibid.*, pp. 111–114

24   GIACOMETTI, A.M., and HARGREAVES, T.F.S.: 'A 140 Mbit/s all regenerative digital transmission system for coaxial cables', *ibid.*, pp. 123–126

25   ROWBOTHAM, T.R., WARING, B., and DOYLE, M.: 'Prospects for digital line links above 140 Mbit/s', *ibid.*, pp. 127–130

26   WEIR, D.A., WILLIAMS, H., USHER, E.S.: 'Digital techniques applied to the subscriber's area', *ibid.*, pp. 53–56

27   TURNER, R.J.: 'Preliminary engineering design of digital transmission systems using optical fibre', *Post Off. Electr. Eng. J.*, 1975, **68**, pp. 7–14

28   BRADLEY, S.D.: 'Crosstalk considerations for a 48-channel p.c.m. repeatered line', *IEEE Trans.*, 1975, **COM-23**, pp. 722–728

Bibliography of some transmission codes

1 Introduction

This Appendix lists a number of codes which have been suggested for the transmission of binary information. It gives a very brief description of each code or coding method and some references to fuller descriptions. No attempt is made to assess the suitability of the codes for the transmission of binary information.

2 The codes

Alternate Digit Inversion
Alternate Mark Inversion (A.M.I.)
B-Code
Binary
Bipolar
Bipolar with Six Zero Substitution (B6ZS)
Bipolar with Zero Extraction
Biternary
Block Coding
Compatible High Density Bipolar (CHDB)
Delay Modulation
Dicode
Digram Coding
Dipulse
Duobinary
Feedback Balanced Codes
High Density Bipolar
Interleaved Bipolar
Low Disparity Binary
Low Disparity Ternary

MS43
Narrow Band Ternary
Paired Selected Ternary
Partial Response Coding
Polybinary
Polybipolar
Quaternary
Time Polarity Control
Twinned Binary
Transparent Interleaved Bipolar
Twinned Binary by Addition
Unit Disparity Binary
VL43
Wolf & Richards Code
Zero Disparity Binary
4B-3T

3 Descriptions

Alternate Digit Inversion
> A 2-level code in which alternate digits in a binary signal are logically comple-
> mented to break up long sequences of marks or spaces. Various extensions of
> this technique exist also.[1-3]

Alternate Mark Inversion (A.M.I.)
> A 3-level code in which one binary level is always represented by the centre
> level of the 3-level code and the other is represented alternately by the two
> extreme levels.[4-9,62]

B-Code
> A 2-level code, basically derived from an unrestricted ternary digit stream, in
> which the 2-level signal is restricted to balance the code i.e. remove its d.c.
> and reduce its low-frequency energy. If the ternary digits are in turn derived
> from a binary information stream the symbol rate of the B-code is 4/3 that of
> the binary source.[10-14]

Binary
> 2-level unrestricted information.[15-17]

Bipolar
> See Alternate Mark Inversion.

Bipolar with Six Zero Substitution (B6ZS)
> A bipolar code which differs only from the basic A.M.I. in that whenever six
> zeros occur without any intervening pulses they are replaced by a code group
> with pulses in it which can be identified at the receive terminal because some
> of the pulses violate the basic A.M.I. rule. This increases the timing content of
> the transmitted 3-level signal.[18,19,56]

Bipolar with Zero Extraction

    A generalised form of B6ZS in which numbers of adjacent zeros are replaced by codes with timing information.[20]

Biternary

    A 3-level signal formed from a binary signal by successive analogue addition of two adjacent binary symbols i.e. digital 'integration' over two symbols. The same conversion can approximately be performed by transmitting the binary signal through a low-pass filter with a Gaussian response.[21-24]

Block Coding

    A sophisticated coding of groups of binary digits into longer groups of binary digits which are interrelated in a way which facilitates some operation such as error correction.[25-28]

Compatible High Density Bipolar (CHDB-n)

    A 3-level code similar to B6ZS except that there are two 'zero-substitution' codes which are used, dependent on the number of binary marks since the previous substitution. CHDB-n replaces groups of n + 1 zeros.[29]

Delay Modulation

    2-level code in which there is always a transition at the end of each bit period. There is an additional transition in the centre of a bit period when the corresponding binary digit is a mark.[30-32]

Dicode

    A 3-level code in which each positive transition in the input binary signal is transformed into a positive rectangular pulse, and each negative transition into a negative pulse.[33-35]

Digram Coding

    The taking of binary symbols in pairs and operating upon the two bits of information in the digram together.[36]

Dipulse

    A 3-level code in which binary spaces are transmitted as the centre 'O' code and binary marks are transmitted as 'dipulses', a + 1 followed immediately by a − 1 within one bit period.

Duobinary

    A 3-level code in which binary spaces are transmitted as the centre 'O' code and binary marks are transmitted as one of the extreme levels. Successive marks are of opposite polarity if the number of intervening spaces is odd, and of the same polarity if the number of intervening spaces is even. Many slight modifications and extensions of this technique have been described.[37-46,55]

Feedback Balanced Codes

    Codes in which the binary information can be coded in two or more ways, or modes; the particular mode chosen at any instant being that which will reduce the accumulated imbalance, or disparity, of the transmitted signal.[47-52]

High Density Bipolar

    A modified bipolar code, again with long runs of zeros replaced by codes with some timing content, identified by 'bipolar violations' in the marks in the inserted codes. This is an earlier form of Compatible High Density Bipolar.[53]

Interleaved Bipolar

The binary information is split into two binary channels at half the rate; each channel is converted to a 2-level bipolar or A.M.I. sequence; and the two bipolar channels so produced are 'bit interleaved'. Thus the bipolar constraint applies to alternate bit periods.[54-56]

Low Disparity Binary

A 2-level code which has redundancy inserted into the binary signal to ensure a low disparity, or imbalance, in the transmitted signal.[49,55,58-62]

Low Disparity Ternary

A 3-level code similar to bipolar but at a slightly reduced symbol rate. The bipolar constraint only applies in this code over intervals of one code 'word'.[63]

MS43

A 3-level code in which three symbols represent four binary digits giving a reduced symbol rate in the ratio 3/4. Mode alternation ensures low overall disparity.[50]

Narrow Band Ternary

A 3-level code similar to duobinary but at a slightly reduced symbol rate.[63]

Paired Selected Ternary

A 3-level code produced by taking binary digits in adjacent pairs and forming pairs of 3-level symbols from them. Of the four possible pairs of binary digits, two result in ternary pairs with no disparity and the other two binary pairs are represented by ternary pairs alternately from two modes such that the overall disparity is minimised. The mode alternation is analogous to that used in A.M.I.[64-69]

Partial Response Coding

The binary signal is passed through some linear device, e.g. a low-pass filter, whose rise time is greater than one bit period, giving rise to a multilevel waveform. Digital equivalents of this also exist e.g. the analogue addition of two adjacent binary digits (Biternary).[23,24,70-72]

Polybinary

An extension of duobinary to generate multilevel codes.[73-75]

Polybipolar

As duobinary or polybinary code, but with symbol inversion around a centre zero to give a balanced, i.e. no d.c., multilevel signal e.g. Duobinary is converted to a 5-level balanced code.[73,74]

Quaternary

4-level transmission at half the binary symbol rate. Variants produce balanced 4-level codes at the original binary symbol rate.[76-78,87]

Time Polarity Control

A 3-level signal in which the binary spaces are represented by the centre zero level and binary marks are represented by one or other of the extreme levels.

The choice of which extreme level is to be used is alternated on a time basis alone.[5,79]

Transparent Interleaved Bipolar (TIBn)

Similar to Interleaved Bipolar except that when a number $n + 1$ of consecutive zeros is found in the composite pulse train, they are replaced by a 'filling' sequence (as in CHDBn) terminated with a bipolar violation in each of the subchannels.[56]

Twinned Binary

A 3-level code formed from the binary signal by delaying the binary signal by one bit and subtracting it algebraically from the original. This can be looked upon as a special case of Dicode where the dicoded pulse width is of exactly one bit duration.[6,55,80,81]

Twinned Binary by Addition

A 3-level signal formed from the binary signal by delaying the binary signal by one bit and adding it algebraically to the original. This is identical to Biternary and is a special case of Partial Response Coding.[21–24,55]

Unit Disparity Binary

A special case of Low Disparity Binary where the imbalance is only one unit over one code word.[55,57–59,61,62]

VL43

A 3-level code of variable word length, an average symbol rate reduction ratio of 3/4 over that of the binary information, and mode alternation to give zero d.c. and small low-frequency energy.[50]

Wolf and Richards Coding

A 3-level code of the partial response type.[24,70,82]

Zero Disparity Binary

A more restricted form of low disparity binary in which individual code words are constrained to have zero disparity.[57]

4B-3T

A 3-level code in which four binary symbols are represented by three ternary symbols, using mode alternation, controlled by the output disparity, to reduce the disparity of the transmitted signal.[47,48,69,83–86]

Note

The above descriptions do not cover all the codes so far encountered. The remaining references[88–107] cover some of those codes less readily classifiable, and several more general papers on code properties which do not easily fit any of the above code categories.

4 References

1   Western Electric Co.: 'Improvement in or relating to apparatus for transmitting and receiving coded information.' British Patent 929, 889

2 FRACASSI, R.D., and FROELICH, F.E.: 'A wideband data station', *IEEE Trans.*, 1966, **COM-14**, p. 648

3 International Standard Electric Co.: 'Digital transmission system.' British Patent 1, 115, 894

4 BARKER, R.H.: 'Electrical signal generating systems.' British Patent 706, 687

5 MAYO, J.S.: 'A bipolar repeater for p.c.m. signals', *Bell Syst. Tech. J.*, 1962, **41**, p. 25

6 AARON, M.R.: 'P.C.M. transmission in the exchange plant', *ibid.*, 1962, **41**, p. 99

7 HOTH: 'The Tl carrier system', *Bell Lab. Rec.*, 1962, **40**, p. 358

8 AARON, M.R., and MAYO, J.: '3-level binary code transmission.' US Patent 3, 149, 323

9 Netherlands Administration: 'A bipolar repeater with the same sensitivity to interference as the binary repeater.' CCITT Com. Sp. D. No. 14-E, July 1969

10 NEU, W.: 'Some techniques of pulse code modulation', *Bulletin des Schweizerischen Electrotechnishen Vereine*, Oct. 1960, **51**, p. 978

11 NEU, W., and KÜNDIG, A.: 'Project for a digital telephone network', *IEEE Trans.*, 1968, **COM-16**, p. 633

12 Swiss Administration: 'Experimental performance of balanced binary vs. bipolar code', CEPT Document Tph/SGT MIC (68) M. Oct., 1968

13 Swiss Administration: 'Comparison of p.c.m. line transmission methods', CCITT Study Group XV Document WP 33/XV-No. 13E. Dec. 1965

14 Swiss Administration: 'On the choice of line signal code', CEPT Document TpL/SGT MIC (68) 16

15 HOUWEN, D.v.d.: 'A binary regenerative repeater for pulse transmission over phantom circuits of low frequency cable', Het PTT-Bedrijf XVI – No. 2, 1969, p. 81

16 DONALDSON, R.W.: 'Optimisation of p.c.m. systems which use natural binary codes', *Proc. IEEE*, 1968, **56**, p. 1252

17 Netherlands Administration: 'Comparison of the binary and bipolar code for two-way transmission via low frequency cables', CCITT Com. Sp. D. No. 15-E, July 1969

18 DAVIS, J.H.: 'A 6·3 Mbit/s digital repeatered line', ICC 69, 1969, 69CP69-COM, p. 34/9

19 AARON, M.R., JOHANNES, V.I., MAYO, J.S., MCCULLOUGH, R.H., and SIPRESS, J.M.: 'Improvements in or relating to pulse transmission apparatus.' British Patent 1, 190, 099

20 JOHANNES, V.I., KAIM, A.G., and WALZMAN, T.: 'Bipolar pulse transmission with zero extraction', *IEEE Trans.*, 1969, **COM-17**, p. 303

21 BROGLE, A.P.: 'A new transmission method for p.c.m. communication systems', *ibid.*, 1960, p. 155

22 RINGLEHAAN, O.E.: 'System for the transmission of binary information at twice the normal rate.' US Patent, 3, 162, 724

23 BECKER, F.K., KRETZMER, E.R., and SHEEHAN, J.R.: 'A new signal format for efficient data transmission', *Bell Syst. Tech. J.*, 1966, **45**, p. 755

24 KRETZMER, E.R.: 'Generalisation of a technique for binary data communication', *IEEE Trans.*, 1966, **COM-14**, p. 67

25 SAVAGE, J.E.: 'A bound on the reliability of block coding with feedback', *Bell Syst. Tech. J.*, 1966, **45**, p. 967

26 PLOUFFE, R.L., and SCHREINER, S.M.: 'Recognising data coded in error correcting codes.' British Patent 921, 946

27 BUCHNER, M.J.: 'The equivalence of certain Harper codes', *Bell Syst. Tech. J.*, 1969, **48**, p. 3113

28 HUFFMAN, D.A.: 'A method for the construction of minimum redundancy codes', *Proc. IRE*, 1952, **40**, p. 1098

29 IBM Europe (CROISIER, A.): 'Proposal for a code allowing unrestricted binary transmission over digital transmission systems.' CCITT Special Study Group D, contribution 33, Sept. 1969

30 HECHT, M., and GUIDA, A.: 'Delay modulation', *Proc. IEEE*, 1969, **57**, p. 1314

31 HANCOCK, P.G., EDLINGTON, M.J.B., and WEST, P.R.: 'Improvements in or relating to data transmission systems.' British Patent 1, 154, 969

32 SCANTLIN, J.R.: 'Information transmission system for use with an a.m. radio transmitter.' British Patent 1, 023, 963

33 Bell Telephone Laboratories: 'Transmission systems for communications.' Western Electric Inc., technical publication, p. 713

34 CROFT, G.F., and DAVIS, J.C.H.: 'Improvements in or relating to electrical signalling systems.' British Patent 1, 023, 621

35 Philips Ltd.: 'Improvements in or relating to transmission systems for pulses.' British Patent 1, 132, 274

36 BYLANSKI, P.: 'Efficient coding of facsimile using an adaptive technique', *Electron. Lett.*, 1965, p. 72

37 CRATER, T.V.: 'Shaping the power density spectra of pulse trains.' US Patent 3, 133, 280

38 LENDER, A.: 'Duobinary coding.' Lenkurt Demodulator **12**, No. 2, Feb. 1963

39 LENDER, A.: 'The duobinary technique for high-speed data transmission', *IEEE Trans.*, 1963, **COM-82**

40 LENDER, A.: 'A synchronous signal with dual properties for digital communications', *ibid.*, 1965, **COM-13**, p. 202

41 Western Electric Co.: 'Improvements in or relating to p.c.m.' British Patent 1, 008, 387

42 Philips Ltd.: 'Telegraphy system.' British Patent 1, 011, 399

43 Philips Ltd.: 'Improvements in or relating to code converters.' British Patent 1, 115, 677

44 TRT Ltd.: 'Improvements in and relating to binary information transmission systems.' British Patent 1, 146, 728

45 TRT Ltd.: 'Encoding binary information signals.' British Patent 1, 179, 307

46 SEKEY, A.: 'An analysis of the duobinary technique', *IEEE Trans.*, 1966, **COM-14**, p. 126

47 KANEKO, H., and SAWAI, A.: 'Feedback balanced codes for multilevel p.c.m. transmission', *ibid.*, 1969, **COM-17**, p. 554

48 WATERS, D.B.: 'Data transmission terminal.' British Patent 1, 156, 279

49 CARTER, R.O.: 'Low disparity binary coding system', *Electron. Lett.*, 1965, **1**, p. 67

50 FRANASZEK, P.A.: 'Sequence state coding for digital transmission', *Bell Syst. Tech. J.*, 1968, **47**, p. 143

51 KAWASHIMA, M. *et al.*: 'On composition of broadband p.c.m. systems', *Fujitsu*, Nov. 1965, p. 123

52 KAWASHIMA, M., HINOSHITA, S., and KATAGIRI, Y.: 'Studies on code balancing and synchronising conversion method for broadband p.c.m. systems', *Electron. & Commun. Japan*, 1966, **49**, p. 392

53 FALCOZ, A., and CROISIER, A.: 'The high-density bipolar code, a method of base band data transmission', International Colloquium on 'Remote data processing', Paris, March 1969

54 Western Electric Co.: Bell System Monograph 4085

55 APPEL, U., and TRONDLE, K.: 'Compilation and classification of codes suitable for digital transmission', *NTZ Commun. J.*, 1970, **23**, p. 11

56 CROISIER, A.: 'Introduction to pseudoternary transmission codes', *IBM J. Res. & Devel.*, 1970, **14**

57 CATTERMOLE; K.W.: 'Low disparity codes and coding for p.c.m.', IEE Conference on Transmission Aspects of Communications Networks, London, Feb. 1964

58 GRIFFITHS, J.M.: 'Binary code suitable for line transmission', *Electron. Lett.*, 1969, **5**, p. 79

59 GRAYSON, H., CATTERMOLE, K.W., and NEU, W.: 'Improvements in or relating to electric p.c.m. systems of communication.' British Patent 849, 891

60 Societe d'Applications Generale d'Elect. et Mech.: 'Improvements to systems for protection against errors in transmission.' British Patent 1, 069, 619

61 VAN DUREN, H.C.A., and DA SILVA, H.: 'System for transmitting digital traffic signals.' British Patent 1, 159, 441

62 CAIN, J.B., and SIMPSON, R.S.: 'A study of major coding techniques for digital communications', US Government report N-70 19873

63 HENRIKSON, U., and INGERMASON, I.: 'Two new pseudo-ternary codes', ESRO Tech. Memo. TN-6, Sept. 1967

64 SIPRESS, J.M.: 'A new class of selected ternary pulse transmission plans for digital transmission lines', *IEEE Trans.*, 1965, **COM-13**, p. 366

65 DORROS, I., SIPRESS, J.M., and WALDHAUER, F.D.: 'An experimental 224 Mbit/s digital repeatered line', *Bell Syst. Tech. J.*, 1966, **45**

66 SIPRESS, J.M.: 'Pulse transmission system.' US Patent 3, 302, 193
67 SIPRESS, J.M.: 'Improvements in or relating to pulse transmission apparatus.' British Patent 1, 087, 860
68 SIPRESS, J.M., and ZARINS, E.. 'Error detection in paired selected ternary code trains.' US Patent 3, 439, 330
69 JESSOP, A.: 'High capacity p.c.m. multiplexing and code translation.' IEE Colloquium on p.c.m., March 1968
70 KRETZMER, E.R.: 'Binary data communication by partial response transmission.' First IEEE Communication Convention, Conference Record, June 1965, paper G1-E
71 GUNN, J.F.: 'Digital transmission on the L-4 coaxial system', *Bell Lab. Rec.*, Feb. 1970, p. 5
72 PIERRET, J.M.: 'Data transmission systems.' British Patent 1, 154, 648
73 LENDER, A.: 'Correlative digital communications techniques.' IEEE International Convention Record, **12**, Part 5, 1964
74 LENDER, A.: 'Method and apparatus for the transmission of intelligence.' British Patent 1, 041, 765
75 HOWSON, R.D.: 'An analysis of the capabilities of polybinary data transmission', *IEEE Trans.*, 1965, **COM-13**, p. 312
76 MELAS, M.: 'Improvements in or relating to data transmission systems.' British Patent 1, 069, 930
77 HOPNER, E., CRITCHLOW, D.L., and DENNARD, R.H.: 'Improvements in or relating to data transmission systems.' British Patent 1, 036, 316
78 RCA Ltd.: 'PM/AM multiplex communication.' British Patent 1, 136, 882
79 THOMAS, L.C.: '3-level binary code transmission.' US Patent 3, 154, 777
80 MEACHAM, L.A.: 'Twinned binary transmission.' US Patent 2, 759, 047
81 AARON, M.R.: '3-level binary code transmission.' US Patent 3, 139, 615
82 WOLF, J.K., and RICHARDS, W.R.: 'Binary to ternary conversion by linear filtering.' Rome Air Div. Center, Report AD278, 264, May 1962
83 JESSOP, A., and WATERS, D.B.: '4B-3T, an efficient code for p.c.m. coaxial line systems.' Proceedings of the 17th International Scientific Congress on Electronics, Rome, March 1970, p. 275
84 KIRTLAND, J.P. *et al.*: 'Digital codes for line transmission.' PO Research Report 118
85 GRIFFITHS, J.M.: 'Method of aligning code translators used in digital transmission', *Electron. Lett.*, 1968, **4**, p. 495
86 BURRELL, E.W., WATERS, D.P., and WILLIAMS, D.A.: 'Choice and performance of codes for p.c.m. systems on coaxial cable including line regeneration.' IEE Colloquium on p.c.m., March 1968
87 SCHERER, E.H.: 'An error correcting code for quaternary data transmission.' Westinghouse Co. Tech. Report AG20
88 FREENY, S.L., KING, B.G., PEDERSEN, J.S., and YOUNG, J.A.: 'High speed hybrid digital transmission.' 1969, ICC Conference Record 69CP-386-COM, p. 38–7
89 BEDROSIAN, E.: 'Weighted p.c.m..' US Government report AD606271
90 VAN GERWEN, P.J.: 'Efficient use of pseudo ternary codes for data transmission', *IEEE Trans.*, 1967, **COM-15**, p. 658
91 BECKER, F.K.: 'Data transmission system.' British Patent 1, 153, 125
92 FILIPOWSKY, R.F.J.: 'Data communication system.' British Patent 1, 120, 975
93 RUMBLE, D.H.: 'Pulse transmission system and apparatus.' British Patent 1, 095, 439
94 Bendix Co. Ltd.: 'Digital data transmission system for increasing information rate without increase of bandwidth.' British Patent 1, 113, 143
95 KAZAKOV, A.A.: 'A method of improving the noise immunity of redundant binary code reception', *Telecommun. Rad. Eng.*, Pt. 1 (USA), 1968, **22**, p. 51
96 MCRAE, D.D.: 'Techniques for the transmission of p.c.m. signals.' Proceedings of the National Telemetry Conference, Chicago, 1961
97 ZLOTNIK, B.M., and LEVIN, B.R.: 'The energy spectra of code sequences having constant weight', *Radiotekh. Elektron.*, 9, No 12
98 TARONI, A.: 'Some stochastic problems relating to p.c.m.', *Onde Electri.*, 1966, **475**, p. 1108

 99 HUGGINS, W.H.: 'Signal flow graphs and random signals', *Proc. IRE*, 1957, **45**, p. 74
100 HOLLOWAY, D.G.: 'Optimum coding for maximum repeater spacing', *ATE J.*, 1954, **10**
101 TSUJII, S.: 'Considerations on optimum waveforms in coaxial cable p.c.m. systems', *Electron. & Commun. Japan*, 1966, **49**, p. 229
102 KIRTLAND, J.P.: 'Line coding methods.' IEE Colloquium on p.c.m. March 1968
103 YOUNG, J.A.: 'Design parameters associated with multilevel transmission systems.' IEEE Northeast Electronics Research & Engineering Meeting, Boston, 1966
104 CATTERMOLE, K.W.: 'Line signals in p.c.m. transmission.' IEEE Northeast Electronics Research & Engineering Meeting, Boston, 1965
105 PIERCE, J.R.: 'Information rate of a coaxial cable with various modulation systems', *Bell Syst. Tech. J.*, 1966, **45**, p. 1197
106 BENNETT, W.R.: 'Statistics of regenerative digital transmission', *ibid.*, 1958, **37**, p. 1501
107 PIERCE, J.R.: 'Some practical aspects of digital transmission', *IEEE Spectrum*, Nov. 1968

Index

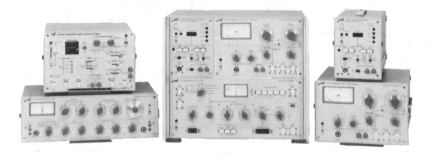